HOMEOSTATIC REGULATORS

HOMEOSTATIC REGULATORS

A Ciba Foundation Symposium held jointly
with the Wellcome Trust

Edited by
G. E. W. WOLSTENHOLME
and
JULIE KNIGHT

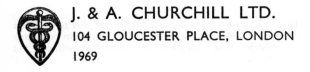

J. & A. CHURCHILL LTD.
104 GLOUCESTER PLACE, LONDON
1969

First published 1969

Containing 84 illustrations

Standard Book Number 7000 1429 2

© J. & A. CHURCHILL LTD. 1969. All Rights Reserved. *No part of this publication may be reproduced, stored in a retrieval system, or transmitted, in any form or by any means, electronic, mechanical, photocopying, recording or otherwise, without the prior permission of the copyright owner.*

Contents

F. Bergel	Chairman's introduction	1
L. F. Lamerton	Cell population kinetics in relation to homeostasis	5
Discussion	A model of homeostasis Regulation of proliferation Allison, Bergel, Elsdale, Gros, Iversen, Lamerton, Leese, Möller, O'Meara, Stoker, Wolpert	15
O. H. Iversen	Chalones of the skin	29
Discussion	Abercrombie, Allison, Bergel, Burke, Iversen, Roe, Stoker, Subak-Sharpe, Vernon, Wolpert	53
C. A. Vernon B. E. C. Banks D. V. Banthorpe A. R. Berry H. ff. S. Davies D. M. Lamont F. L. Pearce K. A. Redding	Nerve growth and epithelial growth factors	57
Discussion	Abercrombie, Allison, Bergel, Burke, Finter, Gros, Miller, Stoker, Vernon	70
J. Mason Em. Wolff Et. Wolff	"Wolff factors" from chick embryo mesonephros and liver or yeast	75
Discussion	Allison, Bergel, Finter, Iversen, Johns, Lamerton, Leese, Mason, O'Meara, Roe, Subak-Sharpe, Vernon	81
R. A. Q. O'Meara	Thromboplastic materials from human tumours and chorion	85
Discussion	Abercrombie, Allison, Jacques, Johns, Mason, Möller, O'Meara, Subak-Sharpe	97
General discussion	Approaches to the study of homeostasis Abercrombie, Allison, Bergel, Iversen, Lamerton, Möller, O'Meara, Stoker, Vernon, Wolpert	101
F. Gros P. Kourilsky L. Marcaud	Pattern of gene transcription during the induction of bacteriophage λ development: a possible model for the control of gene expression in a differentiating system	107
Discussion	Allison, Bergel, Burke, Gros, Möller, Stoker, Wolpert	124
E. W. Johns	The histones, their interactions with DNA, and some aspects of gene control	128
Discussion	Allison, Bergel, Gros, Johns, Möller, Wolpert	140
C. L. Leese	Enzymes and isoenzymes	144
Discussion	Allison, Elsdale, Forrester, Gros, Lamerton, Leese, Stoker, Vernon	161

CONTENTS

N. B. Finter	Interferons as possible regulators	165
D. C. Burke	Interferons as possible regulators—biochemical aspects	171
Discussion	Allison, Bergel, Burke, Finter, Gros, Jacques, Möller, Ormerod, Stoker, Subak-Sharpe	176
P. J. Jacques	Lysosomes and homeostatic regulation	180
Discussion	Allison, Jacques, Möller	193
G. Möller	Regulatory mechanisms in antibody synthesis	197
Discussion	Allison, Bergel, Elsdale, Lamerton, Mason, Möller, Ormerod, Stoker, Vernon, Wolpert	217
General discussion	Immune reactions and homeostasis Allison, Bergel, Johns, Lamerton, Leese, Möller, O'Meara, Roe, Stoker, Subak-Sharpe, Vernon, Wolpert	222
J. A. Forrester	The structure of mammalian cell surfaces	230
L. Wolpert D. Gingell	The cell membrane and contact control	241
Discussion	Allison, Bergel, Forrester, Iversen, Jacques, Möller, O'Meara, Roe, Stoker, Wolpert	259
M. G. P. Stoker	Regulating systems in cell culture	264
Discussion	Abercrombie, Allison, Bergel, Elsdale, Iversen, Jacques, Lamerton, Mason, Möller, O'Meara, Roe, Stoker, Wolpert	271
J. H. Subak-Sharpe	Metabolic cooperation between cells	276
Discussion	Bergel, Burke, Jacques, Leese, Möller, Ormerod, Subak-Sharpe	288
T. Elsdale	Pattern formation and homeostasis	291
Discussion	Elsdale, Lamerton, Stoker, Subak-Sharpe, Vernon, Wolpert	303
General discussion	Abercrombie, Allison, Bergel, Burke, Elsdale, Iversen, Jacques, Lamerton, Möller, O'Meara, Ormerod, Roe, Stoker, Subak-Sharpe, Vernon, Wolpert	307
F. Bergel	Chairman's closing remarks	318
Author Index		320
Subject Index		321

Membership

Symposium on Homeostatic Regulators, held jointly by the Ciba Foundation and the Wellcome Trust, 28th–30th January 1969

F. Bergel (Chairman)	Magnolia Cottage, Bel Royal, Jersey, C.I.
M. Abercrombie	Department of Zoology, University College London
A. C. Allison	Clinical Research Centre Laboratories, National Institute for Medical Research, London
D. C. Burke*	Department of Biological Chemistry, Marischal College, University of Aberdeen
T. R. Elsdale	MRC Clinical and Population Cytogenetics Research Unit, Western General Hospital, Edinburgh
N. B. Finter	Virus Research Department, Pharmaceuticals Division, ICI Limited, Macclesfield
J. A. Forrester	Chester Beatty Research Institute, London
F. Gros	Institut de Biologie Physico-chimique Fondation Edmond de Rothschild, Paris
O. H. Iversen	Institut for Generell og Eksperimentell Patologi, Universitetet i Oslo, Rikshospitalet, Oslo I
P. J. Jacques	Laboratoire de Chimie Physiologique, Université de Louvain
E. W. Johns	Department of Physical Chemistry, Chester Beatty Research Institute, London
L. F. Lamerton	Department of Biophysics, Institute of Cancer Research, Belmont, Sutton, Surrey
C. L. Leese	Chester Beatty Research Institute, London
J. Mason	Laboratoire d'Embryologie Expérimentale, Collège de France, Nogent-sur-Marne
G. Möller	Bakteriologiska Institutionen, Karolinska Institutet, Stockholm
R. A. Q. O'Meara	Department of Experimental Medicine, School of Pathology, University of Dublin
M. G. Ormerod	Chester Beatty Research Institute, Belmont, Sutton, Surrey
F. J. C. Roe	Chester Beatty Research Institute, London
M. G. P. Stoker	Imperial Cancer Research Fund, London
J. H. Subak-Sharpe	Institute of Virology, University of Glasgow
C. A. Vernon	Department of Chemistry, University College London
L. Wolpert	Department of Biology as Applied to Medicine, Middlesex Hospital Medical School, London

* Present address: Dept. of Biological Sciences, University of Warwick.

The Ciba Foundation

The Ciba Foundation was opened in 1949 to promote international cooperation in medical and chemical research. It owes its existence to the generosity of CIBA Ltd, Basle, who, recognizing the obstacles to scientific communication created by war, man's natural secretiveness, disciplinary divisions, academic prejudices, distance, and differences of language, decided to set up a philanthropic institution whose aim would be to overcome such barriers. London was chosen as its site for reasons dictated by the special advantages of English charitable trust law (ensuring the independence of its actions), as well as those of language and geography.

The Foundation's house at 41 Portland Place, London, has become well known to workers in many fields of science. Every year the Foundation organizes six to ten three-day symposia and three to four shorter study groups, all of which are published in book form. Many other scientific meetings are held, organized either by the Foundation or by other groups in need of a meeting place. Accommodation is also provided for scientists visiting London, whether or not they are attending a meeting in the house.

The Foundation's many activities are controlled by a small group of distinguished trustees. Within the general framework of biological science, interpreted in its broadest sense, these activities are well summed up by the motto of the Ciba Foundation: *Consocient Gentes*—let the peoples come together.

CHAIRMAN'S INTRODUCTION

Professor F. Bergel

THERE exist many types of conferences and symposia, from jumbo-sized gatherings via workshops or seminar-like sessions to small discussions. Hardly any organization has contributed so much to the development of the best kinds of meetings, particularly on subjects of medical research, as the Ciba Foundation, and in a more indirect way, the Wellcome Trust, both our hosts for this three-day gathering.

The two, Foundation and Trust, can look back at many meetings and at a prodigious number of topics treated mostly in a retrospective manner. Subjects have ranged from those which invite reviews of all the results so far obtained in laboratories all over the world, to those which like nebulae in space are just beginning to coalesce and require the bringing together of early observations from all points of view. To continue with this simile of heavenly bodies, the nearest to new galaxies and suns are the quasars and pulsars whose origin, composition and properties are even less known, because the signals of their existence are insufficient for drawing any final conclusions from them. There are problems of medical research which are in a comparable state of badly defined existence.

There have not been many conferences on these border-line topics which are, to say the least, of a vague nature. Yet why should not meetings be called from time to time to explore the question of whether something which is still in a conceptual state only, possesses more tangible features than one thought or, alternatively, is rather unreal and will, in all probability, remain so?

One such concept of medical research is that of "cellular homeostatic regulators". Not that the principle of homeostasis *per se* (stability of internal environment, the *milieu intérieur* of Claude Bernard) has not been recognized to play an essential part in living matter and life processes, but the big question is whether it rests on definable entities rather than on a number of inter-linked and patterned happenings with built-in feedback mechanisms. When such a problematic theme is chosen for a symposium it requires an even stronger than usual collective spirit, which from the beginning may have the following aims. If there is anything at all in the role of molecular, though complex, entities as regulators of cell growth, mitosis, development, arrest and cell death—in other words, in all the finer mechanisms of cell kinetics—then a discussion is desirable which without preconceived ideas could investigate the realities of such factors, free or cell-bound (maybe several acting simultaneously), the possibility of a link-up between as yet apparently incongruent research approaches, and if there are encouraging signs, could consider the desirability of collaborative schemes and of a second meeting, say after a couple of years.

After all, one is in good company, at least when considering abnormal cell events or the cancer field which, I should like to underline, is only part of the total territory to be covered; among many others, Farber some years ago postulated the idea of an active principle bringing about maturation of certain cell categories, and Haddow and Van Potter that of deficiencies of special cyto-components.

These ideas may now turn out to be over-simplifications of the true state of affairs. Nevertheless, it should not come as a surprise that when Dr P. O. Williams, the Secretary of the Wellcome Trust, mentioned the rarity of conferences on medical research problems which not only saw the subject through to the very end but also kept it alive *after* the participants had dispersed, I suggested that "cellular homeostatic regulators" could be an experimental discussion topic to prove or disprove the feasibility of such an unorthodox type of symposium (more like a spoken piece of research). As the Wellcome Trust does not possess the necessary machinery and inclination to organize such a meeting on its own, Dr. Williams kindly approached Dr Wolstenholme whose organization has both. By good fortune it so happened that the Ciba Foundation was preparing a meeting on a closely related subject, to be held in Delhi, India in March 1969. The draft programme foresaw papers on growth factors in more complex biological systems (such as tissues and whole organisms), than in our proposal, which was meant to cover research on molecular, subcellular and cellular levels. Thus the two plans not only dovetailed but, even more important, developed the theme of growth regulation from biochemical systems to more and more integrated biological ones; this is a very welcome fact in these days when some scientists seem to live in two different worlds, the molecular and the biological one. Yet only both together make a whole. Another stroke of good luck originated with correspondence between Professor Michael Stoker and Dr Wolstenholme on the possibility of arranging a meeting on cellular regulation, particularly that governing cells in culture. It then became apparent that the way was open to work out a reasonable draft programme. From the final programme you may guess that this was not an easy task.

You will forgive me if I start this meeting by a largely justifiable pretence that we know very little about the whole matter. I recently listened to the Prime Minister of Canada, Mr Pierre Trudeau, and was impressed by his statement that before carrying on old and formulating new policies, he wanted first to pursue a reassessment of the, at present, blindly accepted basis on which they rest. This is how I should like to look at our papers and discussions and at the items on the programme which at first sight may give you the impression of deriving from an exploded Institute of Molecular Biology or Cell Research. I make a plea here for an open-minded and open-ended meeting. Some observations and results, whether molecular, subcellular or cellular, may turn out to be outside our research subject, homeostatic regulators, and to belong to a different system or systems having very little to do with the perhaps multi-membered system of regulation which we think should exist. "Should" and "ought to" are stronger than I dared to use at the start; we might say it more convincingly at the end.

I will bring up briefly only a few points about the meeting which follows. The potential regulating activities of some materials are often demonstrated by testing them on cancers—that is, on abnormal tissues and on certain cells in them; this does not mean that, as indicated before, some of them do not also influence normal events in dividing tissue. You may have noted omissions. It may be asked why we did not include Szent-Györgi's promin and retin. I found that they became more and more ephemeral the longer they were investigated (Eygüd and Szent-Györgi, 1968; Szent-Györgi, 1968). In contrast, the chalones, first discovered by Bullough and by Professor Iversen and his colleagues, and foremost described in their epidermal form, have become better established; we hope to hear from Professor Iversen about those chalones which have now also been described in granulocytes and melanocytes (Bullough and Laurence, 1968a, b; Rytömaa and Kiviniemi, 1968; Mohr et al., 1968).

My remaining point is that a subdivision into molecular entities, subcellular particles and whole cells is at this moment rather artificial. Some of the subcellular components and cells might carry simpler compounds, responsible for the observed cyto-activities of the larger units. On the other hand we should not close our minds to the possibility that regulating principles may exist which are cell bound. Nor should we ignore systems such as endonucleases as restricting enzymes (Meselsen and Yuan, 1968) which may prove to be more important than some which will be the subject of formal papers here.

REFERENCES

BULLOUGH, W. S., and LAURENCE, E. B. (1968a). *Nature, Lond.*, **220**, 134–135; Idem. *Europ. J. Cancer*, **4**, 587–594; 607–615.
BULLOUGH, W. S., and LAURENCE, E. B. (1968b). *Nature, Lond.*, **220**, 137–138.
EYGÜD, L. G., and SZENT-GYÖRGI, A. (1968). *Science*, **160**, 1140.
MESELSEN, M., and YUAN, R. (1968). *Nature, Lond.*, **217**, 1110–1114.
MOHR, U., ALTHOFF, J., KINZEL, V., SUSS, R., and VOLM, M. (1968). *Nature, Lond.*, **220**, 138–139.
RYTÖMAA, T., and KIVINIEMI, K. (1968). *Nature, Lond.*, **220**, 136–137; Idem. *Europ. J. Cancer*, **4**, 595–605.
SZENT-GYÖRGI, A. (1968). *Science*, **161**, 988.

CELL POPULATION KINETICS IN RELATION TO HOMEOSTASIS

L. F. LAMERTON

Department of Biophysics, Institute of Cancer Research, Sutton, Surrey

In this paper I shall be concerned with those aspects of homeostatic regulation which relate to the control of cell proliferation in tissues. At the present time a great deal of work is in progress on the characterization of the pattern of cell proliferation in various tissues, and there is an increasing amount of information on systemic and tissue factors which can affect rates of cell division. However, very little is known so far about the basic mechanisms of control of cell division rate; the advances in molecular biology have, for various reasons, been mainly in the field of control of cell function.

A necessary step in the fundamental attack on the problem is to assemble the quantitative information available on rate of cell division, its capacity for change, the relationship between proliferative state and cell environment and the relationship between proliferative state and cell function. The present paper is a brief review of some of these data.

THE RATES AT WHICH CELLS DIVIDE

In considering rate of cell division it is necessary to distinguish between the intermitotic time of the individual cells in a population and the "doubling" or "turnover" time of the cell population. If all cells are dividing at the same rate, intermitotic time and population turnover time will be equal. If the growth fraction is less than unity, that is, if only a proportion of the cells are dividing, the turnover time will be greater than the cell cycle time. This is an obvious distinction, but one that is not always made in the literature and has led to confusion.

A measurement merely of mitotic index, or labelling index with tritiated thymidine or other DNA precursor, or a mitotic accumulation experiment using colchicine or a similar stathmokinetic agent, will lead to a value only of the turnover time of the population. To determine the intermitotic times of the dividing cells more complex techniques must be used and at present these all require either serial observation or repeated sampling. The most important is the technique of labelled mitoses, described originally by

Quastler and Sherman (1959). Information on the distribution of intermitotic times can be gained from mathematical analysis of the labelled mitoses curves (Barrett, 1966; Takahashi, 1966), and an example of a labelled mitoses curve with the distribution of intermitotic times estimated by Barrett's method is shown in Fig. 1.

To illustrate the range of intermitotic times found in normal unstimulated mammalian tissues a number of values for rats and mice are given in Table I. In normal, unstimulated tissues the shortest intermitotic times

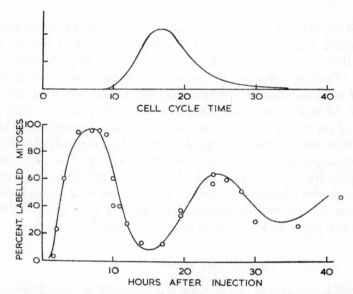

FIG. 1. Labelled mitoses curve (*lower*) of transplantable rat tumour (reproduced from Steel, Adams and Barrett, 1966, with permission), with (*upper*) cell cycle time determined by method of Barrett (1966).

found are 8 to 9 hours in the embryo, but cells in the proliferative zones of bone marrow and small intestine are dividing almost as rapidly. There are reports of shorter intermitotic times in lymphoid tissues, but these may represent the stimulated state. The epidermal tissues of the rat and mouse have intermitotic times of the order of several days. For spermatogonia a value of 13 days has been found. Much longer intermitotic times are found in the so-called conditional renewal tissues, such as the liver and kidney, which normally divide very slowly but can speed up their rate of proliferation dramatically under the appropriate stress. The intermitotic times of liver and kidney in the young adult rat and mouse appear to be of the order of 100 days from the labelling indices, assuming that there is not a

Table I

RANGE OF INTERMITOTIC TIMES

Mouse embryo neural tube	8·5 hr	Kauffman (1966)
tail bud	9·0 hr	Wimber (1963)
Rat bone marrow basophilic erythroblasts		
at 6 weeks	8·8 hr	Roylance (1968)
at 11–12 weeks	9·9 hr	Hanna, Tarbutt and Lamerton (1969)
Rat small intestine		
at 8 weeks	10·4 hr	Lesher et al. (1966)
at 21 weeks	10·7 hr	Lesher et al. (1966)
Mouse small intestine		
at 8 weeks	10·3 hr	Lesher et al. (1966)
Mouse colon	~16 hr	Lipkin and Quastler (1962)
Mouse (hairless) dorsal epithelium	3½ days	Iversen, Bjerknes and Devik (1968)
Rat dorsal epithelium	5–10 days	F. J. Burns (unpublished observations)
Rat seminiferous epithelium	13 days	Clermont and Harvey (1965)
Rat liver and kidney	order of 100 days	Derived from Fabrikant (1968) and Threlfall, Taylor and Buck (1967)

small sub-population of rapidly dividing cells, which is a reasonable assumption on present evidence.

A certain amount of work has been published on the change of intermitotic time with age. For the rapidly dividing tissues, such as bone marrow and small intestine, there is some lengthening with age, though generally it is not very great, as indicated in Table I. For the more slowly dividing tissues there is little information, save for observations of rapid change in the very young individual. There is a dearth of work on differences between species, particularly on the comparison of man with the small rodent. Information available on the gut and skin suggests that intermitotic times in man are longer, perhaps by a factor of about two.

In this short paper it is not possible to go into any detail on the intermitotic time in malignant tumours. In summary, tumours in rats and mice frequently show a wide spread of intermitotic times, with mean values often much longer than those of the more rapidly dividing normal tissues. A very high rate of cell division is not a characteristic of most malignant tumours (Bresciani, 1968); the problem of malignancy resides not in the rate of cell division, but in the failure to stop dividing. In only a few cases has the intermitotic time of human tumours been measured directly. Tubiana and his colleagues (Frindel, Malaise and Tubiana, 1968) have measured the distribution of intermitotic time in some human epithelial tumours and found a wide spread with a mean of 1 to 4 days.

DURATION OF PHASES OF THE CELL CYCLE IN NORMAL MAMMALIAN TISSUES

There have been a number of reports of a remarkable constancy in the duration of the DNA synthesis phase (S) in a given species. It is true that in mice and rats S for a number of tissues has a value of 6 to 8 hours, but there are exceptions. A value of $4\frac{1}{2}$ hours has been reported for embryonic tissue (Kauffman, 1966) and a value of more than 20 hours for alveolar cells in the unstimulated mammary gland of the mouse (Bresciani, 1965). The data available for man suggest a rather longer S period than for the small rodents —perhaps 10–15 hours.

The duration of G_2 is of the order of 1 hour for most tissues of the small rodent. The suggestion that in skin there may be a considerably lengthened G_2 phase (Gelfant, 1963) has yet to be fully confirmed.

The major variation in phase duration, both between cell types and within the same cell population, is found in the G_1 phase, that is, the interval between the end of mitosis and the start of DNA synthesis. In the rapidly dividing tissues of the small rodent, such as bone marrow and small intestine, one finds values for G_1 of 1 to 2 hours. However, there is now evidence that with certain tissues it may be very short, perhaps even non-existent, but one of the difficulties is that the labelled mitoses method does not allow a very accurate measurement of G_1. However, by working with synchronized populations *in vitro*, Robbins and Scharff (1967) showed that G_1 for Chinese hamster lung cells was certainly less than 15 minutes.

In the more slowly dividing tissues G_1 can be very long—of the order of 10 days in the basal layer of the skin in the rat, and of the order of 100 days in the liver. Here, of course, one meets the major problem of how far the interval between mitosis and the start of synthesis should be divided into two main stages, a period of waiting to be triggered into division, and the period following the trigger when the cell is preparing to synthesize DNA. I shall return to this point later.

CAPACITY FOR CHANGE IN DURATION OF CELL CYCLE

In a given cell population the greatest capacity for change is normally shown by the G_1 phase. The following table shows the changes in the cell

	Intermitotic time in hours	Duration of phases in hours		
		S	$G_1 + \frac{1}{2}M$	$G_2 + \frac{1}{2}M$
Control mice	42	8·5	32	1·5
Stimulated mice	26	5	19	1·5

cycle of the uterine epithelium of the castrated mouse when it is stimulated by oestrogens, reported by Epifanova (1966).

Even in tissues that are normally dividing rapidly the value of G_1 may be shortened, as shown by Hanna, Tarbutt and Lamerton (1969) in the basophilic erythroblasts of the bone marrow of the rat kept anaemic by repeated injections of phenylhydrazine:

	Intermitotic time in hours	Duration of phases in hours		
		S	$G_1+\frac{1}{2}M$	$G_2+\frac{1}{2}M$
Control rats	9·9	6·7	1·6	0·7
Anaemic rats	6·9	5·5	0·4	0·9

In both of these examples there is also a shortening of the duration of the S period. Its capacity for change is demonstrated very markedly in the alveolar cells of the mammary gland in mice, where Bresciani (1965) has shown that it shortens from 21 hours to 9 hours on stimulation with 17β-oestradiol and progesterone.

The greatest capacity for change in the G_1 phase is shown by the so-called "conditional renewal" tissues. These tissues, which include the liver, kidney, thyroid and salivary gland among others, normally have a very low rate of cell proliferation, but can increase it dramatically under the appropriate stress. However, one feature of the response of these tissues is the relatively long interval from the application of the stimulus to the first observation of DNA synthesis. For the liver, kidney and salivary gland under appropriate stimulation the interval is found to be 18–20 hours (Baserga, 1968; Threlfall, 1968). For certain other tissues it appears to be rather longer, but of the results reviewed by Baserga the only tissue in which it has been reported as substantially shorter is the skin, and this is a particularly interesting tissue, which will be discussed by Professor Iversen (p. 29).

A reasonable interpretation of the interval between the stimulus and the start of DNA synthesis is that the majority of cells in these tissues are normally not concerned in preparation for division—that is, they are in the "G_0" state—and that the interval of 20 hours or so represents the true G_1; that is, the time needed after the stimulus to complete preparations for DNA synthesis. One may then ask why all cells do not have a G_1 period as long as 20 hours. The answer may be, as has been suggested by Baserga, that the enzymes necessary for the process of DNA synthesis decay with time, so that the length of the true G_1 depends on the interval between the end of the previous division and triggering into preparation for DNA synthesis. It may be that all cells enter the G_0 state immediately after mitosis but that it is of very short duration in the case of the rapidly dividing tissues of the body.

If this concept of G_0 is correct then the proliferative state of all cell populations must depend on the nature and rate of triggering from G_0 into preparation for synthesis. Unfortunately this is all very hypothetical, because at present there is no way of recognizing, or characterizing, a cell in G_0, and information about the presence of G_0 cells can only be obtained by indirect and generally equivocal methods. The term G_0 must be used with great care, particularly with tissues in abnormal conditions. For instance, the term is often applied to tumours but here, if such cells exist, they may well be the result of local nutritional deficiencies, in oxygen or other materials, and both the biochemical state of the cell and the factors governing release from the state of suspended proliferation may be very different from those obtaining in normal tissues.

RELATIONSHIP BETWEEN PROLIFERATIVE STATE AND ENVIRONMENT OF THE CELL

The interrelationships between the cell environment and the proliferative behaviour of the cells represents, I believe, one of the major fields for study in homeostatic mechanisms. The skin demonstrates both the nature and the importance of some of the problems involved. In skin and certain other types of epithelium cell division is confined, normally, to the monolayer of cells in contact with the basement membrane. These cells therefore represent the stem cells of the tissue, and the classical theory of the maintenance of constancy in size of a stem cell population is that asymmetric division occurs, one daughter cell being destined to remain as a stem cell and the other destined to move on to differentiation. Leblond and his colleagues (Marques-Pereira and Leblond, 1965; Leblond, Clermont and Nadler, 1967), working with the oesophageal epithelium of the rat, showed that this was not the process operating. By labelling with tritiated thymidine they demonstrated that sometimes both daughters of a dividing cell in the basal layer moved out, sometimes one and sometimes neither. They concluded from their results that there was a random removal of cells from the basal layer which was determined by the population pressure in the basal layer. Iversen and his colleagues in a recent paper (Iversen, Bjerknes and Devik, 1968) have studied the same problem in considerable detail using the dorsal skin of hairless mice and have arrived at essentially the same conclusion, but with the difference that the chance of a cell leaving the basal layer increases with time since its birth. Their conclusion was that whenever a daughter cell from a mitosis remained in the basal layer and divided again, the mean time from mitosis to mitosis was 84 hours. When the daughter cells differentiated they first remained in the basal layer for an average time of 60 hours.

The conclusion from these experiments is that the property of "stemness" of an epithelial cell is a function of its environment and that the loss of proliferative activity is the result, not the cause, of the cell moving out of the basal layer. One possibility is that the basal layer alone provides the environment for a cell to retain proliferative capacity. The alternative is that the cell can divide outside the basal layer, but is normally prevented from doing so by a high concentration of inhibitor. This inhibitor could also control the rate of division in the basal layer itself. In regeneration from injury there is a marked shortening of the intermitotic time in the basal layer, due presumably to an increased rate of release from the G_0 state, and some tendency for cell division to take place above the basal layer. Also, in certain pathological conditions such as psoriasis (Van Scott and Ekel, 1963), there is considerable cell division above the basal layer. These findings, together with the evidence for chalones, which Professor Iversen will discuss, point to the importance of inhibitory controls in this tissue.

Another tissue in which the relationship between cell proliferation and tissue architecture has been studied in some detail is the small intestine (Cairnie, Lamerton and Steel, 1965). The proliferative zone occupies the lower half of the crypts of Lieberkuhn, whence cells move up the villi and are finally extruded. Over most of the proliferative zone the cells are dividing rapidly, with an intermitotic time of about 10 hours. At the base of the crypt, for the first few cell positions, the intermitotic time is somewhat longer. The change to non-dividing cells occurs over the region of cell positions 10 to 18, numbered from the base of the crypt. The capacity to start new DNA synthesis is lost over this region, but there is no obvious lengthening of the phases of the cell cycle during the process. This process is represented in Fig. 2.

When the small intestine is compensating for loss of cells as, for instance, after exposure to radiation, there is some shortening of the cell cycle, particularly at the base of the crypt (Cairnie, 1967), but the major mechanism of recovery is a temporary extension of the zone of proliferation by the movement of the proliferative boundary up the crypt. In recovery from extensive damage mitotic figures may be found over almost the whole length of the crypt (Lesher, 1967).

At the present time essentially nothing is known about the mechanisms controlling this change in proliferative pattern and how far it is determined by functional demand or size of mature cell population. Nor can we say how far the whole of the proliferative zone should be considered the stem cell compartment; it may be that apart from a few cells at the base of the crypt, the capacity for indefinitely maintained proliferation has already been lost. Here there is a field of great interest for investigation.

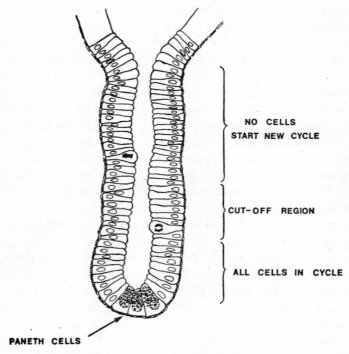

FIG. 2. Diagram of longitudinal section of intestinal crypt of rat, showing spatial pattern of cell proliferation.

RELATIONSHIP BETWEEN CELL FUNCTION AND PROLIFERATIVE STATE

One general problem encountered in any consideration of the cell population kinetics of normal tissues and their homeostatic responses is the relationship between cell division and specialized function and, in particular, how far cell proliferation rate and cell maturation are interdependent.

The liver is a good example of the principle that proliferative capacity is not necessarily lost in cells which carry out highly specialized biochemical functions. Another example is provided by the red cell series where rapidly dividing cells are engaged in haemoglobin synthesis.

Data are now available on the change of pattern of cell proliferation throughout the red cell series which are of considerable interest in the context of homeostatic regulation. Work on rats and mice has shown that the stem cells of the series, which cannot yet be identified morphologically, but can be studied by functional tests, are normally dividing relatively slowly. In the rat the intermitotic time has been estimated at about 30 hours (Blackett, 1968). In the mouse it may be longer (Becker et al., 1965). The start of haemoglobin synthesis is accompanied by an increase in rate of cell proliferation and the recognizable red cell precursors which are dividing have an intermitotic time of the order of 10 hours. Values of the intermitotic times found for the young adult rat (Hanna, Tarbutt and Lamerton, 1969; Hanna, unpublished observations) are shown in Table II. There is only a slight increase in intermitotic time as the cell content of haemoglobin increases from the proerythroblast to the polychromatic erythroblast, at which stage division ceases.

Table II

INTERMITOTIC TIMES OF RAT RED CELL PRECURSORS

	Cell type	Intermitotic time
Unrecognized precursors	"Stem cell"	~ 30 hr
	Erythropoietin-sensitive cell	~ 30 hr
Recognizable precursors	Proerythroblast	9·9 hr
	Basophilic erythroblast	
	Polychromatic erythroblast	10·8 hr
	Orthochromatic erythroblast	—
	Reticulocyte	—
Mature cell	Erythrocyte	—

Until fairly recently it was generally held that this system responded to a demand for increased red cell production in one way only—by increasing the flow of cells into the proerythroblast stage, through the agency of a specific blood-borne factor, erythropoietin. However, there is now evidence that under certain circumstances the kinetic pattern of the recognizable, maturing, precursors can change to provide a measure of homeostatic regulation. For instance, under continuous radiation, where the damage is to the cells of the proliferative compartment and the mature cells are little affected, the cycle time of the recognizable precursors is shortened and their transit time through the various stages lengthened, so providing an increased number of cell generations in the maturing cell stages (Tarbutt, 1969). One possibility is that this effect is related to a reduced cell population in the bone marrow.

It is of interest to note that a shortening of the cell cycle is also a result of bleeding but, if bleeding is maintained, the transit times decrease proportionately, so that the number of cell generations in the maturing stages is not changed appreciably.

This example is a salutary one because it indicates the complexity of homeostatic regulation of cell population size in some tissues and, in particular, that the regulatory mechanisms are not necessarily confined to the stem cells of the system.

CONCLUSION

The attack on the problem of control of cell proliferation has to be carried on at two levels. There is the obvious attack at the cellular and sub-cellular level, but equally important is the attack at the tissue level, where mechanisms will be in operation that will not be apparent from studies on cells in isolation. The characterization of the cell proliferation pattern in various tissues, both normal and under various conditions of stimulation, is setting the stage for this work, but the real advances now depend on the identification of the various regulators present.

SUMMARY

A review is given of the rates at which cells divide in the various tissues of the body, and the capacity for change of the intermitotic time and duration of the various phases of the cell cycle. The relationship between tissue structure and cell proliferation pattern is considered in selected examples. The response to increased cell demand of the normally rapidly dividing tissues, such as gut and bone marrow, is contrasted with that of the "conditional renewal" tissues such as liver, and the concept of the "G_0" cell discussed.

REFERENCES

BARRETT, J. C. (1966). *J. natn. Cancer Inst.*, **37**, 443–450.
BASERGA, R. (1968). *Cell Tissue Kinet.*, **1**, 167–191.
BECKER, A. J., MCCULLOUGH, E. A., SIMINOVITCH, L., and TILL, J. E. (1965). *Blood*, **26**, 296–308.
BLACKETT, N. M. (1968). *J. natn. Cancer Inst.*, **41**, 909–918.
BRESCIANI, F. (1965). In *Cellular Radiation Biology, 18th Annual Symposium on Fundamental Cancer Research*, pp. 547–557. Baltimore: Williams and Wilkins.

BRESCIANI, F. (1968). *Europ. J. Cancer*, **4**, 343–366.
CAIRNIE, A. B. (1967). *Radiat. Res.*, **32**, 240–264.
CAIRNIE, A. B., LAMERTON, L. F., and STEEL, G. G. (1965). *Expl Cell Res.*, **39**, 528–539.
CLERMONT, Y., and HARVEY, S. C. (1965). *Endocrinology*, **76**, 80–89.
EPIFANOVA, O. I. (1966). *Expl Cell Res.*, **42**, 562–577.
FABRIKANT, J. I. (1968). *J. Cell Biol.*, **36**, 551–565.
FRINDEL, E., MALAISE, E., and TUBIANA, M. (1968). *Cancer*, **22**, 611–620.
GELFANT, S. (1963). *Expl Cell Res.*, **32**, 521–528.
HANNA, I. R. A., TARBUTT, R. G., and LAMERTON, L. F. (1969). *Br. J. Haemat.*, **16**, 381–387.
IVERSEN, O. H., BJERKNES, R., and DEVIK, F. (1968). *Cell Tissue Kinet.*, **1**, 351–367.
KAUFFMAN, S. L. (1966). *Expl Cell Res.*, **42**, 67–73.
LEBLOND, C. P., CLERMONT, Y., and NADLER, N. J. (1967). In *Proceedings of the Seventh Canadian Cancer Research Conference*, pp. 3–30, ed. Morgan, J. F. Oxford: Pergamon Press.
LESHER, S. (1967). *Z. Zellforsch. mikrosk. Anat.*, **77**, 144–146.
LESHER, S., LAMERTON, L. F., SACHER, G. A., FRY, R. J. M., STEEL, G. G., and ROYLANCE, P. J. (1966). *Radiat. Res.*, **29**, 57–70.
LIPKIN, M., and QUASTLER, H. (1962). *J. clin. Invest.*, **41**, 141–146.
MARQUES-PEREIRA, J. P., and LEBLOND, C. P. (1965). *Am. J. Anat.*, **117**, 73–87.
QUASTLER, H., and SHERMAN, F. G. (1959). *Expl Cell Res.*, **17**, 420–438.
ROBBINS, E., and SCHARFF, M. D. (1967). *J. Cell Biol.*, **34**, 684–686.
ROYLANCE, P. J. (1968). *Cell Tissue Kinet.*, **1**, 299–308.
STEEL, G. G., ADAMS, K., and BARRETT, J. C. (1966). *Br. J. Cancer*, **20**, 784–800.
TAKAHASHI, M. (1966). *J. theoret. Biol.*, **13**, 202–211.
TARBUTT, R. G. (1969). *Br. J. Haemat.*, **16**, 9–24.
THRELFALL, G. (1968). *Cell Tissue Kinet.*, **1**, 383–392.
THRELFALL, G., TAYLOR, D. M., and BUCK, A. T. (1967). *Am. J. Path.*, **50**, 1–14.
VAN SCOTT, E. J., and EKEL, T. M. (1963). *Archs Derm.*, **88**, 373–381.
WIMBER, D. E. (1963). In *Cell Proliferation*, pp. 1–17, ed. Lamerton, L. F., and Fry, R. J. M. Oxford: Blackwell.

DISCUSSION

A MODEL OF HOMEOSTASIS

Iversen: The first and most important step when one tackles an analysis of biological problems by means of a computer is to construct a model that on the one hand has a reasonable degree of isomorphism with the process one is analysing, and on the other can be transferred to a computer. In mathematics and logics we have a comprehensive set of such models with clear designation rules and with clear and unambiguous interrelations between the elements.

Many biologists of the old school react strongly against the use of mathematical models for interpreting biological phenomena, arguing that all biology is so fantastically complex that any model is bound to be an oversimplification of the grossest kind. On reflection it will be evident, however, that *a good simplification is a big advantage*—as a matter of fact it is a precondition for a model being at all serviceable. If a map included all the details really present, it would be quite useless. There is reason to regard the use of models in an optimistic light. If we succeed in constructing mathematical models that include the main features of the structure of a biological system, such simplification will not rule out the possibility of finding the biological laws. On the contrary, the model offers us the opportunity to understand the main points of what goes on in biology.

By employing mathematical models which have a number of already well-known properties one can hope to learn something new about the quantity under study, in that some of the model's properties may serve to establish isomorphism and others to make new discoveries. It must, however, be possible to keep the model strictly separate from what it is supposed to be a model of. Good models are sufficiently detailed to make it possible to reason and to draw conclusions within the model, and then test these on the observed quantity.

An example is the mouse epidermis, with its basal cells, differentiating cells, and a growth-regulating system. One can build a theoretical model of this hypothetical system, including only known elements and relations. The epidermis of the mouse is complicated and we do not know the details. The theoretical model which we ourselves construct contains nothing but known elements. If it is possible to make a model which has a sufficient degree of isomorphism with the epidermal growth-regulating system, we can use the model to experiment upon and see what the model's answers are to such experiments. Afterwards one can do isomorphic experiments on the mice. Then, if there is similarity between the answer the model gave in advance and the result of the biological experiment, one may draw the following conclusion: *the structure that is built into the model is sufficient to explain the biological reaction*. One cannot prove, however, that what happens biologically necessarily follows the principle vested in the model, as there may be many other principles leading to the same result. This being so, it is important that the model is based on sound biological reasoning and that the hypothesis behind the model is accepted by leading research workers in the field. Preferably the hypothesis should be verifiable by direct biological experiments. In our example this means that the regulating principle should prove to be, for instance, a chemical substance which can be found and characterized.

At the Institute for General and Experimental Pathology in the University of Oslo we have been studying growth regulation problems in epidermis in cooperation with Rolf Bjerknes, research engineer at the Central Institute for Industrial Research, Oslo (Iversen and Bjerknes,

1963). We have tried to analyse the growth-controlling system in the epidermis of the mouse and to study the reaction of the epidermis to single applications of carcinogenic substances. The investigation is based on a combination of experimental studies on mice, the construction of a growth model for the epidermis, and the theoretical analysis of this model, first by means of an electronic analogue computer, later by using a digital computer.

Fig. 1 shows the essential points in our theory. The basal cells move around the circle and each time they reach the branching point a binary

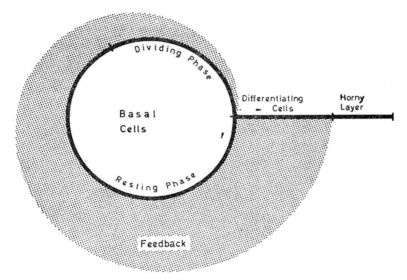

FIG. 1. Deterministic model of the supposed growth regulation system in epidermis.

fission takes place. As a result of each division in a basal cell another cell becomes a differentiated cell and ultimately changes into keratin. The differentiating cells are continuously producing an inhibitory substance which is supposed to diffuse back to the basal cells, where it controls the proliferation rate.

Fig. 2 is a block diagram of the model. The two delays, called the dividing phase and the differentiating phase, correspond to the same phases in Fig. 1.

To simulate the feedback production of inhibitor we must assume a relationship between the rate of production of inhibitor and cell age. A linear relationship, as shown in Fig. 3, is a likely approximation. It

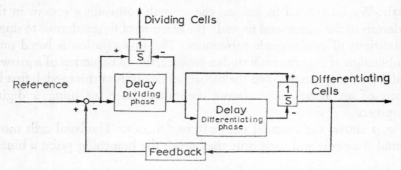

Fig. 2. Block diagram of the model used for simulation on an analogue computer. $\frac{1}{s}$ means that the actual number of cells in this compartment is determined at any given moment.

follows that the mean production rate of the inhibitory substance is proportional to the mean elimination time. If cells are eliminated at a faster speed than normal the mean level of inhibiting signal will decrease.

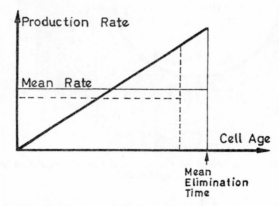

Fig. 3. Relationship between cell age and supposed production rate of the inhibiting signal. It shows how a reduction in the mean elimination time results in a lowering of the *mean* production rate.

It is stressed that this model is constructed according to our theory and is based on reasoning on biological grounds: no "curve fitting" is involved. Assuming that this model was isomorphic with the growth-regulating system in the epidermis, we tested the model against a well-known disturbance of the epidermis: increased wear and tear. It is widely recognized that the epidermis reacts to an increased rate of cell loss by thickening: that is, the number of cells increases. Fig. 4 shows the result: the model reacts to an increased rate of elimination by increasing the number of cells.

DISCUSSION

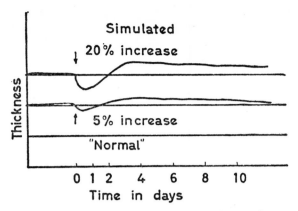

FIG. 4. The model's answer to simulation of increased external wear and tear.

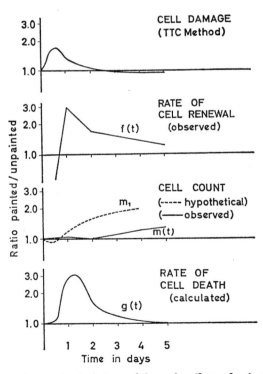

FIG. 5. Observations and calculations of the early effects of a single application of one drop of 1 per cent 3-methylcholanthrene in benzene to skin from a hairless mouse. For explanation, see text. See also p. 33.)

From studies of the effects of a single application of carcinogen to the skin of mice we have demonstrated (Iversen and Evensen, 1962; see also this volume, p. 33) that the following changes in the cell population occur: (1) A very short blocking effect on DNA synthesis and on the process of mitosis. (2) A disturbance of the mitochondria followed by the death of many cells. (3) The relatively sudden increase of rate of cell loss is compensated by an increased rate of cell renewal. (4) After 2 days a transient hyperplasia (thickening) develops. Fig. 5 demonstrates the changes as recorded over the first 5 days after application.

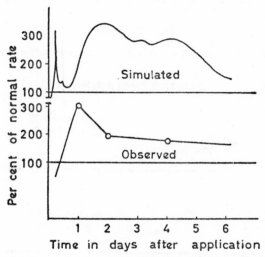

FIG. 6. A comparison of the observed and simulated effects on the rate of cell proliferation.

We then put a disturbance into the model that was aimed to simulate only the destruction of cellular function and cell death. Fig. 6 shows the simulated rate of cell renewal, which is approximately as observed in mice. Fig. 7 demonstrates the change in the number of cells, which is almost identical with the results observed by counting cells in mouse epidermis.

If the model is based on correct principles, and *if* one of the effects of carcinogens is a destruction of cellular function with subsequent death of cells, then the increased rate of cell renewal and the hyperplasia can be explained as a consequence of this destruction in a homeostatic system such as the epidermis.

These theoretical results obtained with a model have given us suggestions for the planning of new biological experiments.

One of the limitations of analogue computers is that only quantitative changes can be tested. It is generally held that the production of a cancer involves a *qualitative* change, initially only in a single or very few cells. Such changes cannot be tested on an analogue computer. Therefore to try to simulate some aspects of carcinogenesis we had to proceed to a digital computer. We have built the model as a population of individual cells. They can be given different properties, and they can be followed from birth to death.

The new model has been used to simulate the outcome of a series of biological experiments. It has also been used to simulate several numerical

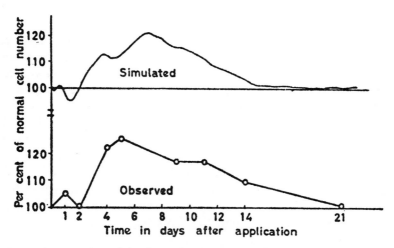

FIG. 7. A comparison of the observed and simulated effects on the number of cells present in epidermis.

aspects of hypothetical tumour growth. The model is especially well suited to studying early changes in the cell population after a carcinogenic influence has been introduced. Work is in progress to test some of the current theories of cancer development. Within the framework of this model there is for instance a high probability that a cell which is given a 10 per cent proliferative advantage over other cells, and which can transfer this property to all its descendants, will take over the whole population after a certain time.

I wish to stress that we do not claim to have *proved* how the control mechanism in the epidermis functions and how it is disturbed in carcinogenesis. Nevertheless, we feel that we have collected some theoretical support for the statement that the main outlines of our theory are correct.

REGULATION OF PROLIFERATION

Bergel: From Professor Iversen's model and Professor Lamerton's review of cell kinetics in different types of cells I gather that there are indications of the existence of regulating mechanisms but not yet of specific factors which partake in this.

Lamerton: This is broadly true, apart from the case of erythropoietin for the red cell series; here is one known specific factor, and we are beginning to learn something about its action. It would appear to act mainly on the cells just before the proerythroblast by increasing their rate of proliferation and starting the process of haem synthesis (Hanna, 1967). So we have at least one factor in one tissue that can be explored.

Iversen: Professor Lamerton commented on the fact that it takes only 3–4 hours after stimulation of the skin before proliferation starts, whereas in many other systems it takes at least 20 hours. He gave Baserga's explanation that this could be due to the time the cells need to go through the last presynthetic part of the G_1 compartment. But if we accept that there are chalones for all organs, such a difference in reaction time could also be due to different half-lives of the chalones. If the half-life of the epidermal chalone is only 3 or 4 hours a reduction in concentration should be felt by the basal cell 3 hours after the removal of cells, whereas if the half-life of the liver chalone is 20 or 40 hours it would be much longer before a reduction was effective, since the supposed liver chalone would remain at a high concentration in the circulation for many hours.

Möller: I have three questions on the labelled mitosis technique. First, how useful is it when only a proportion of the cells divide? Secondly, how useful is it when the generation time varies between different cells? Thirdly, how do you explain the fact that the doubling time may be much shorter than the generation time? This is the case with for example antibody-producing cells which increase with a doubling time of 5 hours although their generation time is 10 hours.

Lamerton: One has to recognize that the labelled mitosis method gives information only on the proliferating population, because the cells followed are those that have taken up the label. It also tends to exaggerate the relative importance of the more rapidly dividing cells; other methods, such as continuous labelling with tritiated thymidine, give more information on the slowly dividing cells.

Möller: If the slowly dividing cells make up a sufficiently large proportion, will they obscure the peaks?

Lamerton: There may be considerable damping out of the peaks, and this in fact is used to estimate the spread of cell-cycle time, but you are quite right that with very slowly dividing cells the labelled mitosis method has severe limitations.

You asked how one could explain why, in the production of antibody, the cells concerned appeared to be dividing once every 10 hours or so whereas the rate of increase of antibody was much faster. I don't know the answer to this. I think one has to look very carefully at the kinetics of the system; one possibility is that cells are being continuously released into the proliferating compartment. The labelled mitosis technique could indicate an average intermitotic time but I cannot see how it could indicate too long an intermitotic time.

Stoker: Professor Lamerton referred to the cells which have a very short G_1. I think it has been suggested that in Ehrlich cells DNA synthesis goes on all the time; that is, there is no G_1.

Lamerton: Yes, it has been suggested that Ehrlich ascites cells have no G_1 (Lala and Patt, 1966), but with the labelled mitosis method it is not possible to distinguish between a non-existent G_1 phase and a very short G_1, of 30 minutes or less. The experiment I mentioned by Robbins and Scharff (1967) used synchronized cells in culture. Cells in mitosis were separated mechanically and measurements then made of tritiated thymidine incorporation. In this experiment a G_1 phase, if it existed, must have been less than 15 minutes.

Stoker: If a break in a block is required to initiate an inevitable progression, is there any evidence that this might occur in the previous G_2, in cells with very short G_1 periods?

Lamerton: This has been suggested.

Stoker: Going one stage further, is there any situation where there might be an initiation of the next mitosis but one?

Lamerton: I don't know any direct evidence for this, but programming ahead may be involved in those situations where the number of divisions a cell can undergo appears to be limited.

Bergel: To approach this question of the control of proliferation in another way, how do the cycling times in bacterial cultures relate to mammalian cells; what kind of differences do we have there, Dr Gros?

Gros: Replication in bacteria is at first sight a rather monotonous phenomenon since during one division cycle there is one unique replicating fork moving along a unique circular DNA entity and usually (that is, in bacteria with an average division time) no new cycle is initiated unless the previous one is completed. In fact the nature of the signal (initiator) which triggers each round of replication, and its precise molecular target, are the subject of much debate. What seems generally agreed is that specific proteins (presumably endonucleases) are needed for reinitiating the cycle and that such initiators do not operate unless specific changes have occurred at the level of the cell wall or the cell membrane. So, at the end of DNA replication the cell surface begins to elongate, which permits the two daughter chromosomes to separate and a new signal, the initiator, restarts the DNA replication cycle.

To come back to the differentiation of red cells, in the erythroblastic system erythropoietin appears to have two functions; on one hand it seems to reduce the generation time of the stem cells and on the other hand it seems also to drive, or gear the system in a particular path that corresponds to the first step in its differentiation. I should like to draw a parallel between these eukaryotic cells and bacteriophages. In the differentiating bacteriophages it is well known that some factors which activate DNA replication must operate before the expression of certain genes, the so-called "late genes", and this constitutes a mandatory time sequence; if there is no DNA replication there is no late-gene expression and therefore the maturation of the phage does not take place. There is in other words some sort of correlation between the rearrangement of the DNA chromosome in bacteriophages and the triggering of certain gene expressions to make new types of specific products. I wonder whether we do not have the same situation in red cell differentiation.

Lamerton: This is very interesting. My colleague I. R. A. Hanna (1967) has shown that within three hours of bleeding there is an increased rate of proliferation in the erythropoietic cells of the rat, mainly in the cells preceeding the proerythroblast. Up to now the generally accepted theory has been that erythropoietin first causes erythropoietin-sensitive cells to differentiate and that cell proliferation follows. Our feeling is that the increase in proliferation comes first.

Gros: So in a sense the situation would resemble the one in phage?

Lamerton: It would indeed.

Allison: There are at least two other vertebrate systems in which this seems to be true. In the mammary gland, synthesis of casein in response to hormonal stimulation seems to depend on DNA synthesis (Lockwood, Stockdale and Topper, 1967) and this seems to be so in the prostate as well; and other examples are being investigated. For example V. M. Ingram, studying the erythroblast of the frog during metamorphosis induced by thyroid hormones, finds that DNA synthesis is required before synthesis of adult-type haemoglobin commences. This is interesting in relation to whether some kinds of reprogramming require DNA synthesis.

O'Meara: In considering growth and differentiation one has to make a clear distinction between the physiological state, which is theoretically one of perfection, and the pathological state. Consider the epidermis; mitosis is taking place in the basal layer with extrusion of cells into the spinous layer higher up, according to Professor Iversen. At this stage the cells switch their metabolism from proliferation to maturation, which in the epidermis takes the form among other things of the synthesis of keratin, and this process continues until they reach the horny layer where the residues of the cells are left as strings of keratin. The important thing about this physiological state is that the end point of maturation is death of the cell, and if death fails to take place at the normal stage of the lifespan or cycle the cells that fail to die accumulate. Hence hyperplasia may be seen in the epidermis as a result of some stimulus which has prevented the cells from dying. When this stage is reached a new state is set up in which messages may go back to the basal layer, perhaps to stimulate more frequent mitosis or mitosis in an uninhibited fashion.

On grounds of observation I have arrived at a somewhat similar conclusion to that which Professor Iversen reached with his computer but for a different reason, namely that in a tissue like epidermis the cells may appear indistinguishable morphologically along the basal layer, or a little higher up, yet we know there are differences. For example, some will show Barr bodies not present in others (Barr and Bertram, 1949); moreover heterochromatin differences are found, reminiscent of Caspersson's A and B cells (Caspersson and Santeson, 1942). Differences in metabolic requirements are probable, such as are found in lymphocytes in relation to asparagine. The preferential action which Professor Iversen has mentioned may quite well act on such a varied population in a selective fashion to eliminate some cells and to allow others to develop.

Wolpert: If we look at growth in developing systems, in general we might have to think of two different sorts of control systems for deciding how big something becomes. If you ask how big a tissue should grow, one way is to produce something like a chalone and measure the dilution against some, perhaps external, space; then depending how big this external space is you can control your growth in relation to an external pool. This does not seem to be what happens in all developing systems. For example, if you look at the early development of the limb bud, this grows out in a very reliable way and to a very reliable sort of length. You can take the limb bud at an early stage, cut away the middle portion and it will still grow out to produce a perfectly normal limb. It is as if the limb could measure its own length.

I want to suggest that spatial factors may be crucial in determining whether cells are going to divide or not. There may be something that we can think of as *positional information*. If you have a line of cells which always grows back to a certain length, which seems to happen in the crypt of Lieberkuhn, as Professor Lamerton pointed out, there may be no diffusing substance that is measured against some external space, but rather the cells have some much cleverer mechanism of measuring the length of the line. The concept of positional information suggests that in such a line the cells perhaps number off from one end and number off from the other end; that is, they measure their distance from two ends of the row. If this sort of mechanism were operative, every cell in that line in principle would have information by which it would compute the length of the line. Then if you changed the length of the line, the line would as a whole grow back to the same length. If this sort of idea is true, one could conceive of quite general mechanisms for specifying positional information, and I shall mention this in my paper (p. 241).

Leese: By studying the electrical communication between normal cells with the aid of microelectrode systems Lowenstein and Kanno (1967) have demonstrated that membranes between cells in contact with each other exhibit a much lower resistance to the flow of ions and substances up to molecular weights of approximately 3000 than was hitherto envisaged on the basis of membrane structures. Through modification of those parts of membranes directly involved in the establishment of cell contacts, information in the form of transport of ions and small molecules can apparently be transmitted over distances of 20–30 cell diameters. Moreover, after wounding of epithelia this form of cell communication is lost by intact cells at the periphery of the wound prior to migration and division of cells populating the site of the wound (Loewenstein and Penn, 1967). Communication is not re-established until the wound surface is repopulated and the epithelial cells regain contact with each other. This type of cell communication would appear to be species specific even between cells which otherwise are phylogenetically related (Loewenstein, 1967).

It is of particular interest that in pathological states this type of intracellular communication is absent, for example between cells in hepatocarcinomas (Loewenstein and Kanno, 1967).

This phenomenon would appear to be a demonstration of one mechanism whereby cells may specify positional information.

Stoker: Loewenstein has pointed out that in an enclosed system with this sort of communication any cell can "know" how big the system is, and how far it is from the edge of the system.

Elsdale: The idea of positional information implies that cells are connected together through their membership of communication networks. If such networks are instrumental in controlling the performances of cells in particular morphological situations, clearly the networks must be closed; that is to say, they must include a finite number of cells and exclude all others in their neighbourhood. Cellular specificity can be involved both in the construction of such networks, by determining membership, and also in the operation of such networks by determining the nature of the messages that pass between members. However, if cellular specificity applies a rigorous membership selection there may be nothing further to be gained by having highly specific messages, if unspecific ones will serve as well to compile addresses in such closed networks. This is one reason for thinking that a positional information controlling system may be essentially different from the chalone controlling system—to take this as an example of a system employing a specific molecular stimulus.

These two methods of control imply different possibilities in the situation where cells are dissociated and placed in culture. When control of cellular performance is provided by chalones or other specific and diffusible substances, then dispersal and culture does no necessary violence to the control system; the cells find themselves in the position of zero chalone present and their behaviour might be expected to be predictable on this basis. Members of a communication network subjected to the same treatment, on the other hand, find themselves in a quite different situation; their control system has been physically destroyed, and the individual cells granted independence. How would we expect them to behave? We could hardly be surprised if they behaved differently from formerly, and a high variability may be a feature of their new behaviour. Later I shall talk more about unreliable cells; the question I want to ask now is: do you find the intermitotic times of your cells highly variable or the opposite?

Lamerton: In general cells which have long intermitotic times show very considerable spreads in the length of G_1. This is true, for instance, of mammary gland, the colon and the skin; spermatogonia are somewhat exceptional in having a long intermitotic time with little spread, but here the population is synchronized.

REFERENCES

BARR, M. L., and BERTRAM, E. G. (1949). *Nature, Lond.*, **163**, 676–677.
CASPERSSON, T., and SANTESON, L. (1942). *Acta radiol.*, Suppl. 46.
HANNA, I. R. A. (1967). *Nature, Lond.*, **214**, 355.

IVERSEN, O. H., and BJERKNES, R. (1963). *Acta path. microbiol. scand.*, Suppl. 165, 1–74.
IVERSEN, O. H., and EVENSEN, A. (1962). *Acta path. microbiol. scand.*, Suppl. 156, 1–184.
LALA, P. K., and PATT, H. M. (1966). *Proc. natn. Acad. Sci. U.S.A.*, **56**, 1735–1742.
LOCKWOOD, D. M., STOCKDALE, F. E., and TOPPER, Y. J. (1967). *Science*, **156**, 945.
LOEWENSTEIN, W. R. (1967). *Develop. Biol.*, **15**, 503–520.
LOEWENSTEIN, W. R., and KANNO, Y. (1967). *J. Cell Biol.*, **33**, 225–234.
LOEWENSTEIN, W. R., and PENN, R. D. (1967). *J. Cell Biol.*, **33**, 235–242.
ROBBINS, E., and SCHARFF, M. D. (1967). *J. Cell Biol.*, **34**, 684–686.

CHALONES OF THE SKIN

Olav Hilmar Iversen

Institutt for Generell og Eksperimentell Patologi, Universitetet i Oslo, Rikshospitalet, Oslo 1

THE skin consists of the corium and the epidermis with its appendages. In this paper a brief description of a locally acting growth-regulating system in the epidermis will be given. The belief that such a system exists is founded partly upon the interpretation of the observed regenerative reactions in the epidermis, partly upon experimental evidence obtained from work with skin extracts *in vivo* and *in vitro*, and partly upon hypothetical speculations within the framework of cybernetics and of the theory of hormone action. The epidermal growth-regulating system maintains a constant thickness of the epidermis under normal conditions. Sudden cell damage and cell death induce changes leading to increased proliferative activity of the cells in the basal layer, followed by a transient period with an augmented number of cells—a hyperplasia.

For many years the main interest in our institute has been directed towards epidermal carcinogenesis (Iversen and Evensen, 1962; Skjaeggestad, 1964; Elgjo, 1968a, b). This has led us to study the cell population kinetics in the epidermis (Iversen, Bjerknes and Devik, 1968). We have tried to find out how the growth regulation functions, and how this system reacts to different traumata of a physical and chemical nature, especially how the system reacts during the early stages of chemical carcinogenesis. We have performed biological experiments on the mouse epidermis, and tried to interpret them by means of a mathematical model that can be realized with an electronic computer (Iversen and Bjerknes, 1963).

Our test animals have been hairless mice. The epidermis of these animals is relatively thin and consists of two cell layers and a horny layer. The basal cells are in contact with the basement membrane. Outside these there are some differentiating cells. The relative numbers are about 60 per cent basal cells and 40 per cent differentiating cells. Most externally one finds the horny layer. The skin of the hairless mouse also contains many keratin-filled crypts and remnants of hair. Deeper in the corium some sebaceous glands and cysts (which may be degenerated sebaceous glands or hair follicles) can be found.

When the cell population kinetics of epidermis are studied it is important to work with a relatively simple model. Our model is the so-called "interfollicular epidermis" (Fig. 1) which is defined as a flat piece of epidermis consisting of basal and differentiating cells with the horny layer; but without crypts, hair roots, sebaceous glands, or cysts. We are aware that this is a simplification in comparison with the real epidermis, in the same way that a map is a simplification of the real landscape. But such a model may be useful as a map to find a route to a better understanding of growth control.

The epidermis is a tissue in dynamic equilibrium. From the surface, horny material and some few cells are continuously lost. This loss is compensated for by mitotic activity in the basal layer and by differentiating cells being transformed to keratin. The classical theory stated that when a basal cell divided, the mitosis was differential and resulted in one basal and one

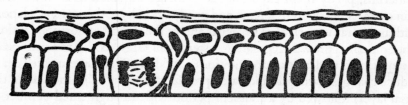

FIG. 1. Schematic drawing of our model of interfollicular hairless mouse epidermis. About 60 per cent of the cells are attached to the basement membrane, about 40 per cent are differentiating cells, and there is a horny layer. A mitosis in a basal cell is seen. This enlarged cell is expelling one of its neighbours out into the differentiating layer.

differentiating cell. Recent investigations (Iversen, Bjerknes and Devik, 1968) have shown that this concept is not true. At least 85 per cent of the cell divisions in the hairless mouse dorsal epidermis lead to two new basal cells. As a secondary result of this a neighbouring cell is expelled out into the differentiating cell layer, possibly because of the increased mechanical pressure among the cells on the basement membrane.

The consequence of this is that in the basal layer there must be two kinds of keratinocytes with different fates. These are the "progenitor cells", which are the cells that after one mitosis will divide again. They belong to the "proliferative pool". In addition there are the cells that after a mitosis will be squeezed out into the differentiating cell layer before having divided. Such cells have been called "non-progenitor" cells. We have found the mean generation time for the progenitor cells in the epidermis of our hairless mice to be 3·5 days. The non-progenitor cells remain on the basal membrane for 2·5 days before they are expelled. The turnover time for all the cells in the epidermis, including the time these cells exist as basal cells, is about 6 days. The turnover time of the horny layer is not fully known, as we do not know how many cell layers it is formed from. Possibly it is about 4–6 days (Downes, Matoltsy and Sweeney, 1967).

ARGUMENTS FOR THE EXISTENCE OF A "CHALONE"

As mentioned, the epidermis is a tissue in dynamic equilibrium. We know that the epidermis is a sensitive tissue, the growth of which is influenced by many factors, namely hormones, nerves, condition of vessels, immunological reactions and external wear and tear. To balance the cell proliferation against the loss under such varying conditions a control system must be operating.

Many theories have been proposed to explain regeneration and wound healing in the epidermis. These theories fall into two large groups: (*i*) theories about possible growth-stimulating factors, so-called "wound hormones"; and (*ii*) theories about growth-inhibiting factors, so-called "chalones". As a result of our experiments we have concluded that the main growth-regulating mechanism must be essentially locally acting, and based upon a feedback principle which is mediated through a signal substance produced in the epidermis itself. A series of observations and theoretical considerations form the basis for this "chalone theory":

(1) *The tendency to hyperplasia.* It is well known that any surface damage of the epidermis leads to a hyperplasia. If one simply increases the outside wear and tear, for instance by rubbing the back of the hairless mouse with sandpaper, the number of cells in the epidermis increases (Iversen and Bjerknes, 1963). Even careful removal of some surface cells by stripping with Scotch tape (Pinkus, 1952) or destroying the upper cell layer with silver nitrate (Oehlert and Block, 1962) leads to an increased proliferation in the basal layer followed by a transient period of hyperplasia. Such an "overshooting" is characteristic of a cybernetic system regulated by negative feedback (inhibition).

(2) *The local type of reaction.* If a child falls and scratches the epidermis on the left knee, the healing processes start only on the left knee; there is no hyperplasia and no rise in mitotic activity on the right knee. The signal for regeneration thus cannot be blood-borne through the general circulation. The proliferative reaction is seen just under the damaged area, and only there. The effect of such superficial wounding, as measured by the proliferative reaction in the basal cell, extends only about 1 mm outside the damage (Bullough, 1962). This is strong support for the theory that the

regulating signal is a locally acting and probably short-lived chemical substance.

(3) *The study of wound healing.* Bullough and Laurence reported in 1960 on the effects of inflicting different sorts of wounds on the skin, in terms of changes in epidermal mitotic activity. The mitotic rate was determined by injecting each mouse, 5 hours before death, with 0·1 mg Colcemid in

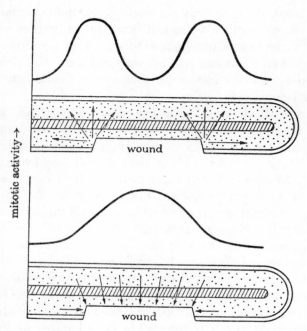

FIG. 2. Diagrams (from Bullough and Laurence, 1960, with permission) to show the regions of high epidermal mitotic activity expected in undamaged ear epidermis opposite an area 3 mm² from which the epidermis and superficial dermis have been removed, on the assumption (*upper diagram*) that a "stimulating wound hormone" is produced by the damaged epidermis, and (*lower diagram*) that the concentration of epidermal inhibitor (chalone) is reduced in the neighbourhood of a wound.

0·25 ml saline. The number of mitoses arrested during 5 hours of Colcemid blockage gives a good estimate of the mitotic rate. One of these experiments was performed with the mouse ear. This is so thin that it could be assumed that a mitotic inhibitor which was freely diffusable would affect even the epidermis on the opposite side of the ear. Fig. 2 shows diagrammatically the situations that would develop after a wound on one side of the mouse ear if (*a*) a stimulating wound hormone is produced by the damaged epidermis (upper diagram); or if (*b*) the concentration of an epidermal mitotic inhibitor is reduced in the neighbourhood of a wound (bottom diagram). The experiments demonstrated that the condition shown in the bottom diagram corresponded best to the facts. From the results of this and other similar experiments the authors proposed the existence of a mitotic

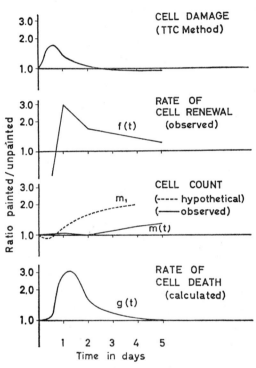

Fig. 3. Observations and calculations of the early effects of a single application of 3-methylcholanthrene to hairless mouse skin. For explanation, see text.

inhibitor substance in the epidermis. Finegold (1965) supported this with his transplantation experiments.

(4) *Observations from the early stages of carcinogenesis.* Experiments in our institute (Iversen and Evensen, 1962) have revealed the following facts (Fig. 3). After a single application of 0·005 ml of a 1 per cent solution of 3-methylcholanthrene (a carcinogen) in benzene to a circumscribed area of the skin of our hairless mice, the following primary reaction occurs: (*a*)

DNA synthesis and the process of mitosis are almost blocked for some hours (Fig. 3, Rate of cell renewal), and (b) the mitochondria of many cells become seriously damaged (Fig. 3, Cell damage). The secondary effects are characterized by the following alterations: (a) The rate of cell proliferation increases 24 hours after application to about three times the normal. It remains high during the next few days, and thereafter returns to normal (Fig. 3, Rate of cell renewal). (b) Many cells which are destroyed by the carcinogen die and are shed during the first, second and third days after application (Fig. 3, Rate of cell death). This is deduced from mathematical calculations based on the experimental findings. Experiments have later demonstrated the correctness of these calculations (Skjaeggestad, 1964). (c) A transient hyperplasia develops after the second day, reaching its peak on the fifth day, after which it slowly dwindles (Fig. 3, Cell count). If the rate of cell renewal is altered as shown in Fig. 3 and the cell loss had remained normal and constant, the hyperplasia would have developed as shown in the hypothetical dotted line labelled m_1 on the third curve from the top. The real hyperplasia, however, followed the curve $m(t)$. This means that the area between the curve m_1 and $m(t)$ is a symbol of the increased cell loss. The rate of this has been calculated in the bottom curve. When we simulated our model on the analogue computer we got good isomorphy between the results obtained on the cybernetic model and what happened with the mouse epidermis (Iversen and Bjerknes, 1963; see also this volume, p. 20). Thus the cybernetic study confirmed and strengthened the theory that the main growth regulation principle in the epidermis was of a cybernetic nature.

In my opinion the theory of a mitosis-inhibiting feedback system within the epidermis itself is the best way to explain these reactions. It is unnecessary to postulate a so-called wound hormone or other stimulating factors, even if we cannot exclude the possibility that such factors also exist.

The local character of the regenerative reaction could theoretically be caused also by a nervous mechanism of the axon-reflex type, but then the borderlines of the reaction would have been irregular like "the red flare" in Lewis's famous "triple reaction" (Lewis, 1927), and have extended much more than 1 mm outside the damaged area.

HISTORICAL BACKGROUND

The theoretical background for this way of thinking was specifically formulated by Weiss and Kavanau (1957) when they published a general theoretical growth model in 1957. The essential features of this model were stated as follows:

"Each specific cell type reproduces its protoplasm by a mechanism in which key compounds ('templates') characteristic of the particular cell type act as catalysts. Each cell also produces specific freely diffusable compounds antagonistic to the former ('antitemplates') which can block and thus inhibit the reproductive activity of the corresponding 'templates'. The 'antitemplate' system acts as a growth regulator by a negative feedback mechanism in which increasing populations of 'antitemplates'

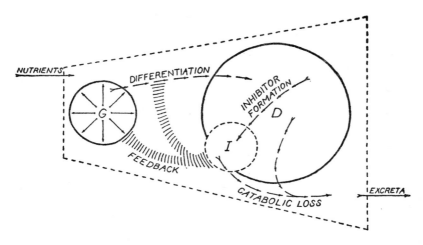

FIG. 4. The model of Weiss and Kavanau (1957, with permission). G = generative mass. D = Differentiated mass. I = inhibiting principle. For further explanation, see text.

render an increasing proportion of the homologous 'templates' ineffective resulting in a corresponding decline of the growth rate. The attainment of terminal size is an expression of a stationary equilibrium between the incremental and decremental growth components and of the equilibrium of the intracellular and extracellular 'antitemplate' concentration."

Fig. 4 shows the main features in the Weiss-Kavanau model. Cells belonging to the generative mass (G) are dividing and some of them are later transferred to the differentiated mass (D), and from there again lost by catabolism or cell death. The differentiated mass produces an inhibitor (I) which, according to Weiss, diffuses back and regulates the rates both of the dividing activity in the generative mass and of differentiation. Weiss and Kavanau used this model to explain the growth curve of chickens and it could also be used to explain the well-known "over-shooting", the hyperplasia when loss of substance is followed by regeneration, already mentioned.

It is interesting to see how time was ripe in the years 1960 and 1961 for applying such a theory to the epidermis. From studies of the changes in cell metabolism in experimental skin carcinogenesis I proposed such a theory for epidermal growth regulation in March 1960 (Iversen, 1960a) and elaborated it at the first International Congress of Cybernetic Medicine in Naples in October 1960 (Iversen, 1960b), as shown in Fig. 5. In the very same year, Bullough and Laurence published their study on wound healing referred to

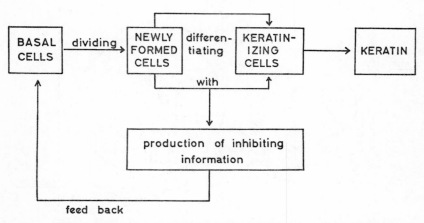

FIG. 5. Simple schematic representation of the model suggested by the present author. The production of inhibiting information is dependent upon the process of differentiation.

earlier (Bullough and Laurence, 1960), and illustrated in Fig. 2. The latter authors concluded: "The evidence shows that, while the mitotic rate of a tissue may be influenced by hormones, it is not controlled by them. The ultimate control of the mitotic rate evidently resides within the tissues themselves. It is obvious that, in appropriate circumstances, the cells of most tissues are capable of indefinite growth and mitosis, and that each tissue specific control mechanism must be antimitotic." These authors also suggested that each tissue produces and contains its own specific inhibitor. In 1961 Mercer proposed a similar theory for epidermis in his book *Keratin and Keratinization*. In 1962 Bullough coined the name "chalone" for substances with the property of acting as tissue-specific mitotic inhibitors. A general definition has been given: "A chalone may be defined as an internal secretion produced by a tissue for the purpose of controlling by inhibition the mitotic activity of that same tissue." The name chalone is derived from a Greek word belonging to the maritime family of words and originally means "to slow down the speed of a boat", "to reef the sails". It must be stressed that this definition is theoretical. The assay systems used to judge a chalone effect can certainly not tell us whether the factor is an "internal secretion" or whether it is there "for the purpose of mitotic inhibition". Our test systems (see below) can only tell us that a certain extract or fraction inhibits proliferation—alone, or in combination with other factors like adrenaline. In the following the practical designation "chalone" will therefore be used for extracts or fractions of extracts of homogenized fresh skin (or other tissues) produced with the methods described. One has to work with the available tests, and from these it is difficult to know whether a certain factor should be called a chalone. The assay systems are also described below.

THE PRODUCTION OF CRUDE EPIDERMAL CHALONE

The most natural next step was to homogenize epidermis, make extracts, and test the effect of these extracts on the mitotic activity of squamous cell epithelium and of other cell types.

Bullough and Laurence published the first results of testing such extracts in 1964. It turned out that simple aqueous extracts of homogenized epidermis contained a mitosis-inhibiting principle. This first extraction was made with mouse epidermis homogenized with an "MSE" glass homogenizer. From our institute, Iversen, Aandahl and Elgjo published a confirmation of these findings in 1965. Our extracts were made from hairless mouse skin. We used the Colcemid technique and an *in vivo* system for determining the mitotic rate. It is seen from Fig. 6 that the mitotic rate is depressed about 50 per cent during the four-hour period which was measured. The chalone solution was injected intraperitoneally.

The extraction of chalone has been performed in different ways and the details of the methods are possibly of importance for interpreting the results. There is therefore need to go into a little detail about the different methods. In our institute chalone-containing crude extracts of skin are now produced in the following way. Hairless mice are killed and the skins immediately flayed off. The skins are then cut into pieces with scissors and mechanically homogenized in a mill which is cooled with liquid nitrogen. The resulting skin powder is extracted in distilled water at $4°$ c. After high-speed centrifuging, all the formed parts are spun down. The clear supernatant contains the mitosis-inhibiting principle. This is lyophilized and it ends up as a

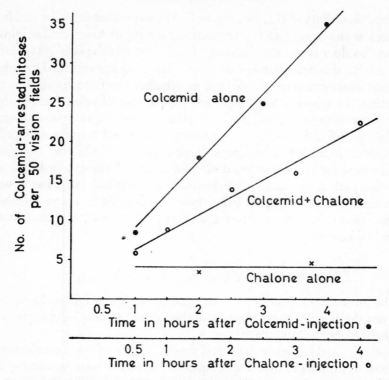

Fig. 6. Mitotic count at different time-intervals after injection with Colcemid alone (*upper curve*), Colcemid+chalone (*middle curve*), and chalone alone (*lower curve*).

powder containing proteins, lipids, carbohydrates, nucleic acids and their derivatives, and salts. This powder is soluble in water and may be injected again into mice. For use in tissue culture it must be filtered in the cold through a Millipore filter, to sterilize it. We have made epidermal chalone in about the same way also from human skin and from codfish skin.

Professor Bullough and his school started cooperation with Dr W. Hondius Boldingh at the Biochemical Research Department, N. V. Organon, Holland. They prepared the chalone-containing crude extract from pig and cod skin using another method (Hondius Boldingh and Laurence, 1968). The main difference is that for most of their preparations they used commercial rind, where the slaughtered pigs were immersed in a water bath at 60°C for about 15 minutes, the hairs were burned off the skin, and the rind (i.e. the epidermis together with some collagenous dermis) was separated. This material was then ground into a fibrous powder after being lyophilized. With this method they found the same effect on mitotic activity as when they (initially) used freshly prepared rind. The cod skin powder was prepared from frozen cod skin, which was processed in a meat cutter, lyophilized and ground as finely as possible, resulting in a coarse powder. From these powders the crude extracts were made by water extraction at 4°C.

PURIFICATION AND CHEMICAL CHARACTERIZATION

Pure chalone has not yet been produced. The most thorough fractionation and purification has been done by W. Hondius Boldingh and Edna Laurence (1968). Various fractionation methods were tried—ethanol fractionation, ion exchange chromatography, gel filtration and column electrophoresis. The most effective method proved to be ethanol fractionation followed by column electrophoresis at pH 3, and then dialysis. Although the biological tests, measuring depression of epidermal mitotic activity, did not give an exact evaluation of the fractions, the results indicate that the active principle of crude skin extracts was purified about 2000 times. Characterization in this way shows that this purified chalone obtained from pig skin is a protein or a glycoprotein which is stable at low pH and has an isoelectric point between 5·2 and 6·8. The most purified material (dialysed) showed a symmetrical peak in the ultracentrifuge corresponding to a molecular weight of 30 000 to 40 000, but it could still be heterogeneous. It is unstable in solution at physiological pH, but more stable at low pH.

We have fractionated the lyophilized skin extracts on Sephadex columns. We have found some activity in the unretarded (high molecular weight?) fraction, but the highest activity in our preparations resides in the retarded fraction and could be a substance of low molecular weight. This corresponds to the characteristics of the granulocytic chalone (Rytömaa and Kiviniemi, 1968a, b).

Many explanations for the discrepancies in behaviour on Sephadex columns are possible. It may be that this type of mitotic control depends upon two factors, a small molecular part (a cofactor?) and a protein part of higher molecular weight which is the basis of the tissue specificity. If so, it could be that through our methods we have managed to separate the two factors. These speculations are at present without any experimental evidence, and many other explanations could of course be given. It is also important that Organon used mannitol during the purification. We have recently found inhibition of mitosis in mouse epidermis by mannitol alone.

But mannitol should not be present in the highly purified (dialysed) Organon preparation. The question of the role of mannitol in these experiments is not yet resolved.

ASSAY SYSTEMS FOR TESTING CHALONE EFFECTS

A most intriguing problem is to find a good assay system for measuring possible effects of chalones. In our institute we have mainly used *in vivo* tests on hairless mice. Chalone solutions of the crude chalone type have mostly been tested, and we have injected into the mouse 0·5 ml of a solution that contains 10 mg protein per ml, or 1 ml of a solution with 5 mg protein per ml. A chalone effect is said to be present if the mitotic rate of the dorsal skin epidermis (measured with the Colcemid method) is reduced by at least 50 per cent by such a procedure. The chalone solutions have been injected either intraperitoneally or subcutaneously.

We have, however, also used small pieces of both mouse skin and human skin *in vitro* in organ cultures (Iversen, 1968; Bullough *et al.*, 1967). It is then necessary to add adrenaline to the culture medium to be able to observe the chalone effect after some hours *in vitro*.

The work in Professor Bullough's laboratory in London has mainly been performed with mouse ear epidermis in short-time organ culture, immediately after the animals have been killed. Small pieces of mouse ear are incubated in a standard buffered solution of saline containing glucose and the test sample. Colcemid is added for the last 4 hours of every experiment. For the routine assay the mitotic rate is measured at the end of a 5-hour period with various concentrations of the preparations to be tested. To ascertain that the inhibition recorded was not due to a non-specific inhibitor Bullough and Laurence used the fact that adrenaline seems to act as a co-factor for the chalone. Every extract that inhibited in the first 5 hours *in vitro* (when endogenous adrenaline is assumed to be present) was retested during a 9-hour period with or without an adrenaline wash after 5 hours. For the latter treatment the pieces of ear were incubated for half an hour in another batch of Warburg flasks in buffered saline solution containing glucose and 0·25 µg per ml adrenaline in the form of the bitartrate. The chalone effect was said to be present in a test sample when the mitotic activity was significantly reduced during the first 5 hours, was normal after 9 hours, but was reduced again if the adrenaline wash was included in the test procedure. A non-specific inhibitor always showed the same inhibition during the whole 9-hour period, and this inhibition was unaffected by adrenaline. This control test for a specific chalone effect can of course not be used in *in vivo* assay systems.

At Birkbeck College in London an *in vivo* test has also been used, very similar to that used in Oslo, but using the ear epidermis instead of the back skin epidermis as the test site (Bullough et al., 1967).

SPECIFICITY OF THE EPIDERMAL CHALONE

The problem of tissue specificity is not easy to decide, because the starting material for all the skin chalone preparations used up till now contained in addition to keratinocytes, melanocytes and Langerhans' cells, also cells from sebaceous glands, hair follicles, connective tissue, endothelium and nerves. But we have started a study comparing extracts of skin with extracts from liver, kidney, lung, heart, brain and intestine, all prepared in the same way.

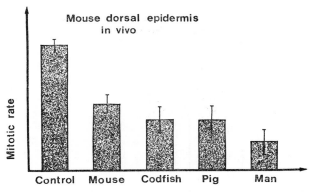

FIG. 7. The effect of skin extracts from mouse, codfish, pig and human on the mitotic rate of hairless mouse dorsal epidermis *in vivo* (Bullough et al., 1967).

These extracts have been tested again on squamous cell epithelium, both in the skin and in the forestomach; and on the intestinal epithelium. The study is not completed, but it seems as if tissue extracts produced similarly to skin chalones, but from liver, kidney, lung, heart, brain and intestine, have no mitosis-inhibiting effect on squamous cell epithelium. On the other side, the skin chalone has no effect on mitotic activity in the small intestine but has an effect on the mouse forestomach squamous cell epithelium. Thus it seems that epidermal chalone is specific for squamous cell epithelium—or that such epithelium is very sensitive to a factor in skin extracts.

Another interesting problem is whether chalones are species specific; that is, whether there is one chalone for human skin, another for mouse skin and a third for pig skin, and so on. Extracts of chalones from human skin, pig skin, mouse skin and from skin of codfish have been tested both *in vivo* in Oslo, and *in vivo* (Fig. 7) and *in vitro* at the Mitosis Research Laboratory in

London (Bullough *et al.*, 1967). We observed that chalones from different sources had a marked mitosis-inhibiting effect on the epidermis. It therefore seems safe to conclude that epidermal chalone is not species specific, nor even class specific, but at least common for man, pig, mouse and codfish.

It is interesting to confirm that chalone, which we know can be produced *from* human skin, has an effect *on* human skin (Iversen, 1968). We were able to get small pieces of fresh human skin from operations and to cultivate these in tissue culture. We had to cultivate the human skin for four days in tissue culture before we reached a stable and countable mitotic rate. We then added chalone, adrenaline and Colcemid. The chalone was partly produced from mouse skin and partly from pig skin. In the controls we

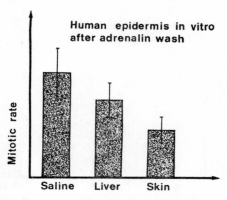

FIG. 8. The effect on the mitotic rate of human epidermis *in vitro* of adding liver extract and skin extract to the culture medium (Iversen, 1968).

added saline or liver extract produced in the same way as we produced chalone from the skin. The results showed that the mitotic rate in human skin can be inhibited *in vitro* by skin chalones from mouse and from pig (Fig. 8). Recently it has been shown that extracts of horny material from human skin inhibit mitosis in mouse epidermis (Born and Bickhardt, 1968).

DOSE–RESPONSE RELATIONSHIP

The question of how to measure the dose of chalone is at present unsolved. In our institute we have used the weight of the lyophilized crude chalone powder as a unit of dose, or the protein content of this powder measured in milligrams. Bullough and Laurence have used a biological definition based on the degree of mitotic inhibition. Thus the unit of dose has hitherto been dependent upon the assay system used.

The dose–response relationship has been tested only preliminarily in our institute on HeLa cells *in vitro*. Fig. 9 shows the results. There is an increase of the response with higher doses, but the slope of the curve is shallow. Similar results (with an even more shallow slope) have been reported by Hondius Boldingh and Laurence (1968) on mouse ear epidermis *in vitro*.

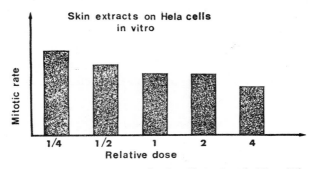

FIG. 9. The effect on the mitotic rate of HeLa cells *in vitro* of adding different doses of skin extract to the culture medium (for details see Amundsen, Elgjo and Iversen, 1970)

POSSIBLE MECHANISM OF ACTION

An interesting question is where in the cell cycle chalone exerts its effect. There seems to be general agreement that the regulation of cellular proliferation in highly differentiated organisms usually takes place in the intermitotic, so-called G_1 phase in the cell cycle (Burns, 1968; Cameron and Cleffmann, 1964). Many believe that the point of regulation is at the so-called "trigger mechanism" for DNA synthesis, even if this most probably is not a definite step in the cell life cycle but comprises many and complicated synthetic processes leading up to DNA synthesis. Such a generalization is supported by observations from many cell types both *in vitro* and *in vivo*.

In tissue culture most cells are arrested in G_1 when the culture stops growing and goes into a stationary phase. This is so if the cause of the halt in cell division is contact inhibition or other unspecific factors. Mature cells that have finished their differentiation are usually diploid (with some exceptions, including many liver cells and some cells in the mouse bladder epithelium); that is, most of them are in the G_1 phase. Variations in cellular proliferation in a tissue or an organ are therefore most often due to prolongation or shortening of the G_1 phase or to alterations in the size of the progenitor cell population.

In early stages of embryogenesis, in some unicellular organisms and in some cell cultures, as for instance of HeLa cells, there is no measurable G_1 phase. A G_1 period is observable parallel to the earliest signs of differentiation to tissues and organs. From a theoretical point of view one may say that this is the time of the earliest need for growth regulation. It seems therefore most probable that the physiological growth regulation mechanism should affect the length of the G_1 phase by affecting the entry of cells into the DNA-synthesizing phase. It is also possible that some growth-inhibiting principles affect G_1 and the dividing phases (DNA synthesis phase, G_2, and mitotic phase) at the same time. There are also observations that some toxic substances have an immediate effect on the mitotic rate

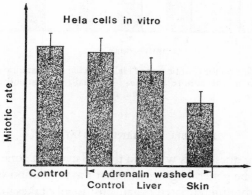

Fig. 10. The effect on the mitotic rate of HeLa cells *in vitro* of an "adrenaline wash", and the addition of liver extracts and skin extracts to the culture medium (for details see Amundsen, Elgjo and Iversen, 1970).

without seriously affecting the DNA-synthesizing phase. Among such substances are the well-known stathmokinetic substances like colchicine and the *Vinca* alkaloids. It is difficult to interpret this other than as a direct action of these substances on mitosis. But if chalone is a physiological growth regulator, it must affect both the rate of DNA synthesis, the transit rate through G_2, and the mitotic rate. If not, a smooth and natural growth regulation cannot occur.

If cells are removed from the epidermis by stripping with Scotch tape there is a lag of 3–4 hours before the increase in proliferative rate is measurable (Iversen and Elgjo, 1967), and in the periphery of a wound the proliferative reaction starts about 24 hours after the wound has been inflicted (Bullough and Laurence, 1960). If epidermal chalone is injected into an animal (Fig. 6) or if it is added to an organ culture of skin, it seems to have an "immediate" effect on the mitotic rate. It is difficult to interpret this other than by assuming that chalone has a direct effect on the entrance of cells into mitosis. But some results, also from our institute (see Fig. 11),

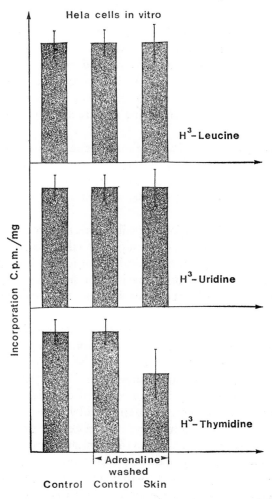

FIG. 11. The effect of skin extract on the incorporation of tritiated leucine, uridine and thymidine into HeLa cells *in vitro*. It is seen that leucine and uridine incorporation is not affected by the skin extract, whereas thymidine incorporation is affected.

show that addition of chalone to tissue cultures of HeLa cells leads immediately to a reduction of the incorporation of [^{3}H]thymidine, whereas the incorporation of tritiated leucine and uridine is not affected.

Rytömaa and Kiviniemi (1968a, b) have demonstrated inhibition of DNA synthesis in cultured granulocytes by adding granulocyte chalone to the medium. Our crude chalone preparations contain small amounts of nucleic acids (of both RNA and DNA nature), but it is unlikely that these low doses of nucleic acids can inhibit DNA synthesis by 50 per cent.

In contrast to these results obtained on cell cultures *in vitro*, Baden and Sviokla (1968) reported that chalone had no effect on DNA synthesis on pieces of epidermis *in vitro*. But there are weak points in this investigation. A chalone effect was found on the mitotic activity of ear epidermis with the Colcemid method but there was no effect on DNA synthesis of newly shaved and plucked dorsal skin, with a supposedly very high proliferative rate.

We have recently found that single and repeated injections of chalone inhibited incorporation of [^{3}H]thymidine into DNA of skin *in vivo* only after a delay of 9–12 hours (Hennings, Elgjo and Iversen, 1969). The chalone level may thus affect the decision of a cell in G_1 to prepare to enter the S phase.

Interesting speculations about possible mechanisms of action at the molecular level can be made in light of modern theories of hormone action (Øye, 1968). According to the classical definition a chalone is not a hormone, but if we use the word hormone in its broadest sense, to mean "intercellular signal molecules", we may include the chalones among the hormones.

About half of the known hormones seem to exert their effect by modifying the protein synthesis of cells by induction or repression. It is now generally accepted that only a small part of the total genetic potency of a cell is active as a template for actual messenger RNA synthesis. Genes may be switched on and switched off as a result of chemical stimuli from the environment (see, for instance, Pitot and Heidelberger, 1963; Tsanev and Sendov, 1966). This is thought to be an effect either on the transcription from DNA to messenger RNA, or on the translation from the nucleic acids to protein. It seems as if some of the sex hormones, corticosteroids, thyroxine and the growth hormone exert their effect in this way. One of the general characteristics of these reactions is that the physiological effect is observable only some time (hours) after the hormone has started to affect the target cells.

If chalone has such an effect, one would expect some hours to elapse before the effect was measurable, and chalone would then have to be either enhancing or activating the physiological repressor to the mitosis operon, or might simply be the repressor of the mitosis operon (Fig. 12).

There are also some indications that chalone may enhance differentiation in cells, and thus may have a blocking or inhibiting effect on the repressor of the functional genes (Fig. 12).

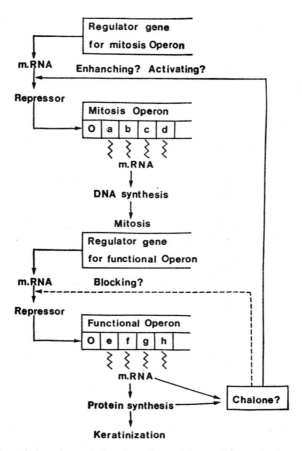

FIG. 12. Speculative schematic drawing of one of the possible mechanisms of action of chalone. It is supposed that the chalone is produced in relation to the process of differentiation (see Fig. 5). If chalone acts by repressing or derepressing operons, it must either enhance or activate the repressor to the mitosis operon, and/or block or inhibit the repressor to the functional operon.

More attractive, however, is the theory that chalone exerts its effect in a similar way to the other group of hormones, namely adrenaline, noradrenaline, serotonin, glucagon, parathyroid hormone, vasopressin, ACTH and the thyrotropic hormone. This was first suggested to me by

Professor F. Seelich of Vienna (personal communication, 1968), and later suggested by Øye (1968). All these hormones are proteins, peptides or derivatives of amino acids, and one of their effects at the molecular level is to activate adenyl cyclase in the cell membrane. Cyclic 3',5'-adenosine monophosphate (cyclic AMP) was discovered by Sutherland and Roll in 1957 (see Sutherland, Øye and Butcher, 1965; Sutherland, Robison and Butcher, 1968). This substance is made from ATP and the process is catalysed by the enzyme adenyl cyclase. Cyclic AMP serves as a mediator or a "signal molecule" in many hormonal mechanisms. Cyclic AMP

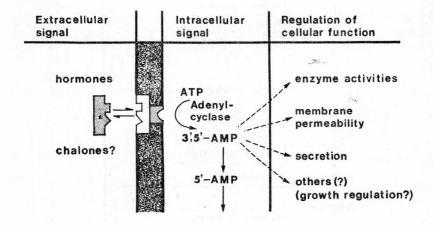

FIG. 13. Schematic drawing of one of the possible mechanisms of action of chalone through the cyclic AMP system (modified from Øye, 1968). For further explanation, see text.

affects many cellular activities and therefore the specificity for these hormones must reside both in the receptor mechanism and in the specific programme of the cells—their differentiation. It is thus the differentiation of the cell and not the hormone itself that determines the physiological effect of these hormones (see Fig. 13). The fact that cell growth *in vitro* can be inhibited by adenosine 3',5'-monophosphate has recently been shown by Ryan and Heidrick (1968).

It is characteristic of this type of action that its physiological results can be seen almost immediately, as is the case with chalones both *in vivo* and *in vitro*. In addition is the fact that chalone needs adrenaline in order to have a measurable effect in long-term skin cultures.

No theories, however, are of value if they cannot form the basis for new experiments. It is possible to design experiments to choose between the two theories mentioned. We have planned such experiments in our institute and shall start them as soon as we find the time ripe; it may be wise, however, to wait until we have a purer and better-defined chalone.

CHALONES IN OTHER SYSTEMS

If a substance like the epidermal chalone exists it is probable that the same sort of substances exist for other, perhaps all, organs and tissues in the body. Saetren (1956) gave some evidence for such substances in liver and kidney. There is a vast amount of recent literature on possible growth-inhibiting factors in liver tissue. Rytömaa and Kiviniemi (1968a, b) have shown that there may be a granulocytic chalone. A similar factor in the rabbit lens has recently been described (Voaden, 1968a, b). Future research will show whether chalones are a group of substances with a mitosis-inhibiting effect and of general occurrence, or whether only a few organs have such systems.

CHALONE AND CANCER

Recently a series of papers discussing the effect of chalone on malignant transplantable tumours and on a transplantable granulocytic leukaemia have been published (Bullough and Laurence, 1968a, b, c, d; Rytömaa and Kiviniemi, 1968c; Mohr et al., 1968). In these papers attention has also been drawn to the possibility of using chalones in treating malignant disease. As a pathologist who has worked with the epidermal chalone for many years I feel that a word of caution must be raised about over-optimistic speculations within this field at the present time. The evidence for such hopeful suggestions is meagre and contradictory and from theoretical considerations it seems unlikely that chalones can be effective in the control of spontaneous tumours.

It is reasonable to assume that the control of growth and differentiation is mediated through many feedback systems consisting of stimulatory, inhibitory and modulating signals that determine the growth rate, the rate of differentiation and the rate of cell loss in the organs. From this point of view carcinogenesis must involve some damage to or destruction of these physiological growth regulation systems, either by the cancer cells losing their ability to respond normally to growth regulation, or by some error in the system itself, for instance in the production or the transfer of the signals.

However, these changes may be secondary and occur as a *result* of the malignant transformation. It is difficult to see theoretically how a high level of chalone can correct this error. Then there is the fact that the growth process as such is only one side of malignancy. The ability of a malignant tumour to infiltrate, to grow destructively and to form metastases is evidently much more important. A therapeutic attack only on the growth process will probably be a failure. The rationale for the use of X-rays, cytostatic agents and hormones in cancer therapy is that the tumour cells are more sensitive than normal cells to these agents. But for chalone the situation is, by definition, quite the opposite. The normal cells retain their normal sensitivity to the growth-controlling signal whereas the cancer cells are less sensitive. Cancer chemotherapy with chalones will theoretically stop the proliferation of the normal cells of that series before it stops the proliferation of the malignant cells. The only possibility for a good effect would be if the mechanism of action of chalones is much more complicated than just inhibition of proliferation, or if it were possible to treat the tumour cells only, a technique that would not be easy to use.

We have as a pilot experiment tested chalone against a highly differentiated transplantable epidermal tumour in hamsters (Chernozemski, 1967). This tumour forms horny pearls, and the desmosomes are easily seen. Four animals with tumours were given small doses of chalone over two weeks, and four other animals were given larger doses for a shorter time. No effect on the tumour growth was observed (Elgjo, personal communication, 1968). We have also tried to use this keratinizing carcinoma as a source of chalone, and so far we have found no effect on the hairless mouse dorsal epidermis *in vivo* with the same doses as used for the skin chalone (Elgjo, personal communication, 1968).

Since we do not know the normal biochemistry and physiology or the chemical constitution of chalones it seems too early to make optimistic statements about chalones and cancer treatment. Every effort should however be directed towards the discovery of the chemical constitution of chalone and its mechanism of action at the cellular and the molecular level. The few experiments published on chalones and cancers warrant no optimistic conclusions about the use of such substances in the treatment of cancer.

There are many interesting perspectives to the chalone work. It is certainly of general interest as a physiological problem of growth regulation, but it may also be relevant to the cancer problem. When it is possible to purify and characterize epidermal chalone and to measure the amount of it per milligram of tissue we shall arrive at a quite new position in epidermal carcinogenesis research.

FACTS AND SPECULATIONS

This paper has given some facts and a lot of speculations. It may be wise to summarize what is fact and what is speculation. I consider it proven beyond doubt that aqueous extracts of homogenized skin contain a factor which, even when given in small doses, has a specific inhibiting effect on the rate of proliferation of squamous cell epithelium. The doses that show mitotic inhibition on squamous cell epithelium have no such effect on intestinal epithelium; and extracts made in the same way from many other organs and used in the same doses have no such effect on the mitotic rate of squamous cell epithelium. The growth inhibition is reversible.

The inhibitory effect can be demonstrated both *in vivo* and *in vitro*. It can be shown directly *in vitro* in short-term cultures, but in long-term cultures adrenaline must be added to the cultures, and hydrocortisone may increase the effect further. This has been used as an argument for a relationship between chalone and adrenaline. This factor found in the supernatant after extraction can be lyophilized and somewhat purified. It seems to contain proteins and sugars but its chemical formula and exact composition are unknown. The factor has an immediate effect on mitosis. It is labile in solution (except at low pH) but may be stored in lyophilized form.

This is what we know. The rest of what I have said consists of more or less well founded speculations, but these may be useful as working hypotheses for further experiments.

SUMMARY

A locally acting growth regulating system in the epidermis is described. This system maintains a constant thickness of the epidermis under normal conditions. Sudden cell damage and cell death introduce changes leading to increased proliferative activity of the cells in the basal layer, followed by a transient period with an augmented number of cells.

Many observations from experimental and clinical pathology point to such a locally acting growth regulating system in the epidermis. These observations include: the tendency to development of hyperplasia, the local character of the reaction to damage, the study of wound healing and observations from the early stages of carcinogenesis.

All these reactions can be explained within the framework of a cybernetic growth-regulating system in the epidermis utilizing a chemical messenger as the inhibiting feedback substance. Experiments have shown that aqueous extracts of skin contain a chemical principle that can be partly purified and which seems to act as a physiological inhibitor of mitosis in squamous cell epithelium, but with no such effect on other tissues.

Biological and chemical work on this substance is reviewed, and some speculations about the possible role of this principle in carcinogenesis are given.

Acknowledgements

The work on skin chalones in our institute has been performed as a team effort with Drs. K. Elgjo, O. Nome, and recently H. Hennings. Dr. E. Amundsen has done the *in vitro* work with HeLa cells. Dr. W. Hondius Boldingh, N. V. Organon, Holland, has kindly provided partially purified pig skin chalone, and has also been helpful with the preparation of this manuscript.

I also thank the other members of our "chalone club" (W. S. Bullough, E. B. Laurence, U. Mohr, T. Rytömaa, and others) for cooperation and valuable discussions during a conference in Oslo in July 1968.

REFERENCES

AMUNDSEN, E., ELGJO, K., and IVERSEN, O. H. (1970). *Cell Tissue Kinet.*, in preparation.
BADEN, H. P., and SVIOKLA, S. (1968). *Expl Cell Res.*, **50**, 644–646.
BORN, W., and BICKHARDT, R. (1968). *Klin. Wschr.*, **46**, 1312–1314.
BULLOUGH, W. S. (1962). *Biol. Rev.*, **37**, 307–342,
BULLOUGH, W. S., and LAURENCE, E. B. (1960). *Proc. R. Soc. B*, **151**, 517–536.
BULLOUGH, W. S., and LAURENCE, E. B. (1964). *Expl Cell Res.*, **33**, 176–194.
BULLOUGH, W. S., and LAURENCE, E. B. (1968*a*). *Nature, Lond.*, **220**, 134–135.
BULLOUGH, W. S., and LAURENCE, E. B. (1968*b*). *Nature, Lond.*, **220**, 137–138.
BULLOUGH, W. S., and LAURENCE, E. B. (1968*c*). *Europ. J. Cancer*, **4**, 587–594.
BULLOUGH, W. S., and LAURENCE, E. B. (1968*d*). *Europ. J. Cancer*, **4**, 607–615.
BULLOUGH, W. S., LAURENCE, E. B., IVERSEN, O. H., and ELGJO, K. (1967). *Nature, Lond.*, **214**, 578–580.
BURNS, R. E. (1968). *Cancer Res.*, **28**, 1191–1196.
CAMERON, I. L., and CLEFFMANN, G. (1964). *J. Cell Biol.*, **21**, 169–174.
CHERNOZEMSKI, I. N. (1967). *J. natn. Cancer Inst.*, **39**, 1081–1087.
DOWNES, A. M., MATOLTSY, A. G., and SWEENEY, T. M. (1967). *J. invest. Derm.*, **49**, 400–405.
ELGJO, K. (1968*a*). *Europ. J. Cancer*, **3**, 519–530.
ELGJO, K. (1968*b*). *Europ. J. Cancer*, **4**, 183–192.
FINEGOLD, M. J. (1965). *Proc. Soc. exp. Biol. Med.*, **119**, 96–100.
HENNINGS, H., ELGJO, K., and IVERSEN, O. H. (1969). *Virchows Arch. path. Anat. Physiol., Abt. B. Zellpathologie*, in press.
HONDIUS BOLDINGH, W., and LAURENCE, E. B. (1968). *Europ. J. Biochem.*, **5**, 191–198.
IVERSEN, O. H. (1960*a*). *Acta path. microbiol. scand.*, **50**, 17–24.
IVERSEN, O. H. (1960*b*). In *Proc. First International Congress of Cybernetic Medicine*, pp. 420–430, ed. Masturzo, A. Naples: Società Internazionale di Medicina Cibernetica.
IVERSEN, O. H. (1968). *Nature, Lond.*, **219**, 75.
IVERSEN, O. H., AANDAHL, E., and ELGJO, K. (1965). *Acta path. microbiol. scand.*, **64**, 506–510.
IVERSEN, O. H., and BJERKNES, R. (1963). *Acta path. microbiol. scand.*, Suppl. 165, 1–74.
IVERSEN, O. H., BJERKNES, R., and DEVIK, F. (1968). *Cell Tissue Kinet.*, **1**, 351–367.
IVERSEN, O. H., and ELGJO, K. (1967). In *Control of Cellular Growth in Adult Organisms*, pp. 83–92, ed. Teir, H., and Rytömaa, T. London and New York: Academic Press.
IVERSEN, O. H., and EVENSEN, A. (1962). *Acta path. microbiol. scand.*, Suppl. 156, 1–184.

Lewis, T. (1927). *The Blood Vessels of the Human Skin and their Responses*, pp. 322. London: Shaw.
Mercer, E. H. (1961). *Keratin and Keratinization*. London: Pergamon Press.
Mohr, U., Althoff, J., Kinzel, V., Süss, R., and Volm, M. (1968). *Nature, Lond.*, **220**, 138–139.
Oehlert, W., and Block, P. (1962). *Verh. dt. Ges. Path.*, **46**, 333–340.
Øye, I. (1968). *Forskningsnytt (Oslo)*, **13**, 59–61.
Pinkus, H. (1952). *J. invest. Derm.*, **19**, 431–446.
Pitot, H. C., and Heidelberger, C. (1963). *Cancer Res.*, **23**, 1694–1700.
Ryan, W. L., and Heidrick, M. L. (1968). *Science*, **162**, 1484–1485.
Rytömaa, T., and Kiviniemi, K. (1968a). *Cell Tissue Kinet.*, **1**, 329–340.
Rytömaa, T., and Kiviniemi, K. (1968b). *Cell Tissue Kinet.*, **1**, 341–350.
Rytömaa, T., and Kiviniemi, K. (1968c). *Nature, Lond.*, **220**, 136–137.
Saetren, H. (1956). *Expl Cell Res.*, **11**, 229–232.
Skjaeggestad, Ø. (1964). *Acta path. microbiol. scand.*, Suppl. 169, 1–126.
Sutherland, E. W., Øye, I., and Butcher, R. W. (1965). *Recent Prog. Horm. Res.*, **21**, 623–646.
Sutherland, E. W., Robison, G. A., and Butcher, R. W. (1968). *Circulation*, **37**, 279–306.
Tsanev, R., and Sendov, B. (1966). *J. theoret. Biol.*, **12**, 327–341.
Voaden, M. (1968a). *Expl Eye Res.*, **7**, 313–325.
Voaden, M. (1968b). *Expl Eye Res.*, **7**, 326–331.
Weiss, P., and Kavanau, J. L. (1957). *J. gen. Physiol.*, **41**, 1–47.

DISCUSSION

Roe: What is the evidence that your material is diffusible?

Iversen: Born and Bickhardt (1968) in Germany have produced chalone from keratinized material only and it affects the basal layer of the cells. Also, in organ cultures it must diffuse into the intact piece of skin to inhibit proliferation. *In vivo* it affects the epidermis after intraperitoneal injection.

Roe: It is clear from what you say that the active principle in extracts can diffuse *into* cells, but is there evidence that under *in vivo* conditions it can diffuse *out of* cells? Perhaps its extraction from tissues is dependent on cells being broken up.

Iversen: We have never really been able to prove that it comes out of cells in your meaning, because to obtain it we have to crush the cells.

Roe: This might be open to investigation by comparing the activity of extracts obtained after different degrees of homogenization.

Iversen: This is an interesting point, but we are now concentrating on trying to fractionate and purify the material before we go further along other lines. We are also working to find a better and less time-consuming assay system that can be used to measure the effect of the fractions.

Bergel: Did the extraction process used at Organon include a dialysis procedure?

Iversen: Yes, in the last step. And then the activity remained in the material inside the dialysing membrane.

Allison: The mannitol would be dialysible.

Iversen: Yes. Organon has produced only very little of the highly purified preparation, and it has been tested only in very few experiments.

Vernon: What is the *minimum* amount of the very highly purified material required in tissue culture to get a response? I gathered that an appreciable fraction of a milligram is necessary; that is a large dose for a highly specific effect. If you are to suppose that chalone is part of a control mechanism, the cells that produce the chalone have to do so at quite a rate, and it has to get across the cell membrane. May I make one more point? I noticed that on the centrifugation plate there was a fast-moving impurity. How do you know that your active material was the major component? Have you attempted to fractionate further?

Iversen: If we can judge molecular weight from the Sephadex column results, our highest activity was found in the region of a 3000 molecular weight molecule only. So one speculation is that chalone is a big protein molecule which may determine the tissue specificity, combined with a small general mitotic inhibitor molecule. So when we use 0·1 mg of the Organon chalone preparation in 4 ml medium we add the whole large molecule, but the real dose of the mitotic inhibitor could be very much less.

Stoker: Professor Iversen has given a very good statement of the situation. One of the most important aspects is the specificity. Does mannitol have any specificity of effect?

Iversen: We came upon the idea that mannitol alone might be inhibiting mitosis in epidermis *in vivo*. High doses did so, but not doses similar to what is added to the Organon chalone preparations. This has also been tested in many early experiments by W. S. Bullough and E. B. Laurence (unpublished results).

Stoker: You seemed to get some effect of the liver extracts on HeLa cells, which I imagine you used as an example of what is presumably an epithelial-derived malignant cell. Have you tried sarcoma cells?

Iversen: We are planning to do this. Dr C. Mittermayer of the Pathologisches Institut in Freiburg has just told me that his preparation of epidermal chalone has an inhibitory effect on fibroblasts in tissue culture. There are, however, small amounts of DNA, RNA and their derivatives in all chalone preparations, and cell cultures are very sensitive to such substances. Cell culture may not be a good assay system at the present time, because it is so sensitive to all types of growth inhibitor.

DISCUSSION

Stoker: Dr R. Bürk (1968) has reported a difference between the adenyl cyclase levels of normal and transformed cells, and this was based on the idea you also showed in your summary hypothesis, namely that cyclic AMP is an important "slowing" substance, and that the difference between normal and transformed cells might be shown in the levels of adenyl cyclase. This was borne out by the observation of a reduction of adenyl cyclase activity in the transformed cells.

Subak-Sharpe: Did I see correctly in your Fig. 7 that mouse chalone has the least effect on mouse dorsal epidermis when compared with pig, human and codfish chalone?

Iversen: This appears to be so because when we did that experiment we had not agreed upon how to determine the dose, and the doses stated are in fact based upon an estimate of how much starting material was involved. I would therefore put no real emphasis on the fact that human skin seems to contain more chalone than mouse skin.

Burke: It may be important how the effects were measured—whether you measured the decrease in the number of mitotic figures or a decrease in incorporation of labelled thymidine. In the latter case one would have to guard against an effect on the cell membrane which prevents thymidine entering the cell. Secondly, have any attempts been made to characterize the active material, even broadly, by the use of specific enzymes—ribonuclease, deoxyribonuclease and proteases—to see whether it falls into any of these major categories of macromolecules?

Iversen: With the exception of the labelling experiment illustrated in Fig. 11 we have used only the Colcemid method for measuring the number of cells arrested in metaphase per hour per 1000 cells. We have just started to use labelling methods.

On the question of specific enzymes, Hondius Boldingh and Laurence (1968) mention that they used two enzymes, pepsin and trypsin, but nothing else has been reported. Chalone was resistant to pepsin but destroyed by trypsin.

Abercrombie: If chalone operates in wound-healing it surely must stop cells entering DNA synthesis, rather than merely stopping them entering mitosis, as has been shown for epidermis.

Iversen: Yes. At first we were unable to show any reduction of [^{3}H]thymidine incorporation into DNA in epidermis *in vivo* by chalone. But more recently we have found that chalone does inhibit incorporation, after a delay of 9–12 hours (see Hennings, Elgjo and Iversen, 1969).

Wolpert: How long does the effect of chalone last when you inject it into the animal? This would give one some idea of the half-life of the chalone *in vivo*. Also, when you inject chalone into the peritoneal cavity and it gets to the skin, the chalone must be diffusing into the skin from the blood, so presumably it must be able to diffuse from the skin into the blood. Is there any evidence for it being present in serum?

Iversen: As far as I know the effect of chalone has not been measured for longer than 5 hours *in vivo* and 9 hours *in vitro*. So it may last longer, but it has not been shown to do so. I think the half-life is only 3–4 hours *in vivo*. In answer to your second question, if chalone can diffuse into skin it must also be able to diffuse out, and then some of it must be present in serum, but this has never been tested, as far as I know.

Bergel: May I raise the question of the role of adrenaline in the action of chalone. You mentioned using adrenaline-washed material. What does adrenaline do? Can it be replaced by other catecholamines, and is it solely in its dihydroxy state? What for instance would adrenochrome, an oxidation product of adrenaline, do, or noradrenaline, or dopa?

Iversen: I don't know. The dose of adrenaline used in the "washing" is really very high, about 100 times the dose you are allowed to give to man. This adrenaline wash therefore implies a very high concentration of adrenaline. As far as I know the other catecholamines have not been tested.

REFERENCES

BADEN, H. P., and SVIOKLA, S. (1968). *Expl Cell Res.*, **50**, 644–646.
BORN, W., and BICKHARDT, R. (1968). *Klin. Wschr.*, **46**, 1312–1314.
BÜRK, R. R. (1968). *Nature, Lond.*, **219**, 1272–1275.
HENNINGS, H., ELGJO, K., and IVERSEN, O. H. (1969). *Virchows Arch. path. Anat. Physiol., Abt. B. Zellpathologie*, in press.
HONDIUS BOLDINGH, W., and LAURENCE, E. B. (1968). *Europ. J. Biochem.*, **5**, 191–198.

NERVE GROWTH AND EPITHELIAL GROWTH FACTORS

C. A. VERNON, BARBARA E. C. BANKS, D. V. BANTHORPE, A. R. BERRY, H. ff. S. DAVIES, D. MARGARET LAMONT, F. L. PEARCE AND KATHARINE A. REDDING

Departments of Chemistry and Physiology, University College, London

INTRODUCTION

NERVE GROWTH FACTOR

Bueker discovered in 1948 that replacement of the limb primordium in three-day chick embryos with a fragment of mouse sarcoma 180 led to invasion of the tumour by sensory nerve fibres and enlargement of the spinal ganglia from which the fibres originated. He concluded that the effect was a consequence of some unknown physicochemical properties of the tumour. Levi-Montalcini and Hamburger (1951) repeated Bueker's experiment and found that both sensory and sympathetic ganglia were involved. They suggested that the sarcoma contained a specific nerve growth-promoting agent, subsequently known as nerve growth factor or NGF. An important development was the discovery that more potent sources of NGF exist, namely a variety of snake venoms (Cohen and Levi-Montalcini, 1956) and certain salivary glands, particularly the submaxillary gland of the male mouse (Cohen, 1960). These discoveries were possible because a relatively simple *in vitro* assay system had been devised. In this system sensory or, less commonly, sympathetic ganglia dissected from chick embryos, usually about seven days old, are cultured in a semisolid medium consisting of chicken plasma and embryo extract. The presence of NGF produces a dense, fibrillar halo of nerve fibres. The assay is difficult to quantify and it is usual to score each particular test on an arbitrary scale from 0 to $+5$, a score of $+3$ being regarded as a strong response.

Over the last decade a large amount of work has been reported, much of it from Levi-Montalcini and her colleagues, concerned with (a) the chemical nature of NGF, (b) its effects *in vitro* and (c) its effects *in vivo*, particularly with chick embryos and neonatal mice and kittens. We have attempted to summarize this work below: for this purpose the views of Levi-Montalcini and her colleagues are taken to be those expressed in a recent review (Levi-Montalcini and Angeletti, 1968).

Chemical nature of NGF

It was early suspected that NGF is protein in nature and this is now firmly established: NGF behaves like a protein in purification procedures and becomes inactive under conditions known to degrade proteins. Most studies have been made with mouse salivary gland as source and it is perhaps surprising that the basic chemical characteristics of the molecule are still a matter of dispute. The earliest purification procedure was devised by Cohen (1960) and, although somewhat ill-defined, apparently led to a protein which sedimented in the ultracentrifuge as a single peak, $S_{20} = 4\cdot 33$ s.* The protein was found to be active in tissue culture (that is, to give a $+3$ response) at a concentration of 10^{-8} g ml^{-1}. It might be supposed that the high level of biological activity and the ultracentrifugation result could be taken as good evidence of homogeneity. However, in the NGF field these criteria have proved inadequate since (a) much higher levels of activity have been reported and (b) NGF appears to form relatively stable aggregates with inert proteins. A modification of Cohen's procedure apparently leads to a protein with a smaller sedimentation coefficient ($S_{20} = 2\cdot 36$ s). Levi-Montalcini and Angeletti (1968) have argued that since the two proteins have the same biological activity they may represent dimeric and monomeric forms. However, more recently (Zanini, Angeletti and Levi-Montalcini, 1968) both forms have been reported to be active at 10^{-10} g ml^{-1}.

The purification of NGF from mouse salivary gland has also been studied by Shooter and his colleagues (Varon, Nomura and Shooter, 1967a, b, 1968; Smith, Varon and Shooter, 1968). They obtained a high molecular weight protein, $S_{20} = 7\cdot 1$ s, which could be dissociated into three subunits ($S_{20} = 2\cdot 6$ s) only one of which possessed biological activity. Since dissociation resulted in loss of activity (by a factor of about 3) which could be restored by reaggregation these workers proposed that the inactive subunits had a regulatory function possibly mediated through allosteric effects. However, the reported changes in activity seem too small to be detected by standard assay methods. Furthermore, Fenton and Edwards (1969) have repeated the purification procedure of Shooter and his colleagues and although they obtain the reported high molecular weight aggregate, they find that it can be fractionated to give a smaller entity of considerably *greater* biological activity.

The situation is further confused by a claim of Schenkein and co-workers

---

* There has been a distressing tendency in this field to suppose that molecular weights can be derived from values of sedimentation coefficients alone. For example, Cohen's protein is said to have a molecular weight of about 44000. We shall place little reliance on such estimates and, for the most part, quote only the measured values of sedimentation coefficients.

(1968) that two different NGF entities can be obtained from mouse salivary gland, one of which, although admittedly inhomogeneous, is active at the incredibly low concentration of 10^{-16} g ml^{-1}. Comment on this particular claim, which entails the view that activity in tissue culture can be produced by only a few hundred molecules of NGF, must wait confirmation by other workers.

The purification of NGF from snake venoms has not been studied extensively. Cohen (1959) obtained a protein from the venom of *Angkistrodon piscivorus* which sedimented as a single component in the ultracentrifuge ($S_{20} = 2 \cdot 2$ s) and which was active at about 10^{-7} g ml^{-1}. A similar preparation has been reported by us (1968) but whether it represents the pure active entity is, for the reasons given above, not known. Multiple forms of NGF have been claimed to be present in the venom of *Crotalus adamanteus* (Angeletti et al., 1967) but full experimental details have not yet been given.

In vitro *effects*

Most of the experiments both *in vitro* and *in vivo* so far reported have been carried out with partially purified NGF from mouse salivary gland. This material when used in the conventional tissue culture assay system (that is, with chick embryonic dorsal root ganglia in a plasma clot) gives a dose–response curve with a well-defined maximum. The decrease in response at high concentration was originally attributed to an inhibitory effect (Levi-Montalcini, 1964) but it now appears that histological examination has shown that with excess NGF the nerve fibres grow in a different way and form a dense fibrillar capsule around the ganglia (Levi-Montalcini and Angeletti, 1968). The response of the dorsal root ganglia to NGF in tissue culture has been reported to vary with the age of the embryo: Levi-Montalcini (1965) writes "*In vitro* as *in vivo*, the growth response is maximal in ganglia from 7 to 9 days incubation, decreases in the following three days and is no longer detectable after the 14th day of incubation".

No very consistent account of the *in vitro* effects of NGF has yet emerged. Levi-Montalcini and Angeletti (1968) appear to favour two main mechanisms by which NGF produces its effects. Firstly, it is supposed that NGF induces differentiation of neuroblasts into neurons. This idea suffers from the disadvantage that differentiation is, in this context, difficult to define. If what is meant is that the neuron differs from the neuroblast in having a fibre, then this constitutes simply a restatement of the experimental observation, namely that NGF induces cells to grow fibres. If, on the other hand, some more complex process is envisaged, the details of it are not yet apparent.

Secondly, it has been claimed that neurons obtained by dispersion of embryonic dorsal root ganglia with trypsin do not survive, even in the presence of serum, unless NGF is added to the culture medium. This suggests that NGF plays a vital role in the survival of the neurons. Levi-Montalcini and Angeletti (1968) comment on this by saying that it argues against the idea that "NGF merely serves as a source of cell proteins and suggests instead that it acts in a catalytic fashion".

In vivo *effects*

Levi-Montalcini and Booker (1960) showed that daily injection of NGF into newborn mice produces a marked increase in the volume of the sympathetic para- and pre-vertebral chain ganglia. The interesting feature of this effect is that it is highly specific; no other cell type and, in particular, no other nerve cells, are affected. Serial sections, particularly of the superior cervical ganglia, are said to show (Levi-Montalcini and Angeletti, 1968) that the effects of NGF are (a) to increase mitotic activity, (b) to increase the number of nerve cells and (c) to increase the size of the nerve cells. An important development in this work came from the use of an antiserum to NGF (Cohen, 1960). It was found that injection of the antiserum to newborn mice and rats caused extensive destruction of the sympathetic ganglia and the term "immunological sympathectomy" was coined to denote a technique for producing animals deprived of a considerable part of their sympathetic system.

Similar findings with rats have been reported by Zaimis and her colleagues (Berk, Filipe and Zaimis, 1965). They found almost complete destruction of the paravertebral ganglia, considerable atrophy of the coeliac ganglia but a rather small effect on the mesenteric ganglia. The subject has been recently reviewed by Zaimis (1967).

EPIDERMAL AND EPITHELIAL GROWTH FACTOR

In the course of experiments designed to test the effects of NGF on neonatal mice Cohen (1962) noticed that, in addition to the characteristic increase in size of the sympathetic ganglia described above, precocious opening of the eyelids and precocious eruption of the incisors occurred. Since these effects tended to disappear as the purity of the NGF increased, Cohen deduced that they were due to some other factor. On the basis of histological evidence Cohen concluded that the primary process is enhancement of keratinization and overall thickness of the epidermis and he accordingly called the factor epidermal growth factor (EGF). Purification studies indicated that EGF is a heat-stable protein with a molecular weight in the region of 15 000. Subsequently Jones (1966) reported that EGF appeared to stimulate the growth of epithelium in a wide variety of organs, including prostate, uterus and vagina, in organ culture, and he suggested that the factor should be renamed epithelial growth factor.

STUDIES ON NGF FROM THE VENOM OF *Vipera russelli*

PURIFICATION OF NGF

Purification of NGF essentially consisted of three steps: (a) passage through G-150 Sephadex, (b) passage through G-75 Sephadex and (c) ion-exchange chromatography on a strongly basic resin (Cellex-T) at high pH (10·8). The overall yield was about 0·2 per cent but the specific activity showed an increase, almost entirely due to step (c), by a factor of about 10^6. This large increase could only be obtained by ion-exchange chromatography under highly basic conditions and would seem to imply that NGF is normally associated with some other protein which masks its activity. The highly basic character of the finally separated material is, of course, consistent with this idea (Banks *et al.*, 1968).

LEVEL OF ACTIVITY

The most active sample prepared by the method described above elicited fibre growth from seven-day-old chick embryonic dorsal root ganglia in hanging drop culture at a concentration of 10^{-14} g ml^{-1}. At this concentration only about 10^4 *molecules* of NGF are present in the assay system and this implies that a response from a given cell can be invoked by an amount of NGF which is of the order of one molecule. We originally reported (Banks *et al.*, 1968) that the very active material was unstable particularly when stored in solution. We are now inclined to believe that this is not the case but that the assay system is intrinsically irreproducible at high levels of activity. Levi-Montalcini and Angeletti (1968) have discussed the difficulties involved in comparing the values found in different laboratories for the activity of NGF *in vitro* assays and, in particular, have pointed out that the common practice of using the same pipette for serial dilutions could give spuriously high levels of activity through a "carry-over" effect.* This, of course, is true but it is also true that the use of a series of different pipettes could give the reverse effect if NGF is readily adsorbed on to glass surfaces. Since NGF is undoubtedly positively charged at near neutral pH this is very

---

* In fact the authors say that when this practice was abandoned "only traces of activity were found in dilutions higher than 10^{-8} μg NGF/ml of medium". This, of course, is the same level of activity as reported by us and is considerably higher than those previously quoted.

likely to be the case and may explain the irreproducibility in measuring the level of activity observed by us with very active samples.

Two other points must be considered before we attempt to compare activity levels from different laboratories. Firstly, the activity recorded may be sensitive to small changes in the assay procedure. This is of particular relevance in our own case since our assay is carried out on collagen and not with the more usual plasma clot. The dose–response curve does not, with the collagen assay, show a well-defined maximum although a rather stunted type of fibre outgrowth is frequently seen at very high concentrations. Secondly, the NGF from mouse salivary glands is unlikely to be chemically identical with that from the venom of *Vipera russelli* and it is possible that the two substances do not produce precisely the same biological effects. There is some evidence (see below) that this is so.

PROPERTIES OF NGF

The protein showed a single peak on sedimentation analysis in the ultracentrifuge, $S_{20} = 2 \cdot 84$ s (mean of seven determinations). Diffusion analysis showed that the areas under the synthetic boundary curves were constant with time and led to a value of the diffusion constant of $D_{20} = 6 \cdot 58 \times 10^{-7}$ cm^2 sec^{-1} (mean of seven determinations). The protein is, therefore, homogeneous with respect to ultracentrifugation analysis and, assuming a partial specific volume of $0 \cdot 735$, its molecular weight is found to be 40000. No significant changes in the values of the sedimentation and diffusion constants were found on dilution (from $1 \cdot 3$ per cent to $0 \cdot 04$ per cent) or on change of pH (from $10 \cdot 3$ to $7 \cdot 5$).

Passage through a calibrated Sephadex column (pH $10 \cdot 8$) gave two active peaks, one, the major component, corresponding to a molecular weight of approximately 24000 and the other to a molecular weight of 38000. The simplest interpretation, namely that two species of NGF are present, is precluded by the ultracentrifugation results. It seems possible that the effect arises from subunit structure but a detailed interpretation is far from obvious. It should be pointed out that the values of the sedimentation and diffusion constants entail that the molecule has an unusual shape—in particular, that it is rod-like rather than spherical. It may be that under these circumstances behaviour on gel filtration is not immediately interpretable.

BIOLOGICAL EFFECTS *in vitro*

Explants of dorsal root ganglia

Experiments with radioactive phosphate. It has been shown (Forbes, 1963) that cytotoxic agents can produce an increased loss of ^{32}P from cells grown in tissue culture and that this increase can provide an index of cell death. Accordingly intact dorsal root (sensory) ganglia from embryos that had been previously injected with [^{32}P]trisodium orthophosphate were grown in glass vials and the loss of radioactivity into the culture medium was measured. The ratio of the activity remaining in the ganglia to the total activity in the culture (ganglia+medium) was compared for cultures treated with NGF and for controls. Results are summarized in Table I.

Table I

LOSS OF RADIOACTIVITY (^{32}P) FROM CULTURES OF SENSORY GANGLIA TREATED WITH NGF *in vitro*

Time (hours)	(G/T)*	
	NGF-treated culture %	Control culture %
0·5	84	83
6	50	50
12	47	45
18	40	34
24	39	30
30	33	23
36	31	22
42	27	14
48	23	14
54	18	13
63	17	10

* G is the percentage radioactivity in the ganglia and T the total radioactivity in the culture.

It is clear that less radioactivity is released into the medium in the presence of NGF and this might be taken as an index of decreased cell death. However, the difference between the control and NGF-containing media, although significant, is not large and the results cannot be used to support the view that the major effect of NGF is to prevent cell death.

Variation of response to NGF with age of ganglia. Ganglia from embryos of age 5 to 16 days were dissected and tested for response to NGF in tissue culture under standard assay conditions. Some of the ganglia from older embryos were cut in half before testing. The results are given in Table II. Each score is the mean of twenty to thirty independent observations.

The results show that all the ganglia responded *in vitro* to NGF. On the differentiation theory the response should be greatest at 5 days and should progressively decrease with age. This is not what is observed and the differentiation theory cannot, consequently, be held as the sole explanation of the mode of action of NGF.

Effect of change of media on the growth of fibres. Ganglia from seven-day-old embryos were maintained in either a control medium (C) or a medium

Table II
VARIATION OF RESPONSE TO NGF WITH AGE OF GANGLIA

Age (days)	Whole ganglia Mean score		Difference
	Control	Plus NGF	
5	1·6	4	2·4
6	1·2	4·6	3·2
7	2·0	7·3	5·3
12	2·9	5·2	2·3
14	1·3	4·9	3·6
15	2·2	4·3	2·1
16	2·7	4·8	2·1
	Half ganglia		
15	3·3	4·7	1·6
16	2·7	4·8	2·1

containing NGF (N) for either 24 (C or N) or 48 (2C or 2N) hours at 37·5° C. It is important to notice that conditions were chosen so that no fibre growth occurred. The ganglia were then cultured in conventional hanging-drop cultures for 24 hours either in a control medium (C) or in a medium containing NGF (N). The fibre growth was scored in the usual way. The experiment involved eight different sets of conditions and within each set twenty-six independent observations were made. The results, which were analysed by a standard statistical procedure (analysis of variance), showed that the order of effectiveness of the various sets in promoting fibre growth was as shown in Table III. Furthermore, the sets could be arranged in groups such that the differences between corresponding members, for example C → N, N → C, were highly significant—that is, $P < 0·001$.

Table III
EFFECT OF CHANGE OF MEDIA ON FIBRE GROWTH

Group	Set
1	2N → N
	N → N
2	C → N
3	2N → C
	2C → N
	N → C
4	C → C
	2C → C

This somewhat complex, but very important experiment, leads to the following conclusions:

(a) Maximum fibre growth occurs when NGF is present throughout the experiment, that is, both in the medium in which no growth can occur and in the medium in which growth does occur. (Group 1 is significantly better than all other groups; group 4 is significantly worse.)

(b) When only one of the media contains NGF, it is better to have the NGF in the second medium when the ganglia are in the first medium for the shorter time (C → N significantly better than N → C) but the advantage disappears when the ganglia are in the first medium for the longer time (2N → C not different from 2C → N).

It is clear that the differentiation theory alone cannot account for these results since differentiation would have been extensive in those ganglia maintained for the longer period in the NGF-containing medium and the treatment 2N → C would have given results differing from those of treatment 2C → N.

Experiments on cell suspensions made from dorsal root ganglia

A method has been developed for dispersing the dorsal root ganglia and thereby obtaining a mixed suspension of fibroblasts, supporting cells and sensory neurons. The following experiment was done with this technique.

Approximately three hundred and fifty spinal ganglia were dissected from 8-day chicken embryos and pooled. They were dispersed into single cells by treatment with pronase and collected in a solution containing 199 medium and buffered saline solution (2:1; total 3 ml). This suspension was made homogeneous by gentle mechanical agitation and a sample withdrawn and counted under a haemocytometer. The cell density was adjusted to be about half that considered to be optimal for the survival of nerve cells. The cell suspension was then dispensed, 100 μl at a time from a micropipette, into glass rings waxed singly on to collagen-coated coverslips contained, in pairs, in sterile Petri dishes. The cell suspension was continually agitated throughout the procedure and the cell density was checked for consistency at the half-way and end stages. Finally serum or serum plus NGF was added to the cultures so that each Petri dish contained a control and a treated culture (199 medium:serum:buffered saline or NGF—2:1:1). In this way 28 identical cultures, save for the addition or omission of NGF, were prepared. The Petri dishes were numbered at random and incubated at $37°c$ in an atmosphere of 5 per cent carbon dioxide in air, saturated with water vapour. At 4 hr, 10 hr, 24 hr, 34 hr, 48 hr, 54 hr and 72 hr, two Petri dishes were selected at random and removed.

The rings were cut free and the cultures were stained with methylene blue vital stain (0·01 per cent w/v in saline), fixed (ammonium molybdate 8 per cent w/v, $4°c$) and mounted.

Accurate cell counts were then made on each culture. Nerve cells were

3*

identified by their characteristic shape and strong affinity for the vital stain. Usually counts were made upon sixty randomly selected fields (400 × final magnification) for each culture. The numbers of nerve cells, and other, supporting cells, per field were recorded separately. The actual area so counted

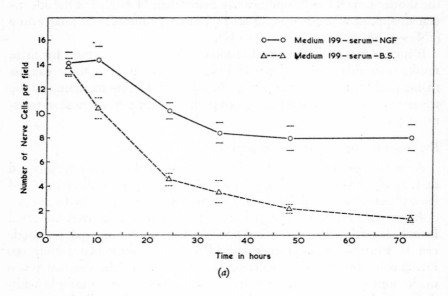

(a)

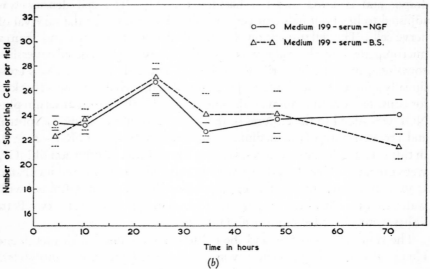

(b)

FIG. 1. Death rate of (a) nerve cells and (b) supporting cells in tissue culture in presence and absence of NGF. For experimental details see text.

was about one-fifth of the total culture area. The counts were statistically interpreted and the mean values of the number of nerve cells and supporting cells per field were calculated, with their standard errors, for each culture. The results are represented graphically in Figs. 1a and b. The error bars represent the composite error of the pipetting and counting and each point is the mean value of two cultures. In this way the behaviour of the nerve cells and the supporting cells with increasing time was studied.

It is apparent that, under the unfavourably low cell density deliberately used in this experiment, extensive death of nerve cells occurs and that death is most rapid between four and twenty-four hours in culture. However, there is markedly less death in cultures treated with nerve growth factor. There are two explanations of this result: the NGF may be acting directly to maintain the viability of the nerve cells, or neuronal death in treated cultures may be balanced by differentiation to recognizable nerve cells from a less differentiated pool. It is seen, however, in Fig. 1b that the behaviour of the supporting cells, particularly in the earlier stages, is independent of added NGF. Thus the large difference in the curves cannot be accounted for by removal and conversion of undifferentiated cells.

The experiment thus strongly indicates that the targets for NGF action are those cells which are sufficiently differentiated to be recognized as nerve cells. These cells have a typical appearance and strong affinity for methylene blue vital stain. They may or may not possess neurites. It has also been shown, by allowing the dispersed neurons to regenerate fibres in plasma clots, that the degree of branching of the fibre is unaffected by the presence of NGF. The average length of fibres is, however, altered by the presence of NGF and may, under suitable conditions, be twice the value observed in control experiments.

Conclusions from the in vitro *experiments*

The following conclusions emerge from the *in vitro* experiments:

(a) The immediate effect of NGF is to trigger and maintain the production of a fibre. The process does not involve cell division (since it is independent of the presence of colchicine) and has no obvious connexion with differentiation. The effect can be produced by a very small amount of NGF but it is not necessarily permanent since in the experiments in which ganglia were successfully maintained in two media more fibres were grown when both media contained NGF.

(b) The presence of NGF *in vitro* decreases the mortality of neurons from dorsal root ganglia. This emerges unambiguously from the results of the experiments with dispersed cultures and is consistent with the results obtained with whole ganglia. It would seem to follow that triggered cells have the greater viability *in vitro* and the observations made with dispersed cultures strongly support this.

(c) It is also necessary to suppose that untriggered cells are most liable to die under conditions where fibre growth is possible (i.e. in a suitable solid support). This follows from the experiment carried out with whole ganglia maintained successfully in two different media where it was found that the treatment C → N was better than N → C. A subsidiary effect, seen with dispersed cells, is that the apparent rate of fibre growth is increased by the presence of NGF. However, this may be a consequence of (a); that is, of the tendency of triggered cells to lose NGF and, thereby, to cease fibre growth.

BIOLOGICAL EFFECTS *in vivo*

Some experiments on kittens have been carried out by Professor Zaimis and her colleagues (1968). In these, 0·1 mg of pure material from the venom of *Vipera russelli* was injected into kittens once per day over a period of three days. Half this amount was then injected daily over the next five days. The kittens were killed on the ninth day and the catecholamine contents of various organs were determined. The appropriate sympathetic ganglia was also examined histologically. A parallel series of experiments at a dose rate ten times greater was simultaneously carried out using mouse salivary gland NGF prepared by the method of Shooter and his colleagues* (Varon, Nomura and Shooter, 1967a,b, 1968; Smith, Varon and Shooter, 1968).

One difference immediately became apparent: the snake venom NGF is much more toxic than that from mouse salivary gland. In fact the difference in the dose given was determined by this factor. Nevertheless, qualitatively, the effects on the catecholamine content of organs and on the sympathetic ganglia were very similar. However, detailed histological examination of the sympathetic ganglia showed that this qualitative similarity may be misleading. Ganglia from animals treated with mouse salivary gland NGF showed considerable hypertrophy of the cells together with an increase in the amount of fibres. With the snake venom NGF the cell hypertrophy was much less pronounced but there appeared to be a very large increase in fibre material. Whether the differences seen reflect some significantly different mode of biological action is, as yet, unknown.

* Material supplied by Dr. D. C. Edwards of the Wellcome Laboratories.

DO THESE FACTORS HAVE A HOMEOSTATIC ROLE?

It is tempting to suppose that NGF and, possibly, EGF, have a normal physiological function. The high specific activity of the purified materials, the selectivity of both the *in vitro* and the *in vivo* effects, and the fact that normal serum exhibits NGF activity all argue in this direction. However, no physiological function has yet been established and it is clear that considerably more work on these substances is necessary.

SUMMARY

Bueker first discovered that certain mouse sarcomas caused a proliferation of sensory but not of motor fibres on implantation into chick embryos. Levi-Montalcini and her co-workers established that the effect is due to the production of a specific protein and that similar proteins are present in larger amount in certain snake venoms and in mouse submaxillary gland. Much work has been published with the aims of establishing (*a*) the nature of the proteins responsible (now invariably called nerve growth factors), (*b*) the mode of action both *in vitro* and *in vivo*, and (*c*) whether or not the nerve growth factor plays any part in a physiological control process. The present status of these problems is reviewed.

Recently nerve growth factor has been obtained from the venom of Russell's Viper in a form which is homogeneous to both sedimentation and diffusion analysis. The substance is a basic protein with a molecular weight of 40000. It is active, *in vitro*, at a concentration of 10^{-14} g ml^{-1} and this corresponds to activity of the order of one molecule per cell. Its effect is apparently to instruct the neuroblast to grow a fibre. This process which, probably incorrectly, might be called differentiation, does not involve cell division. In the absence of NGF the cell tends to die. The relevance of these findings to those of other workers and to function *in vivo* will be discussed.

So-called epidermal growth factor (EGF) was discovered by Cohen who found that a factor, distinct from NGF, was present in mouse salivary gland which caused premature opening of the eyelids and precocious eruption of upper and lower incisors in neonate mice. Cohen claimed that the factor is a heat-stable protein with a molecular weight of approximately 15000. However, the nature of this substance and its relation to epidermal growth has remained obscure. Jones has claimed that Cohen's factor causes the proliferation of a large variety of epithelial tissues: he, therefore, prefers to call it epithelial growth factor. The present position of this subject is reviewed.

Acknowledgements

The authors wish to thank the Whitehall Foundation, New York, for generous financial support, and Miss Janet Cook and Miss Janet Drew for their technical assistance. Cooperation and discussion with Dr. D. C. Edwards and Professor E. Zaimis is also gratefully acknowledged.

REFERENCES

ANGELETTI, P. U., CALISSANO, P., CHEN, J. S., and LEVI-MONTALCINI, R. (1967). *Biochim. biophys. Acta*, **147**, 180–182.
BANKS, B. E. C., BANTHORPE, D. V., BERRY, A. R., DAVIES, H. ff. S., DOONAN, S., LAMONT, D. M., SHIPOLINI, R., and VERNON, C. A. (1968). *Biochem. J.*, **108**, 157–158.
BERK, L., FILIPE, I., and ZAIMIS, E. (1965). *J. Physiol., Lond.*, **177**, 1P–2P.
BUEKER, E. D. (1948). *Anat. Rec.*, **102**, 369–390.
COHEN, S. (1959). *J. biol. Chem.*, **234**, 1129–1137.
COHEN, S. (1960). *Proc. natn. Acad. Sci. U.S.A.*, **46**, 302–311.
COHEN, S. (1962). *J. biol. Chem.*, **237**, 1555–1562.
COHEN, S., and LEVI-MONTALCINI, R. (1956). *Proc. natn. Acad. Sci. U.S.A.*, **42**, 571–574.
FENTON, E. L., and EDWARDS, D. C. (1969). Personal communication.
FORBES, I. J. (1963). *Aust. J. exp. Biol. med. Sci.*, **41**, 255–264.
JONES, R. O. (1966). *Expl Cell Res.*, **43**, 645–656.
LEVI-MONTALCINI, R. (1964). *Ann. N.Y. Acad. Sci.*, **118**, 149–168.
LEVI-MONTALCINI, R. (1965). *Archs, Biol. Liège*, **76**, 387–417.
LEVI-MONTALCINI, R., and ANGELETTI, P. U. (1968). *Physiol. Rev.*, **48**, 534–569.
LEVI-MONTALCINI, R., and BOOKER, B. (1960). *Proc. natn. Acad. Sci. U.S.A.*, **42**, 373–384.
LEVI-MONTALCINI, R., and HAMBURGER, V. (1951). *J. exp. Zool.*, **116**, 321–362.
SCHENKEIN, I., LEVY, M., BUEKER, E. D., and TOKARSKY, E. (1968). *Science*, **159**, 640–643.
SMITH, A. P., VARON, S., and SHOOTER, E. M. (1968). *Biochemistry, N.Y.*, **7**, 3259–3268.
VARON, S., NOMURA, J., and SHOOTER, E. M. (1967a). *Proc. natn. Acad. Sci. U.S.A.*, **57**, 1782–1789.
VARON, S., NOMURA, J., and SHOOTER, E. M. (1967b). *Biochemistry, N.Y.*, **6**, 2202–2209.
VARON, S., NOMURA, J., and SHOOTER, E. M. (1968). *Biochemistry, N.Y.*, **7**, 1296–1303.
ZAIMIS, E. (1967). *Sci. Basis Med. Ann. Rev.*, 59–73.
ZAIMIS, E. (1968). Personal communication.
ZANINI, A., ANGELETTI, P. U., and LEVI-MONTALCINI, R. (1968). *Proc. natn. Acad. Sci. U.S.A.*, **61**, 835–842.

DISCUSSION

Allison: What is the position on the protease effects of these growth factors? Bryn Jones and M. J. Ashwood-Smith tell me that their purest *epithelial* growth factor has protease activity, and I am told the same is true of the Levi-Montalcini group's preparation of NGF.

Vernon: Our NGF does not have protease activity. The preparation of Shooter has protease activity but it is localized in one of the subfractions. It is probably true to say that the amount of work on purification of so-called EGF is rather small, and it is possible that it may prove to be an unspecific substance. The situation is much more confused than with the NGF.

Finter: You said that when injected into the yolk sac, as much as 500 μg of NGF are needed to show activity in the chick embryo. Is this because it must be transported into the embryo itself?

Vernon: This is very likely. I should add that the mortality rate in our eggs is quite high when we inject this amount; we lose about half of them. Obviously there is a transport problem, and of course the egg has proteolytic activity so that the actual concentration that reaches the dorsal root ganglia of the embryo is quite unknown.

Burke: How much NGF is present in snake venom, and is there any significance in its presence there?

Vernon: In our Russell's Viper venom the NGF is about 0·2 per cent of the freeze-dried material. I fancy other snakes have less, but it is difficult to assay crude venom because of the toxins, which interfere with the assay. I don't know its significance, except that I am told by zoologists that the venom sac of a snake is a modified salivary gland.

Gros: I gather that specific anti-NGF antibodies have been used against normally growing nervous tissues; what is the outcome of these experiments?

Vernon: The results were rather confused. You can make an antiserum against the mouse salivary gland NGF and this has an effect *in vivo*, but the effect is entirely on certain parts of the sympathetic nervous system. It was originally implied that you could pretty well destroy the sympathetic system in this way, and the phrase "immunological sympathectomy" was invented. Professor E. Zaimis has since characterized the effects quite carefully, and she finds that damage to the sympathetic system is not so extensive as was originally thought, and that the sympathectomy is only partial (Berk, Filipe and Zaimis, 1965). We don't know what the antiserum to the snake venom NGF will do; we very much hope that Dr D. C. Edwards of the Wellcome Laboratories is going to make some and test it.

Gros: Has anyone attempted to localize the NGF, by say immunofluorescent techniques?

Vernon: People have tried this. We have labelled our NGF preparation with fluorescein isocyanate and it labels without loss of activity. If you now put this into a cell suspension from dorsal root ganglia you can persuade yourself that the fluorescence goes into the neurons and not into the fibroblasts. But I am not very happy about this because these neurons fluoresce strongly anyway. We have also labelled NGF with tritium and tried autoradiography on the grounds that this is a more sensitive technique, but the results are not yet conclusive.

Möller: I would like to ask some general questions. What is the evidence against the possibility that stimulatory agents have a nutritional effect? This possibility is particularly relevant in tissue culture systems, where it is likely that certain factors necessary for cell growth are lacking. Secondly, what evidence is there that the growth factors act specifically on the target cells? Thirdly, what is the evidence that any of these factors has a homeostatic function? Have experiments been done to reduce the production of NGF or to prevent its action, to see whether it has a homeostatic function in any system?

Vernon: It is definitely not a nutritional effect, because the activity is too great. When you are in the region of a few thousand molecules per tissue culture it cannot be a nutritional effect. And remember that when the factor is purified it comes out as pure protein. The activity can be destroyed by treating with trypsin, indicating that the total molecule, with its protein structure, is needed, which argues against it being a nutritional effect.

On the specificity: in fact NGF is specific, and very oddly so. You can induce a very strong response from dorsal root ganglia and sympathetic ganglia but nothing else. We find no other tissue with the same response at all, so NGF is very highly specific.

Whether NGF has any part in homeostasis is quite unknown. One can only say that we know that it is present in normal individuals. It has a specific effect and is highly active, so one would suspect that it has a regulatory function, but to prove it would be very difficult.

Abercrombie: I have a suggestion as to how NGF might act which is purely speculative. Assume that NGF is produced by fibroblasts; hence its presence in sarcomas and in granulation tissue (Levi-Montalcini, 1964). If NGF is consumed by sympathetic nerve fibres, then in normally innervated connective tissue its concentration would be kept low whereas in an injured area, filled with fibroblasts and blood-vessel sprouts, but lacking nerve fibres, the NGF concentration will rise. Hence growth of surrounding nerve fibres into the area will follow. This would not be exactly a homeostatic mechanism but rather a mechanism by which one component of the tissue is adjusted to match other components.

Bergel: This seems to be more a kind of anatomical or histological homeostatic device.

Stoker: Dr Möller, what do you really mean by a nutritional effect?

Möller: If a factor which is necessary for cell growth is lacking in a system, the addition of this factor will stimulate growth. Thus, the factor appears to be growth stimulating but is actually only *necessary* for growth; this is a nutritional effect.

Vernon: This is very true, but on the other hand most vitamins act as cofactors for enzymes and they are required at a much higher level.

Bergel: There exists a case of molecule-to-molecule interaction; the anti-eggwhite-injury vitamin biotin, and avidin, a protein of eggwhite, combine to form a stable complex which prevents the biotin from exercising its growth-promoting activity.

One thing is rather strange and it could be a coincidence: growth factors such as NGF have something to do with nerve tissue in which adrenergic, and, if acting as a receptor site, sympathomimetic action in the widest sense, takes place. Since chalones are seemingly linked up with adrenaline, which is one of the sympathetic effector agents, some tenuous relation between NGF and chalones may exist.

Möller: May I add a point which would serve to distinguish nutrition from specific induction. Antibody production is an excellent example of specific induction. The responding cells are equipped with receptors for the antigen. Upon contact with the receptor, the antigen initiates antibody synthesis and division. If the receptor is first blocked by a hapten, which by itself cannot stimulate the cell, the complete antigen is no longer capable of stimulation. A nutritional effect could not be blocked, however, because it lacks specificity.

Bergel: I recall that when you attempted to block the immune receptors of leukaemic cells you had to have proteins of a given size present in order to bridge the two points of attachment.

Möller: That is correct.

Burke: Another way in which a nutritional effect could be distinguished from specific induction would be to see whether it was restricted to differentiated cells or not.

Vernon: I don't really know what differentiation means! But we have been worried about this. We see mitotic figures in the cultures of ganglia for the most part only in cells which we classify as fibroblasts. We have blocked this mitosis by appropriate doses of substances like colchicine, and the response to NGF is precisely the same as before—nerve fibres grow out. You may want to call "differentiation" the change in a cell in which the information which instructs it to grow fibres is blocked, to a state in which it is apparently unblocked and grows a fibre. But it doesn't seem to be a very good word for it because the change is much more like the phenomenon of induction of enzymes in bacteria, whereas differentiation describes a total process and has the implication that cell division comes in, which the NGF effect does not require, as far as we can see.

REFERENCES

BERK, L., FILIPE, I., and ZAIMIS, E. (1965). *J. Physiol., Lond.*, **177**, 1P–2P.
LEVI-MONTALCINI, R. (1964). *Ann. N.Y. Acad. Sci.*, **118**, 149.

"WOLFF FACTORS" FROM CHICK EMBRYO MESONEPHROS AND LIVER OR YEAST

J. Mason, Em. Wolff and Et. Wolff

Laboratoire d'Embryologie Expérimentale, Collège de France, Nogent-sur-Marne

Two lines of human malignant tumours have been maintained in organotypic culture for some years in the Laboratoire d'Embryologie Expérimentale of the Collège de France by Professor Etienne Wolff and Dr. Emilienne Wolff (Et. Wolff and Em. Wolff, 1961, 1964, 1965).

The first, Z 200, originated from a hepatic metastasis of a tumour of gastric origin. Macroscopically the line presents a translucent homogeneous appearance; in section the compact structure with numerous mitoses is seen. This line has now been in organotypic culture for 7 years (Fig. 1a).

The other line, Z 516, came from an epithelioma of the descending colon and has a quite different macroscopic aspect. Deep convolutions give a superficial cerebral appearance and mucus is secreted which distends the whole explant. This line has been cultured organotypically *in vitro* for $5\frac{1}{2}$ years (Fig. 1b).

The fragments were originally taken from patients during surgical interventions and put straight into organotypic culture, using the following technique. Fragments of tumour were closely associated with pieces of mesonephros about 0·2 mm in diameter on the vitelline membrane from a non-incubated hen's egg. The vitelline membrane was then placed on a semi-solid agar medium composed of 1 per cent agar in Gey's solution, $8\frac{1}{2}$-day chick embryo juice and horse serum, in the proportions 10:4:4. The vitelline membrane was then folded over the mosaic of tissues. In these conditions the explants soon fuse and form a continuous plate, the cancer cells growing at the expense of the mesonephros. The cultures are transferred to new media weekly, the explants being cut into several pieces at each transfer and reassociated with new mesonephros fragments.

Cell-to-cell contact is in fact unnecessary for survival and growth, since it was found that the tumour explants could be separated from the mesonephros fragments by a dialysing membrane, vitelline membrane of non-incubated eggs or by cellophan membranes. In these conditions the tumour explants continue to proliferate normally, doubling or trebling their size between each weekly transfer. This enables them to be cut into 2–3 fragments at each transfer.

The stock is now maintained by this technique, the tumour explants and the mesonephros being separated in successive folds of the same vitelline membrane.

It should be noted at this point that the basic medium alone (embryo extract and horse serum) is not sufficient to support the growth of the two lines. On basic medium the explants seldom live beyond 2–3 transfers, without increasing in size, and this despite the fact that the medium has been shown to be sufficient for the survival and differentiation of embryonic organs in the same conditions.

It seemed therefore at this point that the embryonic organs associated with the tumour explants were elaborating substances essential for the survival and proliferation of the cancer cells but not present or present in insufficient quantities in the basic medium of embryo extract and horse serum. It should be noted that other embryonic organs such as gonads, liver and metanephros are able to support the survival of the tumours *in vitro* but are less favourable to their proliferation than mesonephros (Et. Wolff and Em. Wolff, 1962).

It was evident that the factors elaborated by the embryonic tissues were able to pass the vitelline membrane in sufficient quantities to support the high growth rate of the explants. It was therefore decided to attempt the isolation of these factors by making crude extracts of $8\frac{1}{2}$-day chick mesonephros or embryonic liver. The biochemical analysis of these factors was done in collaboration with Y. Croisille and J. Mason. The extracts were added to the basic medium, replacing the living embryonic organs that had previously been associated with the fragments; the new proportions of the media were 11:4:4:1 (agar: embryo juice: horse serum: embryo extract). These media proved to be toxic but an extract of an acetone powder of brewer's yeast (*Saccharomyces cerevisiae*) was able to stimulate growth indefinitely (Et. Wolff, Em. Wolff and Croisille, 1965).

Subsequently, it was shown that dialysates of the crude organ extracts lacked this toxicity. Both these dialysates and also the dialysate of the yeast extract were extremely favourable to the growth of the cultures. The lines were maintained on the dialysates for 41 weeks and 47 weeks (embryonic liver and mesonephros) and 180 weeks (yeast).

The Visking tubing used for dialysis has a permeability limit of about 15 000 molecular weight, so we could assume that the active factors were below this weight, which excluded complex proteins. Further proof of the

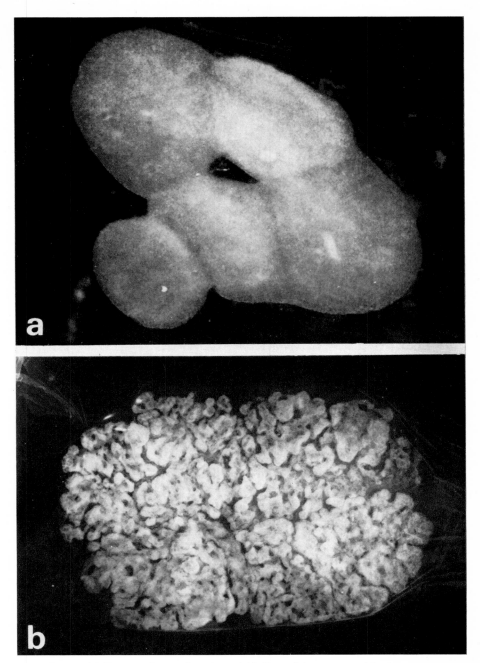

FIG. 1. (a) Z 200 explants cultured on chick liver dialysate. Magnification ×22. (b) Z 516 explants cultured on chick liver dialysate. The enveloping vitelline membrane can be seen. Magnification ×10.

To face page 76

relatively simple nature of the activity came from the fact that boiling the extracts for 1 hour at pH 7·2 in no way decreased their activity.

Since the supply of material for the embryonic liver and mesonephros extracts posed a difficult problem we decided to continue fractionation of the yeast extract and of a dialysate of 3-month-old chick liver, which was found also to be highly active. The tumour explants have now proliferated for 140 weeks (transfers) on this dialysate of chick liver. The fractionation procedures adopted will be described only for the 3-month chick liver dialysates, although as far as has been tested the activity of the yeast extract is of the same order.

As a first step the dialysate was adsorbed on to a column of the cation-exchange resin Amberlite 120 in the uncharged form. The unattached "acidic" and "neutral" compounds were washed off with distilled water; the adsorbed compounds were then displaced with 10 per cent ammonia solution. Both fractions were concentrated under reduced pressure and the ammonia removed in this way. The water-eluted fraction was shown to be completely inactive while the ammonium-displaced fraction supported growth for 36 weeks before the cultures were abandoned, still growing well.

As a next step the fractions were adsorbed on to a column of Amberlite 400 (an anion exchanger) in an uncharged form. The unattached ions were again washed off with distilled water and the adsorbed ions displaced with 2N-acetic acid. The fractions were concentrated under reduced pressure and the acetic acid eliminated. Fig. 2 shows the procedure. It can be seen that only the "ampholyte" fraction (so named because of its attachment to both cation and anion exchangers) proved active and supported full proliferation for 36 weeks before we stopped the series. All the other fractions behaved in a similar manner to the controls on basal medium; that is, they supported mere survival for a few transfers (Croisille *et al.*, 1967).

It is immediately evident that the amino acids are important constituents of the active fraction. However, synthetic media containing amino acids, such as medium C (Et. Wolff *et al.*, 1953) or more complete media such as Parker's medium 199, have up to now failed to promote growth in the same way when added to the basic medium at various concentrations.

As a further step the "ampholyte" fraction was separated on a column of Sephadex G-10 using water as an eluent (Fig. 3). The fraction labelled Dis A was shown to be active and supported full growth and proliferation for 33 weeks before the culture series was stopped. In the conditions in which Sephadex was used, aromatic compounds tend to be adsorbed on to the column and are eluted at stages later than would have been expected from their molecular weight. All the fractions other than Dis A were inactive, but it should be noted that all the ninhydrin-positive material fell in this Dis A fraction.

It was decided to use differential solvent fractionation as the next step in the separation of the active constituent(s). Twenty ml of Dis A were dried under reduced pressure to give a well-dispersed film on a 500 ml round-bottomed flask. The residue was then extracted (at 55°C) with twenty

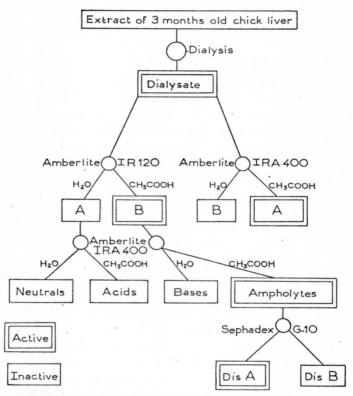

FIG. 2. Procedure for purification of the 3-month-old chick liver dialysate on ion-exchange resins and Sephadex columns.

volumes of acetone containing 9 ml of concentrated hydrochloric acid per litre. This was repeated three times for 20 minutes each time, the flask being constantly rotated. After washing with pure acetone and drying, the residue was extracted with 20 volumes of butanol at 70°C for 20 minutes. This was repeated once. The residue was then dried and taken up in water.

A first series of experiments, which should be regarded as preliminary, has given the following results. The final water-soluble residue proved inactive, the cultures surviving only 8 weeks, while fractions soluble in butanol and acetone-HCl contained the activity. The butanol-soluble substances were able to maintain the cultures for 28 weeks while the acetone-soluble fraction maintained cultures for 17 weeks. Thin-layer and paper chromatography showed that the acetone-soluble fraction contains the bulk of the amino acids, whereas the only amino acid present in important quantities in the butanol-soluble fraction is lysine, although small traces of other amino acids also present in the acetone-HCl fraction were noted.

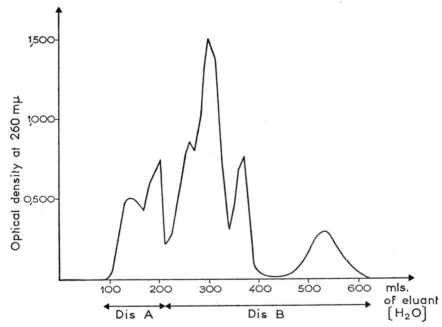

FIG. 3. Separation of the ampholyte fraction on Sephadex G-10.

The addition of lysine to the basic medium in various concentrations has never resulted in the continued growth of the tumour explants. Thus the additional amino acids provided by embryo extracts do not seem essential for growth, although a complete separation of growth-promoting activity from the presence of amino acids has not been achieved. Although the active factors may not be amino acids, an adjuvant role cannot be eliminated since neither of the two extracts supported the indefinite growth normally seen with extracts of embryonic issues.

In an attempt to define the chemical nature of the active principles, with the protein nature of some known growth regulators in mind, a hydrolysate of the "ampholyte" fraction (see Fig. 2) was prepared. The fraction was heated in 6N-hydrochloric acid in a sealed tube for 24 hours and the hydrochloric acid was then removed by repeated evaporation to dryness under reduced pressure. The residue was taken up in water and the pH adjusted to 7·2. This fraction proved to be highly active, the cultures being stopped in a healthy condition after 28 transfers. It thus seems that a protein or polypeptide active principle can be discounted.

The hydrolysate was analysed for amino acids and the concentrations were determined. No unusual ninhydrin-positive peaks were noted and, as can be seen from Table I, the hydrolysate contains no unusual amino acids.

Table I

COMPARISON OF THE AMINO ACID CONCENTRATION OF THE AMPHOLYTE FRACTION OBTAINED FROM CHICK LIVER AND THE SYNTHETIC MEDIUM TESTED

	Concentrations of amino acids	
	In synthetic medium mg/100 ml	In hydrolysate m-mole/100 ml
L-lysine HCl	159·0	0·87
L-arginine	8·71	0·05
DL-methionine	80·6	0·27
DL-histidine	118·0	0·31
L-glutamic acid	268·0	1·82
DL-aspartic acid	343·6	1·29
L-proline	99·0	0·86
L-cysteine	15·8	0·09
DL-phenylalanine	115·6	0·36
L-alanine	143·8	1·61
L-leucine	152·0	1·16
DL-isoleucine	147·0	0·56
DL-threonine	181·0	0·76
DL-valine	300·0	1·28
L-glycine	112·5	1·50
L-tyrosine	37·1	0·20
L-serine	223·0	1·06
L-ornithine	44·0	0·21

A synthetic medium was prepared to contain the same amino acids in the same quantities (Table I) and this was compared in culture with the hydrolysate. It should be noted that where only DL-amino acids were available, as shown in Table I, the concentrations were doubled to bring the final L-amino acid concentration to the level found in the hydrolysate.

The results again showed that the hydrolysate was highly active, whereas cultures on basic medium plus the corresponding synthetic amino acids did not survive more than six transfers.

CONCLUSIONS AND SUMMARY

Despite considerable purification, the active fractions in embryonic organs are still complex and the nature of the active principles still unknown. However a certain number of substances can be discounted, including nucleotides, nucleosides, free bases, thiamine, riboflavin, pyridoxine and folic acid, all these being eliminated during the fractionation procedures. The activity must be due to a relatively simple molecule or molecules resisting acid hydrolysis at 105°C. Even if this represents only a small part of the original molecule the results show that this carrier molecule must have a molecular weight of less than 15000. Finally, the behaviour on ion-exchange columns would seem to indicate that the molecule or molecules possess several functional groups.

The "Wolff factors" responsible for growth regulation and the maintenance of differentiation *in vitro* of organized tumours therefore remain mysterious. The evidence is still mainly negative and we can exclude large numbers of substances without being able to identify the active factors. They are probably low molecular weight compounds, possibly new and still unknown ones; however, we think this unlikely and are attracted by the amino acids, although there are already numerous amino acids in the basic medium of embryo extract and horse serum. On the other hand a fraction which contains few amino acids is active, the main amino acid remaining being lysine, which is inactive when added to the basic medium.

What then are the "Wolff factors" which Professor Bergel has kindly dedicated to us? It is possible that there may be no Wolff factors in the strict sense. They may be a mixture of common substances in definite proportions or an equilibrium between known components which has yet to be defined.

REFERENCES

CROISILLE, Y., MASON, J., WOLFF, Em., and WOLFF, Et. (1967). *Europ. J. Cancer*, **3**, 371–379.
WOLFF, Et., HAFFEN, K., KIENY, M., and WOLFF, Em. (1953). *J. Embryol. exp. Morph.*, **1**, 55–84.
WOLFF, Et., and WOLFF, Em. (1961). *Presse méd.*, **69**, 1123.
WOLFF, Et., and WOLFF, Em. (1962). *Natn. Cancer Inst. Monogr.*, **11**, 180.
WOLFF, Et., and WOLFF, Em. (1964). *C.r. hebd. Séanc. Acad. Sci., Paris*, **258**, 2439.
WOLFF, Et., and WOLFF, Em. (1965). *Presse méd.*, **73**, 1154.
WOLFF, Et., WOLFF, Em., and CROISILLE, Y. (1965). *C.r. hebd. Séanc. Acad. Sci., Paris*, **260**, 2359–2363.

DISCUSSION

Bergel: We still have here a very complex situation, although chemically speaking it has become simpler. The biological activity, which is always tested by Mme Wolff on her organ cultures, has both a growth-maintaining and possibly a differentiation-protecting (anti-dedifferentiating?) effect.

Allison: Dr Mason, have you looked at the effects of asparagine in your cultures?

Mason: We have tried asparagine added to the basic medium in various concentrations and the survival of the explants is not greater than that of the control explants on basic medium alone—this is mere survival with no growth for 3–4 weeks. Probably these two lines have the biosynthetic powers which are lacking in the leukaemic lines.

Lamerton: How do the cultures die in inadequate media; does cell proliferation slow up?

Mason: Yes. On the basic medium the mitoses gradually decrease and there is no growth at all.

Leese: The Wolffs (1966) have reported that explants of human malignant gastrointestinal tissues (Z200 and Z516) grown in the presence of chick mesonephros extracts lose normal tissue elements and that the tumour cultures attain morphological and histological characteristics typical of those of their primary growth structure. Do you find the same degree of organization when explants are grown on cultures containing the fractionated extract? And does the degree of organization increase with the number of times that explants are passaged?

Mason: Yes. Tumours maintained on extracts show the same histology, with organotypic development. The degree of organization remains about the same. I think one can hardly say that the organization is due to the extracts; it is more due to the semi-solid nature of the medium.

Roe: Is there any explanation of why small explants won't grow?

Mason: This also could be connected with the basic medium and the surface properties. The vitelline membrane seems to fall between agar gel and the glass cover slip in the kind of growth that it allows. The cells can spread, but they spread less well on the vitelline membrane than on glass and on agar gel they don't spread at all; hence organotypic growth is possible.

Lamerton: Do only relatively few tumours grow in this system? It is not yet, I imagine, a routine method of growing human tumour biopsies.

Mason: The Wolffs have a number of lines besides these two which grow, for instance HeLa cells and KB cells. In this system they regain their original structure.

Human tumours have been cultured by this method but it is still rather a delicate system and one cannot get all explants to grow in it indefinitely. Several teams, including Röller, Owen and Heidelberger (1966) and Tchao and co-workers (1968) among others, have attempted to develop routine methods of culturing biopsy samples but the techniques seem to be limited by the poor survival of the explants and their variability.

Iversen: You say that HeLa cells differentiated to a squamous cell carcinoma. Was that with keratinization?

Mason: There are epithelia of several layers at the upper and lower surfaces of the explant, but we have never seen any keratinization.

Bergel: This method could have enormous practical use for culturing biopsy samples from patients to test therapeutic procedures *in vitro*. If I remember rightly, out of 20 to 25 different samples that Mme Wolff tried, 14 or 15 were successfully maintained, but there was a selection beforehand.

Mason: That is true. We have indeed selected tumours of epithelial origin, such as cancers of the uterus, stomach, intestine and lung.

One also wonders how much these fragments are representative of the original tumour, because both lines Z200 and Z516 came from one or two fragments of the original tumour. However, they do maintain their original specific histological structure and as far as can be ascertained their physiological and biochemical characteristics. For instance the Z516 continues to secrete copious quantities of mucus.

Lamerton: One could only expect to preserve the organization if the proliferative and differentiation characteristics of the tumour were not affected in any way by systemic factors. For instance, if hormonal influences play a part *in vivo*, the behaviour *in vitro* is bound to be different, however good the culture system.

Bergel: Has Mme Wolff tried to grow normal tissue under exactly the same conditions as these pieces?

Mason: Normal adult tissues lack the capacity for proliferation of the cancer tissues and, as in other systems of culture, fail to survive. Embryonic tissues differentiate very well on the basic medium but there is no continued proliferation.

Vernon: Is it possible that the critical factors are some trace elements that the tumour cell requires over and above the normal cell? Many metal cations will migrate on columns with amino acids and cannot be separated easily. Have you tried ashing your factors and testing the residues?

Mason: No. But would one not expect these elements to be present in the embryo juice?

Vernon: Not necessarily, if it was an unusual metal, and after all there are a lot of elements. You could eliminate the possibility by ashing the factors. If they are still active, it must be a simple element.

Johns: You said that the extract was highly active after acid hydrolysis in 6N-hydrochloric acid for 24 hours. Have you tried any longer periods of hydrolysis? You would then start selectively to destroy the amino acids.

Mason: No; I have not tried longer times.

O'Meara: Has a lipid element in these mixtures been excluded? Lipid solvents have been used quite a lot in the extractions. Did anything of a lipid nature come through?

Mason: Compounds of a lipid nature would probably be eliminated on the ion-exchange resins during the separation procedures. They would either pass along with the neutral compounds or even stick on the resins and fail to be removed by the subsequent elutions.

Bergel: The strange thing is that horse serum is absolutely essential, and therefore you have a two-fold system. You have something in the horse serum which is necessary for growth of these organ cultures, and you have something which you are studying at this moment. So it is a multi-membered system.

Mason: This is the difficulty, that there are so many variable components. There are high molecular weight components which are essential in the embryo juice and also in the horse serum, since if one uses dialysates of them with one of our active extracts the cultures die off.

Finter: Did you consider the possibility that you are neutralizing something in the horse serum or embryo extract that is toxic for these particular tumour cells?

Mason: We have no evidence of a toxic effect of the serum, but this is a possibility.

Subak-Sharpe: Was the very low amount of arginine in your mixture simply due to its destruction in your extraction procedure?

Mason: Arginine is probably destroyed. There is also a very low concentration of cysteine which must be due to destruction during hydrolysis.

Allison: Have you tried replacing horse serum with another macromolecule, such as methyl cellulose?

Mason: No, I have not, but it would be very interesting to try say dextran or Ficoll for the same reasons. But possibly on a semi-solid medium physical factors are less important.

REFERENCES

RÖLLER, M. R., OWEN, S. P., and HEIDELBERGER, C. (1966). *Cancer Res.*, **26**, 626–637.
TCHAO, R., EASTY, G. C., AMBROSE, E. J., RAVEN, R. W., and BLOOM, H. J. G. (1968). *Europ. J. Cancer*, **4**, 39–44.
WOLFF, Et., and WOLFF, Em. (1966). *Europ. J. Cancer*, **2**, 93–103.

THROMBOPLASTIC MATERIALS FROM HUMAN TUMOURS AND CHORION

R. A. Q. O'MEARA

Department of Experimental Medicine, School of Pathology, University of Dublin

THIS paper deals essentially with the participation of those agents in human tumours which promote clotting and their relationship to the changes in homeostasis of tissues which form part of the growth pattern of human malignant disease. Since the altered homeostatic state in this particular instance is of a special character it is first desirable to place it in proper relationship to homeostasis in the tissues in general. Normal tissues, of which the epidermis provides a good example, are capable of maturation growth only (O'Meara, 1960); that is to say, as they grow they ripen or mature to a definite and stable end point. They can only depart from this pattern of growth as a result of the action of trauma or other causal agents of disease. The various possibilities are set out in Table I.

Table I

RELATIONSHIP OF HOMEOSTASIS TO TYPE OF GROWTH OF TISSUE CELLS IN VARIOUS CIRCUMSTANCES

Tissue state	*Type of growth*
Homeostatic normal	Maturation to definable limits
Homeostatic traumatized	Regeneration to definable limits
Homeostatic transplanted	If viable, maturation limited or unlimited
Homeostasis altered by various causes	Maturation impaired, manifestations of disease
Homeostasis selectively lost through action of carcinogens, viruses, hereditary factors, radiations, hormonal imbalance and other undetermined causes of neoplasia	Neoplastic

In this table the first column gives the tissue state in the various circumstances of physiological or pathological proliferation and the second column the corresponding types of growth encountered as changes from the physiological to the pathological take place. Thus in normal homeostatic tissue the only form of growth possible is proliferation of cells accompanied by maturation changes (such as keratin formation in the epidermis) as the cells grow. This maturation growth takes place to clearly defined limits which are only exceeded in the event of disease.

If the tissue is traumatized regenerative growth which is again of maturation type takes place and ceases as soon as regeneration is complete, leading to the restoration of the normal homeostatic state.

Should the cells of the tissue be transplanted and survive in their new site, the growth which they exhibit is once more of simple maturation type but is not necessarily subject to the same restraints as those found with simple regeneration *in situ*. In consequence the maturation growth which they display may be either limited or for practical purposes unlimited. A number of illustrative instances from human pathology come to mind. Thus with implantation of cells from the epidermis into the dermis a cyst is formed lined by stratified squamous epithelium with the basal layer in contact with the connective tissue of the dermis, forming a junction identical in character to that of normal epidermis with the dermis. The interior of the cyst is filled with keratinous debris formed from maturing cells of the epidermis which behave in the same fashion as cells of normal epidermis, forming a keratinous layer on the inner surface of the cyst. In the experimental field a number of workers (e.g. McLoughlin, 1961) have made interesting observations on the influence of the type of mesenchyme into which the epithelial cells are implanted and it has been shown that in fact the form of maturation may be altered by implantation in mesenchyme of a different type to that with which the implanted epithelium is normally associated. Other instances may be provided from disease in man of implantation of normal epithelium leading to unlimited proliferation, for example, when an appendix or gall-bladder is blocked by a concretion or other cause of obstruction and dilates to form a sterile mucocele which bursts into the peritoneal cavity, on the surface of which the mucin-producing epithelial cells become implanted and continue to proliferate in an uninhibited fashion. In such circumstances the epithelium matures normally to provide a mucous secretion and the permanent state of *pseudomyxoma peritonei* is established. There is no invasion of the underlying tissues by the transplanted epithelium, which is not neoplastic: rather its state is that of unlimited maturation growth, and in consequence homeostasis is maintained.

In many diseases the normal homeostatic state of tissues is lost to a greater or lesser extent, leading to a failure of normal maturation growth in one or more parts of the organ altered by disease. In psoriasis, for example, the normal homeostatic state of the epidermis is partially lost with failure of normal maturation of the cells as they grow. They retain residues of the DNA of their nuclei much later than normal, as shown by Dixon, Gresham and Whittle (1962), so that the cells of the horny layer contain nuclear material and the epidermal cells, failing to mature normally, persist in protein synthesis to an abnormal degree. This change in the epidermis is accompanied by parakeratosis and other departures from normal such as irregular hyperplasia together with focal loss or hyperplasia of the normal stratum granulosum and by changes in the associated dermal tissues as well. The aetiology of psoriasis is still undetermined but it may well belong to the group of auto-immune diseases, which contains many instances of altered homeostasis associated with manifestation of disease.

By comparison the agents which are responsible for the development of malignant neoplasia are of a most varied character. Cancer of the skin and other organs may be induced by physical agents such as radiations over a wide range of wave lengths; by organic chemicals of a most diverse character from the point of view of structure; in certain instances by inorganic particles, by hereditary factors, pre-existing benign tumours and various other causal agents. While the diversity of causal agents is almost unlimited, the end result of their operation leads to disturbance of homeostasis in a very specific fashion in so far as some of the body cells cease to remain in homeostasis with their neighbours and develop an independence of growth, proliferating on their own. The two outstanding changes in their morphology and behaviour which mark them as neoplastic are failure to mature normally during growth and an ability to invade the tissues in their neighbourhood and later to enter lymphatics and blood vessels and thus to travel to other sites in the body. In cancers, maturation growth is impaired or wholly absent, while invasive growth develops as maturation growth wanes.

The development of independence of cellular proliferation with loss of homeostasis and the simultaneous onset of invasive growth is closely bound up with the development of increased clotting activity by the site which becomes cancerous. Since I first showed in 1958 that human cancers produced a labile, freely diffusible clotting factor which causes the formation of fibrin in their environment or in certain circumstances in their interior, confirmation has come from many sources as well as from my own laboratories to indicate the importance of this change in tissue behaviour as a characteristic property of malignant neoplasia. The evidence which has accumulated is derived from both human and animal sources, since similar or supportive findings to those in human tumours have been obtained with spontaneous and experimentally induced animal cancers as well.

Thus Hiramoto and co-workers (1960) demonstrated fibrin in a variety of human carcinomas and sarcomas using the fluorescent antibody technique, while Day, Planinsek and Pressman (1959) found radio-iodinated rat fibrinogen preferentially localized in three different transplantable rat tumours. In the dibenzanthracene-derived Murphy-Sturm lymphosarcoma, 40 per cent of intravenously introduced fibrinogen localized preferentially in the tumour during a period of 6 hours at the height of its growth. These authors also showed localization in the tumours of fibrin and fibrinogen antibodies given intravenously. Their findings were confirmed and extended by Spar, Goodland and Bale (1959) who showed high localization of labelled antibody to rat fibrin in rat lymphosarcoma. This technique has now been used for the diagnosis and localization of human tumours using the appropriate labelled antibodies. Spar and co-workers (1967) found anti-fibrin antibody to localize preferentially in neoplasms both in human subjects and in the dog, while Marrach and co-workers (1967) found that ^{131}I-labelled antibody to fibrin localized selectively in a group of intracranial tumours. These results all point to a concentration of fibrinogen and fibrin in tumours, and the uptake of anti-fibrin antibody is selective to such an extent that it has been suggested as a possible vector of therapeutic agents to the tumours.

Another approach approximating more closely to that which I and my colleagues have used consists in the study of the clotting properties of spontaneous tumours in laboratory animals. Holyoke and Ichihashi (1966) studied the thromboplastic activities of homogenates and extracts of spontaneous mammary tumours in mice and also from T-241 Lewis sarcoma and found a marked increase in thromboplastin levels in the tumours by comparison with normal tissue. Frank and Holyoke (1968) made a further study of the T-241 Lewis sarcoma, confirming its high thromboplasticity. They found that interstitial fluid taken from near the tumours had greater clot-promoting power than that taken at a distance from the tumours. Fluid from near the tumour had a greater tendency to undergo spontaneous clotting as well. The authors refer to the thermostability of the thromboplastin derived from their tumour. This question has been clarified by Boggust and O'Meara (1966) who showed that the thermolability and thermostability of thromboplastins of both human and animal origin are reversible, thermolability depending upon the —SH state of the thromboplastin and its environment during the thermolability test. As far as the formation of fibrin by tumours is concerned, no special importance can therefore be attached to the heat lability or stability of the thromboplastin of the tumour. Likewise the availability of fibrinogen in tissue fluid has presented no difficulties since it is well known that oedema fluid clots and, as has been shown by applying to fibrinogen the method used by Sterling (1951) for albumin, total body fibrinogen is divided between the intravascular and extravascular pool, approximately 20 per cent being extra-

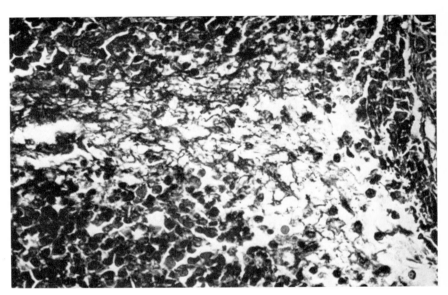

FIG. 1. Fibrin network in human plasmacytoma. Phosphotungstic acid haematoxylin. ×450.

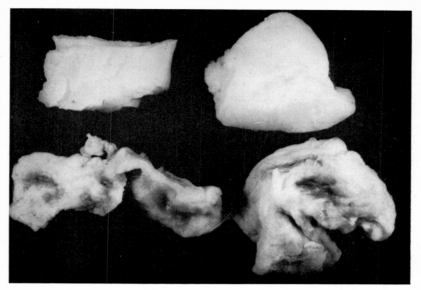

FIG. 2. *Above:* tissue from site of subcutaneous injection of neutral olive oil at 14 days. No vascularization. *Below:* tissue from site of subcutaneous injection of lignoceric acid in olive oil at 14 days. Newly formed vessels apparent in tissue.

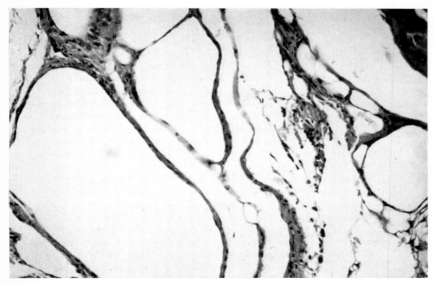

FIG. 3. Microscopic appearances at 14 days at site of subcutaneous injection of neutral olive oil. Haematoxylin and eosin. × 150.

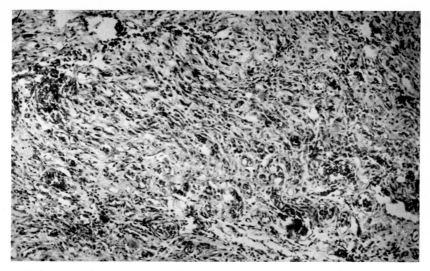

FIG. 4. Granulation tissue at 14 days after subcutaneous injection of arachidic acid in olive oil. Haematoxylin and eosin. × 150.

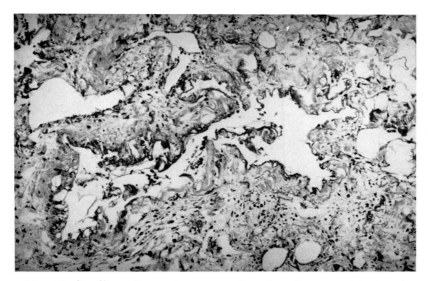

FIG. 5. Hyaline fibrous tissue and spaces at 14 days after subcutaneous injection of arachidic acid in olive oil. Haematoxylin and eosin. × 150.

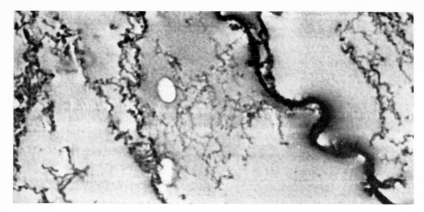

Fig. 6. Acidophilic exudate with much fibrin at site of subcutaneous injection of linoleic acid in rat. Haematoxylin and eosin. ×250. By courtesy of Professor C. W. M. Adams.

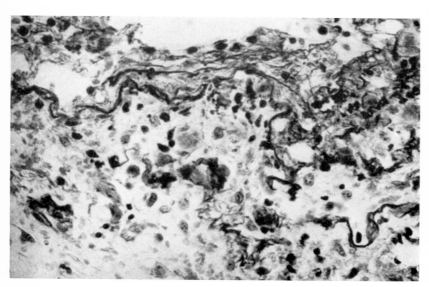

Fig. 7. Networks of fibrin in granulation tissues at 14 days after subcutaneous injection of arachidic acid in olive oil. Phosphotungstic acid haematoxylin. ×450.

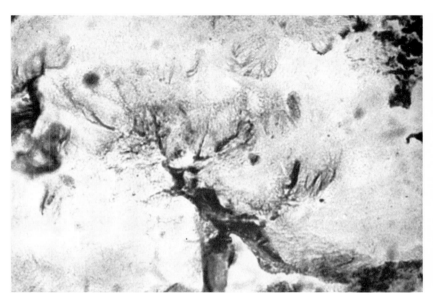

FIG. 8. Fine fibrin threads on connective tissue spur projecting into space (see Fig. 5). Phosphotungstic acid haematoxylin. × 500.

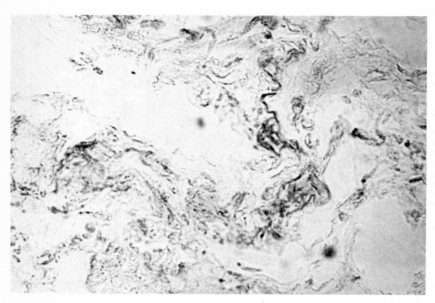

Fig. 9. Same as Fig. 7 but stained to show tryptophan. *p*-Dimethylaminobenzaldehyde-nitrite. ×450.

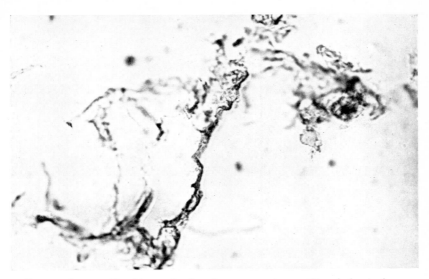

FIG. 10. Same as Fig. 8 but stained to show tryptophan. p-Dimethylaminobenzaldehyde-nitrite. ×450.

FIG. 11. Collagen control in the same tissue, negative for tryptophan. p-Dimethylaminobenzaldehyde-nitrite. ×450.

vascular. It has also been shown by Lewis, Ferguson and Schoenfeld (1961) that fibrinogen migrates more readily than albumin from the circulation in spite of its much greater molecular weight.

Among the most important and interesting results with spontaneous animal tumours is the finding by Laki and Yancey (1968) that a spontaneous plasmacytoma of the mouse, YPC-I, first observed by Yancey (1964), forms networks of fibrin as it grows. This agrees with the finding of similar networks of fibrin in a slide of human plasmacytoma that I made in 1959 (Fig. 1). As pointed out by Laki, Tyler and Yancey (1966), their clotting agent, which corresponds in its properties to that found by me (1958) in human tumours, is thrombin, which has been isolated from the tumour by Laki and Suba-Claus (1966) and shown to act specifically on fibrinogen with the removal from it of peptides A and B. Laki, Tyler and Yancey (1966) made the further interesting discovery that their tumour also produced a fibrin-stabilizing factor conferring stability on the clot. In the formation of thrombin by tumours, tissue thromboplastins derived from the tumours play an essential role and we have found that thromboplastic materials are present in excess in tumours by comparison with the normal tissues from which the tumours arise. Moreover, like thrombin they diffuse from the tumours into their environment.

The relationship between cancer thromboplastin formation and thrombin formation by cancers was determined by Hannigan, Boggust and O'Meara (1966). It was found that the diffusible thromboplastin in cancers and chorion reacted with a factor present in plasma and serum to give a complete thromboplastin capable of converting prothrombin to thrombin.

Much of the work in my laboratories has been planned to determine the exact nature of the thromboplastic properties of tumours and to define their intracellular source. In this work part of the investigation has been made using human chorion as a prototype material for study, as an alternative to biopsy specimens of human tumours.

Finkelstein (1945) and Schneider (1947) had shown that watery extracts of placenta possessed high thromboplastic activity. The properties which they described were very similar to those which I had encountered with human tumour extracts. Human chorion provided extracts of a similar character and possessed the advantage of being much more uniform in composition than placenta. Unlike placenta, chorion is, moreover, devoid of blood vessels. Analysis of watery extracts of human chorion and human cancers as well as normal tissue by Boggust, O'Meara and Thornes (1961) and Boggust and co-workers (1963), employing DEAE cellulose ion-exchange column chromatography with stepwise elution using ammonium acetate adjusted to pH 9·0, followed by 0·5 N-NaOH, has shown close similarities between cancers and chorion, though they are not identical. Human cancers have four distinct groups of clot-promoting activity, arbitrarily labelled A, B, C, D, corresponding respectively to the 0·2 M, 0·4 M and 2·0 M-ammonium acetate and the 0·5 N-NaOH fractions eluted from the column; whereas chorion shows only B, C, and D with little or negligible activity in A. By comparison normal tissue shows no activity at A, C and D and only very little activity corresponding to B. Benign tumour extracts were found virtually inert at all points. The results of these experiments fully justified the use of chorion as prototype material for the study of the thromboplastic activities of cancers.

In the course of these experiments it was noticed that the nitrogen content of the thromboplastically active fractions of chorion and cancer extracts, as determined by nitrogen estimations, was much lower than might be expected from the ultraviolet absorptions of the eluates, if these were attributable to their protein or nucleoprotein content. Moreover, by starch gel electrophoresis it was seen that each fraction is extremely heterogeneous in its protein components.

In the study of the intracellular location of the clot-promoting agents in chorion low nitrogen values were also found by Clarke and O'Meara (1966) relative to the thromboplastic activity shown by the most active fractions, in particular those derived from microsomes. From all these data it was concluded that the thromboplastic activity of chorion and hence of cancers was attributable to factors carried by protein rather than proteins themselves. Confirmatory evidence for this view was derived from the finding by Boggust, O'Meara and Fullerton (1968), using Sephadex G-200 gel filtration applied to thromboplastic chorion extracts, that the thromboplastic activity of the fractions from the column did not correspond to the emergence of the protein fractions but was distributed across the entire fractionation even in places where no ultraviolet absorption occurred. Three minor phases of increased clot-promoting activity can be seen in such fractionations. These phases can be reproduced artificially by combining fatty acids with albumin to give fatty acid–protein adducts which correspond chromatographically in ultraviolet absorption and in the location of thromboplastic activity with the extracts of cancer or chorion. A similar result is obtained by extracting the fatty acids from chorion and reconstituting the mixed fatty acids so obtained with albumin, thus again reproducing chromatographically and by the results of clotting tests findings similar to those with chorion extract.

It should be noted in conjunction with these results that the main intracellular source of the thromboplastic material emerging from the cancer or chorion cell is the endoplasmic reticulum, with less activity demonstrable in the mitochondria and nuclei (Clarke and O'Meara, 1966). The form in which the fatty acids are present in the cell may be either in combination with albumin and other proteins or as part of the phospholipid complex. Probably little free fatty acid is present at any given time within the cell. Bermingham, Boggust and O'Meara (1968) have shown that chloroform extracts carrying the phospholipid fraction of endoplasmic reticulum correspond in their properties to those made from conventional sources such as acetone-dried brain. These extracts possess a built-in hydrolytic system, probably enzymic in character, which renders them unstable and prone to release free fatty acids. Simultaneously they develop procoagulant properties. The endoplasmic reticulum forms part of the cell membrane system and it would seem that in cancers some instability has developed or alternatively that the energy mechanisms necessary for the transport of thromboplastic or fatty acid molecules across the membrane have undergone alteration. It is recognized that fatty acids have a high affinity for albumin and after their release from the cell they may be expected to combine with albumin. Although, depending upon their composition and physical state, they may exhibit either antithromboplastic or thromboplastic activity, as shown by Fullerton, Boggust and O'Meara (1967), the general effect of their release is procoagulant. The clot-promoting effect is much enhanced by factors present in serum and plasma (Fullerton, 1968; see also Hannigan, Boggust and O'Meara, 1966).

Accurate qualitative and quantitative information about the fatty acids released from cancers is not yet available but is engaging our attention. Meanwhile, since it is apparent that the thromboplastic activity of cancer and chorion extracts depends upon the fatty acids which they carry, it has been considered desirable to study the effects of introducing fatty acids into animal tissues in such a way as to permit as far as possible their slow escape from the site of inoculation, thus simulating the conditions found in cancers. Studies in the rat have been made on the tissue changes induced by subcutaneous inoculation of fatty acids by Professor Adams and his colleagues (Abdulla, Adams and Morgan, 1967) and he has kindly permitted me to use one of his preparations to illustrate the early stages of tissue transformation induced by their introduction (see Fig. 6).

My own studies, as yet incomplete, have been carried out in the rabbit. The fatty acid (100 mg) has been injected subcutaneously in the rabbit in 3 ml of neutral olive oil to allow of slow absorption, a corresponding amount of the neutral oil constituting the control. The acids used so far were between C_{16} and C_{24}, namely palmitic, linoleic, oleic, stearic, arachidic, behenic and lignoceric acids. After the injections no appreciable reaction is apparent for 7 days, after which some thickening can be felt at the site of injection of the higher fatty acids. This continues to increase up to 14 days, beyond which no observations have yet been made. The process is silent and at no stage is evidence of irritation shown by the animal. When the animals are examined macroscopically at 14 days no reaction is visible at the site of injection of the control oil but for acids from C_{16} upwards it is apparent that vascularization of the tissue, most pronounced with the higher acids, has taken place at the site of injection. An example of the macroscopic appearances is shown in Fig. 2 (p. 86). The control site with oil only shows no abnormal vascularity such as is seen when oil and fatty acid are inoculated together.

Microscopically the site of injection of the control shows minimal reaction, principally by way of dilated fat spaces, as shown in Fig. 3. There is no new vessel formation. By contrast at sites where the higher acids have been inoculated together with the oil there is a marked tissue reaction, as shown by granulation tissue formation with many new capillary vessels (Fig. 4), and the laying down of hyaline fibrous tissue around spaces containing amorphous material which is partly protein in character (Fig. 5). These spaces are also found at an early stage filled with acidophilic exudate in which abundant strands of fibrin are apparent, as seen in Fig. 6 in a specimen taken 3 days after inoculation of 20 mg linoleic acid in the rat. It was considered unlikely that fibrin would be found in the preparations in the rabbit at 14 days but nevertheless it was sought and, surprisingly, found after suitable fixation and staining for fibrin with phosphotungstic acid haematoxylin. Networks of fibrin were seen to be present in the granulation tissues, as shown in Fig. 7. Even more interesting, fibrin was also found to be forming at the margin of the spaces referred to above. It projected into the spaces in the form of fine threads which were in the process of undergoing fibrous organization from the surrounding hyaline tissue, as shown in Fig. 8. That the networks and fibrils are composed of fibrin was further demonstrated by the p-dimethylaminobenzaldehyde-nitrite reaction of Adams (1957) which revealed their high tryptophan content, characteristic of fibrin and fibrinogen, as shown respectively in Figs. 9 and 10. For comparison a zone of collagen fibres in the same section is shown in Fig. 11. The collagen shows no reaction for tryptophan.

These illustrations are characteristic of what was found with the higher saturated acids in the series studied and represent an approximation to what may be found in a tumour stroma with formation of vascular granulation tissue, fibroblastic proliferation and hyaline tissue formation as prominent features. The fibrin formation also occurs in a fashion similar to that observed in association with tumours. In one respect the response to the acids was not fully typical of tumours; foreign body giant cells were more numerous than commonly found in association with tumours. Their formation in these experiments was due to difficulty in bringing the higher acids into solution. It is anticipated that when more information is available about the exact composition of the mixed fatty acids which emerge from tumours it will be found that the acids of lower melting point will hold those of higher melting point in solution and will provide an even closer approximation to the tumour stroma. Meanwhile it is interesting that lignoceric acid should provide so satisfactory a result. It is the major fatty acid component of cytolipin H, isolated from human epidermoid carcinoma and analysed by Rapport, Skipski and Sweeley (1961).

DISCUSSION

Some of the implications of fibrin formation by tumours have already been discussed by me in previous publications (O'Meara, 1958, 1960, 1962, 1964). They may now be discussed with greater precision in relation to loss of homeostasis when a tissue becomes malignant. The early stages of the development of malignancy at the primary site are of special significance. When malignant transformation of a tissue occurs, the cells of the affected site undergo a change which leads to the formation of thrombin in and around the malignant focus. The formation of thrombin is set in motion by materials derived from the cell organelles which come to the cell surface. These materials, which are normally found within the cell cytoplasm, especially in the endoplasmic reticulum and also possibly to some extent in the mitochondria and nuclei, possess thromboplastic properties. There is at present no conclusive evidence that the thromboplastic agents which reach the surface of malignant cells differ from those found in the interior of normal cells. They behave in the same fashion in their physical, chemical and biological properties as those found by Erwin Chargaff in his classical studies on tissue thromboplastins (for review see Chargaff, 1945). Nevertheless, their emergence to the surface of the cell and their escape beyond points either to some modification in the permeability of the cell membrane, or to a change in its composition, or to some alteration in cell metabolism accompanied by a redistribution of energy utilization. Since we have determined that the active components of the thromboplastins relevant to our problem are the fatty acids which they carry, it may be expected that accurate qualitative and quantitative information about these acids, now being sought, will provide answers to some of these essentially theoretical considerations.

More important from the practical point of view is the fact that leakage of fatty acids from the cancers is capable of causing tissue reaction in their environment comparable to that found in cancer stroma and in the granulomas. This finding provides experimental evidence to support the view that in cancer the definitive parasites responsible for the pathological changes found in the disease are the proliferating cells of the tissue which has become malignant. In common with all other parasites whether viral, bacterial or protozoal they share the ability to proliferate and their pathogenic properties like those of other parasites are dependent upon their cellular products escaping into the body fluids and tissues around them. The fatty acids which we find responsible for one aspect of the pathogenic properties of malignant cells are not necessarily exclusive, since it is obvious that changes in cellular metabolism and alterations in cell membrane permeability may lead to the diffusion from the cancer cells of other cellular components. These, while they are held within the cell, cause no tissue reaction. When, however, they come to the surface or escape into the tissues they may set up a reaction in relation to the site of their escape or further afield. The loss of homeostasis in the tissues which occurs when cancer begins to grow is, therefore, attributable to the pathogenic properties which the cells of the cancer develop at about the same time. The pathological manifestation of this pathogenicity is the formation of a granulation tissue in which the cancer grows. Consequently one may reasonably suggest that, for the further study of its pathological relationships, cancer should be classified among the granulomas.

To classify it in this fashion greatly simplifies the planning of the experimental attack on the disease since much is already known and understood about the granulomas, their pathological relationships and immunological behaviour. Likewise a rational approach is made possible on the basis of hypotheses which can be developed for experimental test. It may be permissible to suggest that cancer research has suffered from too rigid an adherence to preconceived ideas related predominantly to causation, with consequent neglect of other important aspects. When theories of causation are analysed they are seen to be formulated primarily in the belief that the mode of action of the causal agents responsible for inducing cancer has particular reference to an alleged capacity on the part of these agents to induce uninhibited proliferation of cells. While in established cancers the cells commonly multiply at an exceptionally rapid rate, there are two other properties of neoplastic cells which would be capable of inducing tumour formation independently of rate of multiplication. These are, firstly, the failure of cancer cells to mature normally, and secondly the fact that the altered cells permit the escape into the extracellular spaces of cellular components capable of inducing pathological change in the surrounding tissues. Cells which fail to mature normally fail to undergo necrosis at the appointed time in their life cycle. In consequence they must accumulate. In malignant disease the correction of this failure of maturation in hormone-sensitive tumours such as those of the prostate and breast by appropriate hormonal treatment leads to a rapid, and for a time sustained, regression of the tumour. Hormonal treatment fails to eradicate the growth completely in the majority of cases because it fails to cope with the pathogenic properties which some of the tumour cells have developed.

Table II
CONVENTIONAL VIEW OF CARCINOGENESIS AND AN ALTERNATIVE

Conventional view of carcinogenesis

Carcinogen $\xrightarrow{\text{acts on}}$ Cell $\xrightarrow{\text{producing}}$ Cell with modified nuclear material (mutation) $\xrightarrow{\text{giving}}$ Cell transformed to rapid uninhibited growth

Alternative view of carcinogenesis

Carcinogen $\xrightarrow{\text{acts on}}$ Tissue $\xrightarrow{\text{giving}}$ Tissue modified in a selectively specific fashion $\xrightarrow{\text{so that}}$ Cells of the affected tissue
 (1) Fail to mature normally
 (2) Develop pathogenic properties against the tissues around them
 (3) Become autonomous as a result of (2)
 (4) Show accelerated growth and dedifferentiation as autonomy increases

The currently accepted hypothesis of the action of carcinogens on a purely cellular basis may be contrasted, as shown in Table II, with an alternative based on human tumour behaviour as observed clinically and by pathological study. It will be seen from this table that the conventional view of the action of carcinogens tends to limit consideration of the problems of malignancy to one aspect only of malignant growth. By enlarging the field of vision and adjusting theoretical considerations to agree more closely with observed facts a wider view of carcinogenesis becomes possible, leading to a fuller recognition of the various properties which delineate a malignant tissue and cause it to depart from the homeostatic state.

SUMMARY

Normal adult tissues are strictly limited to maturation growth only, maintaining homeostasis by passing through orderly stages of growth accompanied by maturation. Death of the cells at their appointed time is the end point. This cycle can only be upset by disease and in a tissue reacting to a carcinogenic agent the cycle is affected in a specifically selective fashion which leads to the development of malignancy.

When cancer begins to develop maturation during growth is impaired or wholly lost. As maturation growth wanes the cells fail to undergo necrosis at their appointed time; autonomy of cell proliferation and invasive growth supervene. As a prelude to invasion the tissue which is becoming malignant acquires pathogenic properties leading to the formation of a granulation tissue which is later invaded by the tumour.

As one manifestation of the pathogenic properties developed by the tissue, it is shown that fibrin is produced by the action of thrombin. Thrombin formation is induced by the tumour, which liberates fatty acids with thromboplastic properties. Fatty acids inoculated into the tissues evoke a response resembling that given by tumours. The successive changes through which a tissue passes to reach the malignant state at the primary site in man are tabulated.

Acknowledgement

My thanks are due to Mr F. A. Murray for the photographs.

REFERENCES

ABDULLA, Y. H., ADAMS, C. W. M., and MORGAN, R. S. (1967). *J. Path. Bact.*, **94,** 63–71.
ADAMS, C. W. M. (1957). *J. clin. Path.*, **10,** 56–62.
BERMINGHAM, M. A. C., BOGGUST, W. A., and O'MEARA, R. A. Q. (1968). *Nature, Lond.*, **218,** 695–696.
BOGGUST, W. A., O'BRIEN, D. J., O'MEARA, R. A. Q., and THORNES, R. D. (1963). *Ir. J. med. Sci.*, **447,** 131–144.
BOGGUST, W. A., and O'MEARA, R. A. Q. (1966). *Ir. J. med. Sci.*, **481,** 11–21.
BOGGUST, W. A., O'MEARA, R. A. Q., and FULLERTON, W. W. (1968). *Europ. J. Cancer*, **3,** 467–473.
BOGGUST, W. A., O'MEARA, R. A. Q., and THORNES, R. D. (1961). *Biochem. J.*, **80,** 32P.
CHARGAFF, E. (1945). *Adv. Enzymol.*, **5,** 31–65.
CLARKE, N., and O'MEARA, R. A. Q. (1966). *Br. J. Haemat.*, **12,** 536–545.
DAY, E. D., PLANINSEK, J. A., and PRESSMAN, D. (1959). *J. natn. Cancer Inst.*, **22,** 413–426.
DIXON, K. C., GRESHAM, G. A., and WHITTLE, C. H. (1962). *Br. J. Derm.*, **74,** 283.
FINKELSTEIN, M. (1945). *Nature, Lond.*, **155,** 202–203.
FRANK, A. L., and HOLYOKE, E. D. (1968). *Int. J. Cancer*, **3,** 677–681.
FULLERTON, W. W. (1968). Ph.D. Thesis, Dublin University.
FULLERTON, W. W., BOGGUST, W. A., and O'MEARA, R. A. Q. (1967). *J. clin. Path.*, **20,** 624–628.
HANNIGAN, F., BOGGUST, W. A., and O'MEARA, R. A. Q. (1966). *Europ. J. Cancer*, **2,** 325–331.
HIRAMOTO, R., BERNECKY, J., JURANDOWSKY, J., and PRESSMAN, D. (1960). *Cancer Res.*, **20,** 592–593.

HOLYOKE, E. D., and ICHIHASHI, H. (1966). *J. natn. Cancer Inst.*, **36**, 1049–1056.
LAKI, K., and SUBA-CLAUS, E. (1966). Quoted by Laki, K., and Yancey, S. T., in *Fibrinogen* (1968). p. 360, ed. Laki, K. London: Arnold; New York: Dekker.
LAKI, K., TYLER, H. M., and YANCEY, S. T. (1966). *Biochem. biophys. Res. Commun.*, **24**, 776–781.
LAKI, K., and YANCEY, S. T. (1968). In *Fibrinogen*, pp. 359–367, ed. Laki, K. London: Arnold; New York: Dekker.
LEWIS, J. H., FERGUSON, E. E., and SCHOENFELD, C. (1961). *J. Lab. clin. Med.*, **58**, 247–258.
MCLOUGHLIN, B. (1961). In *Biological Approaches to Cancer Chemotherapy*, pp. 371–385, ed. Harris, R. J. C. London and New York: Academic Press.
MARRACH, D., KUBALA, M., CORRY, P., LEAVENS, M., HOWZE, J., DEWEY, W., BALE, W. F., and SPAR, I. L. (1967). *Cancer*, **20**, 751–755.
O'MEARA, R. A. Q. (1958). *Ir. J. med. Sci.*, **394**, 474–479.
O'MEARA, R. A. Q. (1960). *Archo De Vecchi*, **31**, 365–384.
O'MEARA, R. A. Q. (1962). In *The Morphological Precursors of Cancer*, pp. 21–34, ed. Severi, L. Perugia: Division of Cancer Research.
O'MEARA, R. A. Q. (1964). *Bull. Soc. int. Chir.*, **23**, 30–35.
RAPPORT, M. M., SKIPSKI, V. P., and SWEELEY, C. C. (1961). *J. Lipid Res.*, **2**, 148–151.
SCHNEIDER, C. L. (1947). *Am. J. Physiol.*, **149**, 123–129.
SPAR, I. L., BALE, W. F., MARRACH, D., DEWEY, W. C., MCCARDLE, R. J., and HARPER, P. V. (1967). *Cancer*, **20**, 865–870.
SPAR, I. L., GOODLAND, R. L., and BALE, W. F. (1959). *Proc. Soc. exp. Biol. Med.*, **100**, 259–262.
STERLING, K. (1951). *J. clin. Invest.*, **30**, 1228–1237.
YANCEY, S. T. (1964). *J. natn. Cancer Inst.*, **33**, 373–382.

DISCUSSION

Subak-Sharpe: The view of carcinogenesis presented by Professor O'Meara does not seem to fit the known facts of viral carcinogenesis *in vitro*; for example polyoma virus can transform single cells suspended in agar that have no contacts with other cells. One can then derive clones of tumour cells from these single cells, implant them into animals and obtain characteristic tumours. There is no reaction with other tissues here of the type which Professor O'Meara postulates.

O'Meara: I would agree with Professor Subak-Sharpe for the transformation of an individual cell by a virus, but this is something quite distinct; in human cancer the tumour is commonly multicentric; it is not started in any particular cell but is very often found to be occurring at a number of separate sites simultaneously. In human tumours the viral origin is still very questionable, I think.

Möller: Is it possible that the fibrin coat could protect the tumour cells against certain surveillance mechanisms in the host? Have you studied whether reduced formation of fibrin would increase the vulnerability of the cells to either a homograft reaction or a tumour-specific rejection?

O'Meara: A considerable amount of work has been done on this, for example by Clifton (1966). Fibrinolytic agents have been used to keep the cells mobile so that they should remain in the circulation and eventually be phagocytosed or disposed of in some other way rather than settle down in capillaries and give rise to tumours. We have approached the problem of tumour formation in another way and have used agents which neutralize the cellular thromboplastins, to see if we could stop the reaction at an earlier stage, before fibrin forms at all. With protamine we got a certain amount of temporary regression of tumours, in no sense curative (O'Halloran and O'Meara, 1964). This may seem contradictory because protamine is commonly used to neutralize heparin. But protamine is very basic and fatty acids are the active factors in the thromboplastins, which may be the secret of their neutralization. Nevertheless there is some specificity in the reaction since other bases even stronger than protamine may fail to neutralize thromboplastin. As regards graft rejection, I haven't done anything and fail to recall any work on it.

Johns: You said that as the cells become autonomous this leads to dedifferentiation. Do you envisage any change in the chromatin?

O'Meara: Dedifferentiation to my mind is a departure from maturation growth to growth for the sake of growth—that is, growth which is determined primarily by the availability of nutrients, especially suitable energy sources, such as occurs during the logarithmic phase in a bacterial culture. The cells are growing at the maximum possible rate, and this is what tends to happen when cells become dedifferentiated. In human tumours as growth progresses it becomes less and less possible to distinguish one tumour from another because the cells originally from different organs all come to resemble one another closely (which takes one back to the original theories of cancers, that they are a reversion to the embryonic type of growth). This is what I mean by dedifferentiation; if you give cells the opportunity to grow freely they will dedifferentiate. Given autonomy of growth the cells show striking abnormalities of chromatin when any restraint is imposed on their proliferation.

Mason: What is the fibrinolytic activity of malignant tissues, and what is the position on mast cell tumours?

O'Meara: The fibrinolytic activity of malignant tumours is in general less than that of the normal tissues around them. This largely depends on the fact that the principal activator of fibrinolysis in the tissues is located in the venous endothelium, as shown by Todd (1960), but the vessels which form in response to fatty acids or thromboplastic agents or tumour exudates are of capillary type and don't carry this activator, so that in general there is a reduction, but it is not absolutely uniform throughout. Certain tumours have built-in fibrinolytic activity but they also have built-in thromboplastic activity—prostatic tumours, for example.

The mast cell tumour does not occur to any extent in man; I have seen it only in the dog, and I haven't investigated it.

Abercrombie: I wonder whether the increased thromboplastic activity of the tumours compared with the tissue of origin is on a per cell basis. Could it be accounted for by the fact that the tumour is more cellular than the tissue of origin? And secondly, could the liberation of the thromboplastic substances be due to cell death?

O'Meara: An estimation on a weight for weight basis is difficult but I think one can say that the thromboplasticity of the tumour is considerably higher than that of the normal tissue from which it springs. This was demonstrated for animals tumours by Holyoke and Ichihashi (1966).

The escape of the thromboplastin is something to which we can't give a complete answer yet. Normally the thromboplastins are held in the cell organelles and in the normal cell they tend to remain within the cell, as Dr Clarke and I showed (Clarke and O'Meara, 1966). This also was shown to a certain extent by Barnhart (1957); Chargaff's work might be interpreted to indicate the same sort of thing (Chargaff, Moore and Bendich, 1942). The chief site is the endoplasmic reticulum. There are a number of ways in which the thromboplastin held in the endoplasmic reticulum could come to the surface of the cells and escape: one way is by the endoplasmic reticulum joining with the cell membrane, and it has been suggested that endoplasmic reticulum does come to the surface in this way in cancer cells. Another way is by an alteration in the energy relationships. There was reference earlier (p. 48) to cyclic AMP in relation to homeostasis; we are rather interested in this. The transfer of molecules across the cell membrane is usually held to require expenditure of energy and there may be changes in the energy relationships which will enable what is probably a relatively large molecule to be transferred across a membrane such as the cell membrane in the cancer cell but not in a normal cell. The ATP system may be involved. With massive cell death, either spontaneous or following radiation therapy, thromboplastins are released and fibrin forms, as one would expect.

Jacques: The crossing of the cell membrane by individual molecules (permeation) would indeed be determined by the permeability properties of that membrane. This would not be true however, if the thromboplastic agents are packaged into secretory granules and become discharged in bulk (exocytosis) without actually crossing the plasma membrane.

O'Meara: The materials responsible for the thromboplastic activity escape into the surrounding medium very readily without requiring any homogenization of the cells, if one merely suspends the tissues in water or isotonic saline solution. When this extract is freeze-dried it contains quite a lot of material of high molecular weight (from 30 000 to 200 000 or more), corresponding to the sort of thing you get from the endoplasmic reticulum after homogenization; so it doesn't seem likely that it would escape in the form of particles from the endoplasmic reticulum, which is too stable a system to give rise to particles in this fashion. On the other hand much of the material is of high molecular weight.

Allison: Several of us have been interested in the possibility that some deviation of the cyclic AMP controlling mechanism might be found in cancer cells. There are indications that cyclic AMP synthesis may be related to growth in slime moulds and even in bacteria, and Professor Stoker mentioned Dr Bürk's work, where the transformed cells had a lower adenyl cyclase activity than normal cells and so would be expected to produce less cyclic AMP (see p. 55). But unfortunately this doesn't seem to be true of all tumours. For example Brown and co-workers (1968) have found that hepatomas, including the so-called minimum deviation hepatomas, have an increased adenyl cyclase activity, so it seems that there is no simple generalization about the activity of the enzyme in tumour cells. The fact that the adenyl cyclase of tumours was less responsive than the normal enzyme to adrenaline may be interesting.

REFERENCES

BARNHART, M. I. (1957). *J. Mich. St. med. Soc.*, **56**, 1450–51.
BROWN, H. D., MORRIS, H. P., CHATTOPADYAY, S. K., PATEL, A. B., and PENNINGTON, S. N. (1968). *J. Cell Biol.*, **39**, 163a. (Abstract of paper presented at 7th annual meeting of American Society for Cell Biology.)
CHARGAFF, E., MOORE, D. H., and BENDICH, A. (1942). *J. biol. Chem.*, **145**, 593–603.
CLARKE, N., and O'MEARA, R. A. Q. (1966). *Br. J. Haemat.*, **12**, 536–545.
CLIFTON, E. E. (1966). *Fedn Proc. Fedn Am. Socs exp. Biol.*, **25**, 89–93.
HOLYOKE, E. D., and ICHIHASHI, H. (1966). *J. natn. Cancer Inst.*, **36**, 1049–1056.
O'HALLORAN, M. J., and O'MEARA, R. A. Q. (1964). *Bull. Soc. int. Chir.*, **23**, 30–35.
TODD, A. S. (1960). In *Thrombosis and Anticoagulant Therapy*, pp. 25–31, ed. Walker, W. Edinburgh: Livingstone.

GENERAL DISCUSSION

APPROACHES TO THE STUDY OF HOMEOSTASIS

Iversen: We had two very exciting weeks in Oslo six months ago when we conceived the idea that chalone was possibly a large protein molecule coupled to a small general mitotic inhibitor, which we thought might be the ketoaldehyde retin. Dr Jellum at the Institute of Clinical Biochemistry started to investigate all our chalone preparations with the arsenite method of determining 2-keto-3-deoxyglucose. All chalone preparations contained large amounts of retin! Then Otsuka and Egyud's paper (1968) appeared showing that the method used to find 2-keto-3-deoxyglucose is in itself a very good producer of this substance. Dr Jellum tried with only simple carbohydrates and produced a positive ketoaldehyde reaction simply from glucose (Jellum, 1968). So it seems that the retin work was based on a method that in itself produced this substance.

Bergel: It is very disturbing that Szent-Györgi and his collaborators did not approach their work in a more critical manner; he started with an assumption and stuck to it. He altered only the tissue from which he claimed to have obtained retin, and the chemical property of the active principle. At one time carbon suboxide was the choice.

Another somewhat confusing piece of research was that on h_2, a protein fraction found in rat liver and which Sorof and Freed (1966) claimed to be not only a kind of homeostatic regulator but also involved in liver carcinogenesis, because when the animal was fed a liver carcinogen the h_2 factor disappeared. It looked as if the carcinogen acted by causing the disappearance of the antimitotic regulator. A little later it was put forward that h_2 was arginase. It is difficult to see at the moment how all this fits together with homeostatic regulation.

Stoker: I would like to ask how in general one ought to study this whole business of homeostasis. We have heard so far about three different classes of compounds which may be involved in homeostasis, some of which may show specificity, and it is a reasonable guess that many more substances will be isolated which will have effects on growth, inhibitory or stimulatory, in intact tissues. If all or many of the possible growth-controlling substances are involved in the growth of each cell in the body, it is likely that the homeostasis that we observe is the end result of an incredibly complex balance, too difficult to analyse at present. And if so, one should perhaps conclude that it is at present no use studying organs and tissues *in vivo*, including even skin; that one should turn instead to single cell systems or even bacteria. On the other hand if there are single overriding and fairly simple mechanisms involving only one or very few classes of molecule in intact tissues and organs, it is worth looking at *in vivo* situations. It is the old problem of whether to choose the complex situation which is nearer to real life or to study a very simple model.

Bergel: This is what I felt when we began to plan this meeting, that the whole subject needs sorting out. Maybe we shall conclude that homeostatic regulators do not exist as characterizable entities at all.

Möller: Controlling mechanisms are obviously needed at the intracellular level in bacteria as well as in mammals. However, in multicellular organisms there is a unique necessity for intercellular control in order to regulate the number of cells in different parts of the body. It is particularly important to have some surveillance against the major threat to multicellular organisms, which is aberrant cell differentiation, leading either to cells which do not function properly, and if not eliminated would accumulate, or to cells which do not obey control and can multiply and kill the organism. As Burnet (1967) has pointed out, the basic regulating mechanism that we should look for in multicellular organisms is one which tends to eliminate neoplasia and other forms of aberrant cell division. He considered the cellular immune response to be particularly important in this respect. Another surveillance mechanism, the allogeneic inhibition phenomenon, which is a cell-to-cell contact phenomenon leading to the elimination of cells which are different in cell surface structure, has been suggested (Hellström and Möller, 1965).

Both cellular immunity and allogeneic inhibition are phenomena of the recognition of cell surface differences. It may be speculated that histocompatibility surface antigens are important markers in this context. Their complexity and widespread chromosomal location makes them useful as detectors of genetic changes, leading to stimulation of surveillance mechanisms.

Lamerton: We may be making the situation too complicated, particularly with regard to the number of homeostatic mechanisms at work in regenerative processes. The problem may not be so much the complication of the chemistry as the problem of perturbing the system so that we can observe the operation of the homeostatic agents. One can learn very little about a system in dynamic equilibrium without perturbing it, and the number of ways in which a specific and quantitative perturbation can be given to the tissues of the body is very limited. The red cell system has the advantage that the mature product can be removed without damaging the proliferative compartment; this represents a physiological type of perturbation which has proved of the greatest importance in the identification and isolation of the specific controlling factor, erythropoietin. I suspect that the problem of identifying specific controlling factors of similar significance in other systems depends to an important extent on finding ways in which to perturb the systems so that we can observe in a fairly pure way the operation of mechanisms and their response.

Abercrombie: One can add another system to Professor Lamerton's list; the control mechanism of the thyroid is fairly well established.

Iversen: It is possible to study homeostasis in the way described in cybernetic theory, either by causing known perturbations in the system and "listening" to the answer, or by "opening up the loop" and seeing what happens. It should not be considered strange for biologists to go to the engineers and to look at cellular homeostatic mechanisms in the light of cybernetic theory. There exists a well-developed theory of how to investigate cybernetic systems with negative and positive feedbacks, which can be applied to a biological system (see Ross Ashby, 1958).

Vernon: I sympathize with Professor Stoker, because I think that within the next ten years we shall find many more proteins which affect the behaviour of cells in tissue culture. The only reason why more haven't been properly characterized is because the combination of a difficult biological assay and a rather small amount of material catches one both ways. This is why NGF came out relatively easily. You need an assay which is rapid and specific and a source of material such that you won't finish up with less than about 10 milligrams of pure protein. That is the limit of the techniques available at the moment. There must be many more of these factors, because tissue culturists need to add serum and embryo juice to cultures, and if you fractionate these and leave components out you get bizarre effects. Thus for example serum contains a substance which makes cells stick together; and it also contains a substance which makes them *not* stick together. The proteins already known are the tip of the iceberg and we shall discover many more when techniques are available.

Wolpert: I am a little bit nervous that we are talking in terms of "how many factors". This presupposes a "factor" type of organization, and what perturbs *me*, if we take the chalones as an example, is that a great deal more effort is not being devoted to designing experiments which might reveal whether there is a diffusible substance in skin which has any relevance to its normal regulation, rather than asking what the substance is. We should first design experiments which will tell us about the control mechanism at a higher level of organization: is it diffusible or not diffusible? Is cell contact important or not important? Biochemists tend to live in a space-free world, but there are many other levels involved beyond the level of testing substances.

Bergel: I did start off by asking the meeting whether we know that such regulating agents exist at all, or alternatively whether there are signs of a regulatory mechanism based on a pattern of interlinked events.

Allison: There is a lesson to be learned from the history of endocrinology, in which it seemed impossibly difficult at first to isolate the many interacting hormones. Later they were found to be heterogeneous, with very different structures, but only a limited number turned out to be interesting and significant. All were assayed in the whole body system and not in individual cell systems; if you try to do assays of hormones or chalones in artificially simple cellular systems you may run into difficulties. I disagree with Professor Wolpert's approach, because the few topics about which there has been really useful discussion at this meeting so far were those where we have something in a test-tube in reasonably pure form, which can be added to a biological system to give a clear-cut result—the NGF and erythropoietin. Given some kind of biological interacting system, I think that it is right first to isolate and purify the components, say chalones, to find out what they are. Then one can go into the kind of problems which Professor Wolpert mentioned. With impure preparations you never know whether an effect is specific or due to a contaminant.

Bergel: But there might be a third possibility. Some factors or agents which have this regulatory activity could be cell- or particle-bound. There, one cannot study excretion or humoral distribution, but one would have to work like the immunologists, who happen to have both circulating immunobodies and cell-bound antibodies.

Wolpert: Dr Allison is not being fair! Historically, nerve growth factor came from carefully designed biological experiments in which tumours were implanted close to the central nervous system, and it was on this basis that people began to suspect that there may be a diffusible substance. I am sure the same happened with erythropoietin. When you have that evidence, it becomes worth looking for a factor, but until then, you may be on totally the wrong track.

O'Meara: Consider also the history of gastrin; it had a curious development from its descriptions at the beginning of the century as a product of the antral mucosa of the stomach to its very recent identification and synthesis.

Bergel: I agree. I also agree with Professor Lamerton's point about perturbation. One may have to stir up something to discover that, and how, it works. Dr Allison and others will remember the symposium on the interaction of drugs and subcellular components (Campbell, 1968) where quite unexpected and interesting changes were reported in mitochondria and lysosomes after drugs had been applied. It may be that apart from interfering with growth processes, drugs exist which tend to upset the pattern.

Möller: Homeostatic mechanisms are disturbing because they tend to obscure biological principles. The study of a particular biological phenomenon is made difficult by feedback systems, since they automatically correct any experimentally induced change. This will result in poor quantitative data. It is of course very important to study the homeostatic mechanisms in order to eliminate their influence and to be able to reach the basic biological principles.

Bergel: We have a pharmacological example of an ideal homeostatic regulating mechanism in the autonomic nervous system* together with acetylcholine and adrenaline and noradrenaline which act on it. In cellular homeostatic mechanisms we are, I believe, at a stage comparable to the period 1907–1910 when Henry Dale and George Barger started on the catecholamines; one could hardly have foreseen the function of acetylcholine with its hydrolysis and its resynthesis by enzymes. In fact the cholinergic and adrenergic systems regulate in a stabilizing sense all our autonomic nervous activities. The problems of growth, at tissue, cellular and subcellular levels, may be no less insoluble.

Lamerton: We have, I think, to distinguish between regulators from outside the cell and the buffering mechanisms operating within the cell. It is the external factors that I would regard as homeostatic regulators in the context of this meeting.

Wolpert: Dr B. Goodwin's concept (1963) of relaxation time may be useful here. Relaxation time is that time required after you have given a system a small perturbation before it gets back to a steady state again. He talks about two types of biological system, the metabolic one which has a very short relaxation time, in the order of seconds, and the epigenetic

---

* "The stability of the internal environment (the *milieu interne* of Claude Bernard) which is so characteristic of the healthy body, is spoken of by Cannon as *homeostasis*. According to Cannon, the essential and particular function of the autonomic system is to bring about the internal adjustments upon which this constant state depends. He therefore refers to the autonomic nerves as the *interofective* system. He speaks of the voluntary system (i.e., the central nervous system and the somatic nerves) as the *exterofective* system, since through its exteroceptors and effectors a direct relationship is established with the external environment."
(Extract from *The Physiological Basis of Medical Practice*, 1950, 5th edn, p. 1079. Best, C. H., and Taylor, N. B. London: Baillière, Tindall and Cox.)

one in which relaxation time is in the order of hours. This may be the distinction that Professor Lamerton is seeking. The sort of homeostatic mechanisms we are looking for are, I think, of the epigenetic type, in which the time-scale is of the order of minutes, or I would have said hours. Whereas the metabolic system, which has a much shorter time-scale, probably will not have much to do with homeostatic regulation.

Lamerton: This is a very useful distinction.

Möller: In a two-component system characteristic of feedback systems there are two important variables for regulation: the intensity of the correcting force and the latency period between perturbation of the system and the initiation of the correcting force. Furthermore, all such systems are characterized by cycling effects. The system may reach a steady state if the correcting force is weak in comparison with the change in the other variable and when the latency period is of a certain magnitude compared to the frequency of cycles in the system. On the other hand, the cycling of the system may increase in amplitude under different conditions (strong correcting force and a certain latency period in relation to the cycling frequency). In certain systems, such as particular cases of antibody synthesis, these variables can be estimated and the result of perturbation of the system can be predicted.

REFERENCES

BURNET, F. M. (1967). *Lancet*, **1**, 1171.
CAMPBELL, P. N. (ed.) (1968). *The Interaction of Drugs and the Subcellular Components in Animal Cells.* Co-ordinating Committee for Symposia on Drug Action of the Biological Council. London: Churchill.
GOODWIN, B. (1963). *Temporal Organization in Cells.* New York: Academic Press.
HELLSTRÖM, K. E., and MÖLLER, G. (1965). *Prog. Allergy*, **9**, 158.
JELLUM, E. (1968). *Biochim. biophys. Acta*, **170**, 430–431.
OTSUKA, H., and EGYUD, L. G. (1968). *Biochim. biophys. Acta*, **165**, 172.
ROSS ASHBY, W. (1958). *An Introduction to Cybernetics.* London: Chapman and Hall.
SOROF, S., and FREED, J. J. (1966). *IX Int. Cancer Res. Congr., Tokyo*, Abstracts, S.O. 315.

PATTERN OF GENE TRANSCRIPTION DURING THE INDUCTION OF BACTERIOPHAGE LAMBDA DEVELOPMENT: A POSSIBLE MODEL FOR THE CONTROL OF GENE EXPRESSION IN A DIFFERENTIATING SYSTEM

F. Gros, P. Kourilsky and L. Marcaud

Institut de Biologie Physico-chimique, Fondation Edmond de Rothschild, Paris

The study of regulatory mechanisms in bacteria has unravelled the multiplicity of ways that can be used by unicellular systems to adapt their physiology to changes in the environment. Homeostatic regulation can be achieved at different levels of the cell organization, as shown by the remarkable network of feedback inhibitory mechanisms on biosynthetic enzymes, as well as by changes in gene activity during enzyme induction or repression. Genetic regulatory circuits in bacteria are so designed as to permit a perfectly coordinated expression of adaptive mechanisms, since the gene units involved in the same physiological function are often clustered into specific operons, the overall activity of which is under the command of a unique DNA sequence: the operator. Repressors as internal inducers produced by regulatory genes are protein signals that convey to the operator the appropriate orders.

The teleological value of these genetic circuits is of such an appeal that many biologists interested in cell differentiation have suggested that they could also constitute the key elements that ensure the programme of gene expression in differentiating systems.

Unfortunately, the genetic organization of eukaryotic systems is too complex to permit such an exploration at the present time, and many laboratories, including our own, have been seeking systems in which the pattern of differentiation would be easier to study at the molecular level. One such system is that of bacteriophages multiplying in bacteria, since the replication and the maturation of the viruses obviously obey a very specific and precise chronological control.

We shall therefore devote this paper to the pattern of gene *transcription* during the development of the DNA-containing bacteriophage lambda.

We shall then discuss our results in terms of a possible model for the control of gene expression in differentiating systems.

DESCRIPTION OF THE SYSTEM

Bacteriophage lambda (λ) is, at present, the most extensively studied of the so-called *temperate* bacteriophages. Temperate phages possess the remarkable property that, upon infection, one part of the cell population can undergo lysis, while the other part can survive and become lysogenic. In a lysogenic bacterium the genetic structure of the phage—now called *prophage*—has become integrated linearly inside the bacterial chromosome. The different phage functions are repressed by a phage *repressor*, sometimes called "immunity substance", because it makes the cell immune to superinfecting phages of the same specificity as the prophage. The prophage is replicated once per cell generation without expression of any other viral function than the phage repressor itself.

Lysogenic *induction* is the phenomenon whereby the prophage is converted back to the stage of an infectious episome. This results from inactivation of the phage repressor by any indirect (ultraviolet irradiation) or direct mechanism (a temperature shift inactivating a thermosensitive repressor). The scheme of phage development may then be described as follows:

(1) Expression of *"early"*, i.e. prereplicative functions
 Prophage excision out of the bacterial chromosome
 Synthesis of the phage enzymes needed for replication
(2) Phage DNA* replication
(3) Expression of *"late"*, i.e. post-replicative functions
 Synthesis of head and tail proteins
 DNA maturation and phage assembly
 Host lysis

Such a temporal pattern is formally similar to the evolution of a differentiating system since a single event, namely the loss of the immunity substance, triggers an orderly sequence of biochemical changes.

We have chosen to study this temporal sequence at the *transcription* level, by characterizing, as a function of time after induction, the various messenger RNA (mRNA) species synthesized by the phage. However, before the relevant results are described it is appropriate to give a brief outline of the genetic structure of the λ chromosome.

---

* The abbreviations used are: DNA, deoxyribonucleic acid; RNA, ribonucleic acid; mRNA, messenger RNA; GC content, percentage of guanosine+cytosine among the four DNA bases; S, sedimentation coefficient.

STRUCTURE OF THE λ CHROMOSOME

Lambda DNA has a molecular weight of $32 \cdot 10^6$. It contains enough information to code for about 40 distinct proteins with an average molecular weight of 40000. A schematic map of the vegetative chromosome is shown in Fig. 1. In the prophage state, the gene order is circularly permuted (Campbell, 1962).

As is the rule in bacteria (but not in bacteriophages) many genes involved in the same function are clustered together. This is the case, for instance, for functions such as head formation, tail formation, integration, excision and recombination. The existence of these large viral operons is supported by some genetic evidence, and biochemical proofs which will be given later.

Most of the late genes, except S and R, which are involved in cell lysis, are located on the left half of the map, while all the early ones are clustered on the right half, including the known regulatory genes: C_I, C_{II}, C_{III}, N and Q. Three of these, C_I, N and Q, deserve special attention.

C_I is the gene that codes for the λ repressor, which has been characterized by Ptashne (Ptashne, 1967a). This C_I product is being purified in several laboratories. As shown by *in vitro* studies, it is a protein with a molecular weight of about 30000 which binds to λ DNA at two distinct sites. These two "operators" lie on each side of the C_I gene and can be independently mutated (Ptashne, 1967b; Ptashne and Hopkins, 1968).

The products of genes N and Q behave as triggers in the temporal control of gene expression. Although their mode of action is still unknown they appear to be diffusible inducers of early and late genes (Thomas, 1966; Dove, 1966).

Different subfractions of the λ chromosome can be separated by a number of techniques. For instance, the two DNA halves, obtained by appropriate mechanical shearing of intact DNA molecules, can be fractionated on the basis of their GC content (45 per cent and 54 per cent for the right and left halves respectively) (Hogness and Simmons, 1964; Nandi, Wang and Davidson, 1965). The two λ DNA strands obtained after heat denaturation of phage particles in the presence of a detergent differ in their ability to bind ribopolymers, and can therefore be separated by caesium chloride density centrifugation (Hradecna and Szybalski, 1967; Cohen and Hurwitz, 1967). Measurements of the information content of DNA fragments of known size (Hogness, 1966) as well as electron microscopic analyses of heteroduplexes (Davis and Davidson, 1968; Westmoreland, Szybalski and Ris, 1969) have allowed the physical distances between a number of genes to be

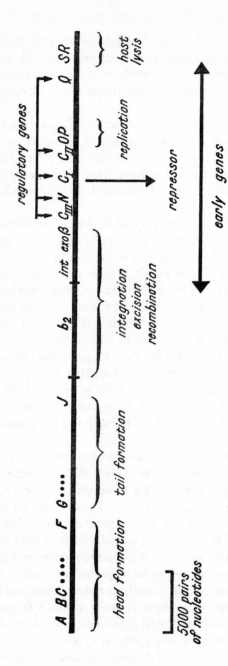

Fig. 1. Genetic structure of the λ chromosome.

estimated. The map in Fig. 1 has been drawn to scale, the unit of length representing five thousand nucleotide pairs.

CHARACTERIZATION OF λ-SPECIFIC mRNA's

It is beyond the scope of this paper to describe in detail all the experimental approaches. We shall emphasize only the guiding principles.

(1) *Recognition of λ-specific mRNA*

The main tool for recognizing λ mRNA among the bulk of bacterial RNA's is DNA-RNA hybridization. This technique is highly specific.

It is possible to use in the annealing reaction not only λ DNA as a whole but also purified DNA pieces, such as the DNA halves or the DNA strands. The direction of transcription, for one particular mRNA species, can be inferred from that DNA strand to which it anneals specifically. The direction of transcription, with reference to the map of Fig. 1, is from right to left for an mRNA complementary to the so-called L strand (L standing for light strand), and from left to right for an mRNA complementary to the H strand (H standing for heavy strand).

(2) *Identification of mRNA species*

The various mRNA species found upon induction can be distinguished on the basis of their molecular weights, determined by either sucrose gradient sedimentation (Kourilsky and Luzzati, 1967; Kourilsky, Luzzati and Gros, 1967) or gel electrophoresis (Kourilsky *et al.*, 1968). Presence of λ-specific RNA in the different fractions of the gradient or the gels is monitored by hybridization with an excess of denatured λ DNA (or the DNA halves, or the DNA strands). The existence of one reproducible peak in a pattern is taken as an indication that one species at least of λ mRNA has the corresponding molecular weight.

(3) *Mapping of the various mRNA species*

Localization on the genetic map is obtained by hybridizing one particular species of mRNA isolated from a sucrose gradient or an acrylamide gel with different DNA's extracted from phages carrying deletions. If a given mRNA species hybridizes with one deleted DNA, but not with another, it can be inferred to originate from a region of the chromosome which lies between the extremities of the two deletions.

When defective prophages carried by lysogenic bacteria could not be purified, a more elaborate technique was used: a given mRNA species was, as above, hybridized with λ DNA. The hybrid was purified and λ mRNA eluted. The (almost pure) λ mRNA thus obtained could then be hybridized with *bacterial* DNA's carrying various deleted prophages, since bacterial-specific RNA had been previously eliminated by the first step of hybridization.

The originality of the methods that we have developed depends on our ability to characterize and isolate *single* messenger RNA species. The results that we shall present are in good agreement with, or confirm, data obtained by other groups by direct hybridization of non-fractionated RNA preparations (Skalka, 1966; Joyner et al., 1966; Skalka, Butler and Echols, 1967; Taylor, Hradecna and Szybalski, 1967; Cohen and Hurwitz, 1967 and 1968; Bøvre, Iyer and Szybalski, 1968).

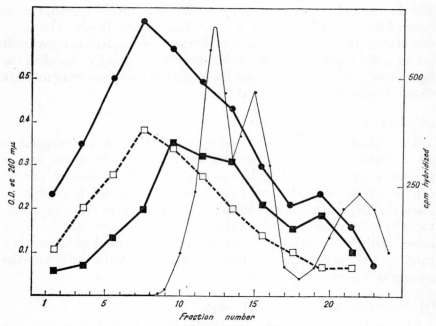

FIG. 2. Sedimentation pattern of late λ mRNA.

$$\frac{\% \text{ hybridized with the left half}}{\% \text{ hybridized with the right half}} = 1 \cdot 2$$

$$\frac{\% \text{ hybridized with the H strand}}{\% \text{ hybridized with the L strand}} = 5$$

——————— optical density (O.D.) at 260 mµ (nm).
——●——●—— counts/min (cpm) hybridized with total λ DNA.
- - -□- - -□- - - counts/min hybridized with the left half of λ DNA.
——■——■—— counts/min hybridized with the right half of λ DNA.

VARIOUS mRNA SPECIES FOUND AFTER INDUCTION OF THE PROPHAGE λ

Three experiments will be taken as examples to illustrate the use of the previously described techniques.

(1) *High molecular weight mRNA found at early and late times*

A sucrose gradient of late mRNA, pulse-labelled for one minute, 15 minutes after heating a thermoinducible prophage λ C_{I857} at 42°C, is shown in Fig. 2. Most of the material hybridizable to λ DNA sediments in the

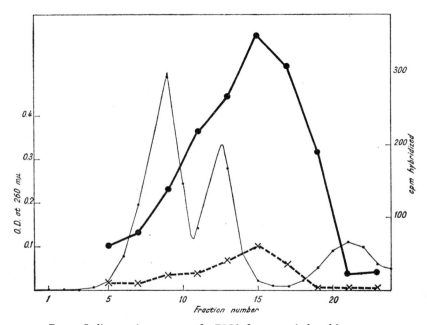

FIG. 3. Sedimentation pattern of mRNA from non-induced lysogens.

$$\frac{\% \text{ hybridized with the L strand}}{\% \text{ hybridized with the H strand}} = 7$$

——————— optical density at 260 mμ (nm).
———●———●——— counts/min hybridized with λ DNA.
– – – ×– – – – ×– – counts/min hybridized with $\lambda_{i\,434}$ DNA.
The background has been subtracted.

30–35 s region of the sucrose gradient. It hybridizes mainly with the left half and the H strand. Assuming that it does not represent aggregates, as supported by many control experiments, we can estimate the molecular weight of the corresponding mRNA molecules to be about $2 \cdot 10^6 – 2 \cdot 5 \cdot 10^6$. Messenger RNA of such a high molecular weight has been found at early as well as at late times (Kourilsky et al., 1968); it is extremely likely to be polycistronic. We therefore conclude that there exist *several large operons* in λ, a fact which warrants our approach to the problem: since the number of mRNA species of such large molecular weight is necessarily restricted, their identification is much easier.

(2) *mRNA found in non-induced lysogens*

A low, but detectable level of hybridization with λ DNA is exhibited by RNA preparations from non-induced lysogens. In terms of the percentage of the input counts hybridized, this level is 0·15 per cent, a hundred times less than the level observed for preparations of late mRNA. Since in our hands the background noise of the annealing technique is equivalent to 0·02 per cent, the hybridization figure of the non-induced lysogens can be considered as significant. The sedimentation pattern of this RNA is shown in Fig. 3. The molecular weight estimated from the sedimentation coefficient of the peak is about 300 000. Hybridization with the DNA from $\lambda_{i\,434}$, a phage in which the λ immunity region is deleted and replaced by that of phage 434 (see Fig. 8), is very low. Therefore, the mRNA species observed originates mostly from the λ immunity region. It hybridizes mainly with the L strand (Taylor, Hradecna and Szybalski, 1967; Cohen and Hurwitz, 1968).

(3) *mRNA found in lysogens induced in the presence of chloramphenicol*

Fig. 4 shows the pattern obtained after electrophoresis of an RNA preparation extracted from cells induced in the presence of chloramphenicol (200 μg/ml). Three species of mRNA can be detected, one complementary to the H strand, two to the L strand. They are easily located on the λ map since they hybridize poorly with the DNA of phage λ_{i21}, in which genes N, C_I and C_{II} from λ are replaced by the immunity region of phage 21. The mRNA species which migrates together with the bacterial transfer RNA's is being extensively studied. It has a molecular weight of 30000 and does not possess any of the characteristics of the known transfer RNA's (L. Marcaud, unpublished results).

The three mRNA species described above are presumably directly controlled by the λ repressor, since they are synthesized in the absence of any inducer protein. The fact that a pattern similar to the one just described is observed after induction of an N mutant indicates that the N protein is probably the inducer which allows the transcription of early species of mRNA of high molecular weight (Kourilsky et al., 1968).

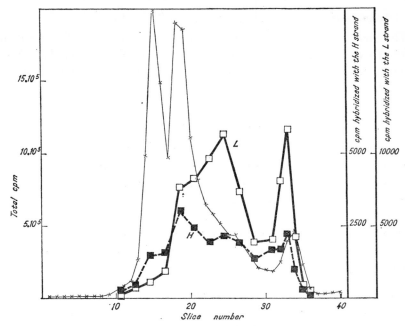

FIG. 4. Electrophoretic pattern of λ mRNA obtained after heat induction of a λ C_{I857} prophage in the presence of chloramphenicol.

The chloramphenicol (200 μg/ml) was added 5 minutes before heating for 3 minutes and pulse-labelling for 1 minute.

$$\frac{\%\text{ hybridized with the H strand}}{\%\text{ hybridized with the L strand}} = 3$$

——×——×—— total counts/min.
——□——□—— counts/min hybridized with the L strand.
----■----■-- counts/min hybridized with the H strand.

Table I

VARIOUS MESSENGER RNA SPECIES OF LAMBDA

Denomination	Molecular weight	Complementary to		Genetic location
		DNA half	DNA strand	
l_0	300000	right	L	Immunity region
l_1	30000	right	L	Region of gene N
l_2	250000	right	L	Region of gene N
l_3	$1 \cdot 8 \cdot 10^6$	right	L	Region of genes β, *exo*, *int*
h_1	$0 \cdot 9 \cdot 10^6$	right	H	Region of genes C_{II}, O (and P?)
h_2	$2 \cdot 10^6$	right	H	Region of gene Q
h_3	unknown	right	H	Unknown—presumably S and R
h_4, h_5	$2-2 \cdot 5 \cdot 10^6$	left	H	Number of species unknown—size compatible with 3 large operons for head and for tail formation

(4) *Characteristics of the various species of λ mRNA identified*

The size, genetic location and direction of synthesis of the various mRNA's which we have identified up to now are listed in Table I. More precise data on the genetic location of these species are given in Fig. 8 (p. 120).

KINETICS OF λ INDUCTION

In order to correlate the results mentioned above with the development of a normal phage, it is now necessary to study the kinetics of induction of a non-mutated phage. The experiments are performed in the following way. The lysogenic cells, carrying a λ C_{I857} prophage, are grown at 33°C and induced by heating at 42°C. Samples are pulse-labelled at different times after heating and the extracted RNA's are hybridized with an excess of λ DNA so as to measure the rate of synthesis of λ mRNA as a function of time.

(a) *Hybridization with the two DNA halves*

As shown in Fig. 5, λ mRNA synthesized during the first five minutes at 42°C hybridizes mainly with the right half of the DNA. The RNA's formed later hybridize with both halves (Naono and Gros, 1969). Since phage DNA synthesis starts five minutes after the onset of thermal induction (Naono and Gros, 1967), it follows that prereplicative RNA's possess the characteristic of being essentially complementary to the right half of λ DNA where all the early genes are clustered. This experiment demonstrates (*i*) that the definition given for early and late mRNA is consistent; (*ii*) that phage transcription is not initiated at random, but starts on the right half and proceeds on both halves.

(b) *Hybridization with the two DNA strands*

More information can be obtained by eluting the mRNA's specifically hybridized with each of the DNA halves at different times of the inductive cycle, and by rehybridizing these purified mRNA's with each of the DNA strands. In this way it is possible to check the rate of synthesis of λ mRNA on each of the strands of the two DNA halves. The results of such an experiment (Sheldrick, 1969) are shown in Fig. 6.

The data clearly indicate (*i*) that the temporal pattern varies for each of the four sectors (I, II, III and IV) of DNA tested; and (*ii*) that early mRNA is synthesized from the right half and from both strands, while late mRNA is synthesized from both halves and mainly from the H strand.

(c) Very early kinetics of induction

In order to analyse further the very early kinetics of prophage induction, short pulses (10 seconds) of a radioactive precursor were administered every five to ten seconds after a very rapid thermal transition (less than

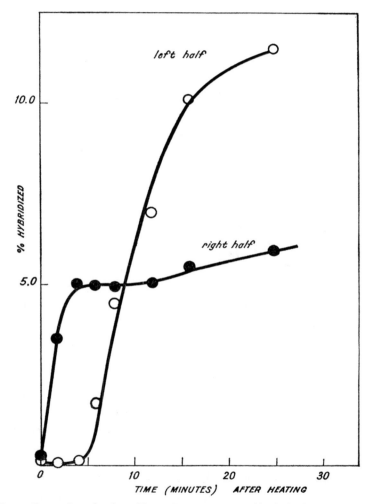

FIG. 5. Formation of early and late mRNA as measured by hybridization with the right and left halves of λ DNA.

———●———●——— percentage of input RNA counts/min hybridized with the right half.

——— ○ ——— ○ ——— percentage of input RNA counts/min hybridized with the left half.

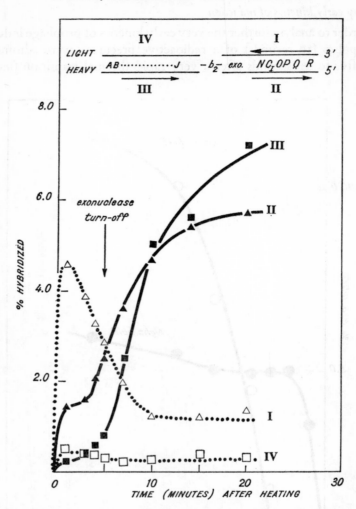

FIG. 6. Formation of early and late mRNA as measured by hybridization with λ DNA halves and λ DNA strands.

5 seconds). In Fig. 7 the percentage of hybridization, which measures the rate of λ mRNA synthesis, is plotted against the time of heating. One observes a lag of about 25 seconds, during which the "induced" mRNA, as well as the mRNA extracted from non-induced lysogens, hybridizes poorly with λ_{i434} DNA. This mRNA is therefore transcribed from the immunity region.

After the lag period and during the subsequent 25 seconds the rate of λ mRNA synthesis increases abruptly, when measured both with λ and λ_{i434} DNA's, but not with λ_{i21} DNA. Thus the mRNA synthesized between the twenty-fifth and the fiftieth second of induction is transcribed from genes situated outside the immunity region, and absent from λ_{i21}

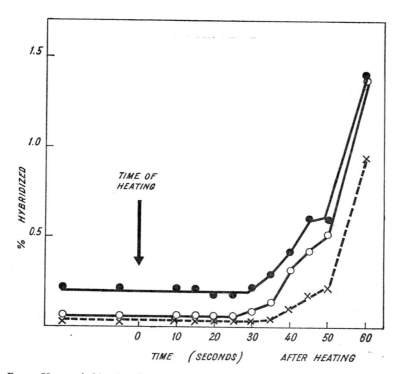

FIG. 7. Very early kinetics of λ mRNA synthesis after induction by heating to 42°C.

———●———●——— percentage of input RNA counts/min hybridized with λ DNA.
———○———○——— percentage of input RNA counts/min hybridized with λ_{i434} DNA.
---×----×--- percentage of input RNA counts/min hybridized with λ_{i21} DNA.

DNA; that is, N, C_I, C_{II} (see Fig. 8). Since RNA's formed during that period hybridize with both DNA strands, it can be concluded that after a short lag of about 25 seconds, *transcription starts on both sides of C_I and on each of the DNA strands simultaneously.*

Results from Fig. 7 contain additional information. It appears from this figure that by 50 to 60 seconds after induction the rate of synthesis of λ mRNA is the *same* when measured with λ and λ_{i434} DNA's. This means

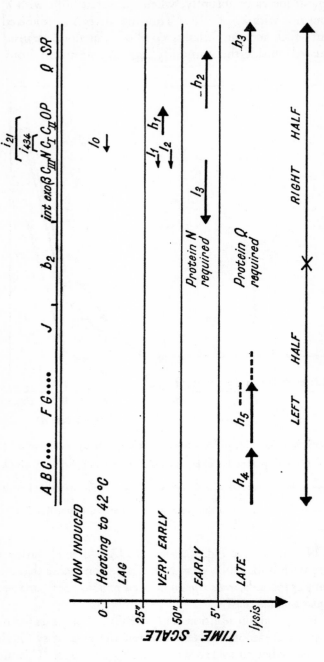

FIG. 8. Stepwise transcription of the λ chromosome.

The various mRNA species present at different times have been approximately drawn to scale, and a number of genes positioned on the map according to physical data, as explained in the text. There is good correspondence between the size of the mRNA species, their genetic location, and the physical distances between the genes.

that the rate of synthesis of the mRNA complementary to the immunity region falls to zero. More elaborate experiments, which confirm this observation, are in progress. We shall tentatively conclude that, as already shown by genetic studies on certain λ mutants (Eisen, Pereira da Silva and Jacob, 1968), the synthesis of the λ repressor is "turned off" after induction.

TEMPORAL CONTROL OF PHAGE RNA SYNTHESIS AFTER INDUCTION

A large number of experiments have been designed to reconstitute the stepwise pattern of RNA transcription following induction of a lysogenic system. Data from the kinetic experiments outlined above have been correlated with those derived from the use of specific prophage mutants. Most of our results so far are summarized in Table II and Fig. 8.

An examination of these schemes allows us to describe λ development after heat induction in the following way:

(1) In non-induced lysogens, the immunity region is the only part of the λ chromosome to be transcribed.

(2) After a thermal shock, a lag of about 25 seconds elapses, during which the immunity region continues to be transcribed to the exclusion of other regions of the λ chromosome.

(3) Between 25 and 50 seconds after the thermal shock, three species of mRNA are synthesized *de novo*. One of these (h_1) probably directs the synthesis of one (or two) replication enzyme(s). Another one (l_2) probably codes for the N protein. The role of l_1 is unknown.

(4) By 50 to 60 seconds, two additional events take place. The synthesis of the C_I mRNA is probably shut off; and protein N is functional and allows the transcription of two large operons (species l_3 and h_2). One of these (l_3) is probably responsible for the expression of those functions involved in the excision of the prophage out of the bacterial chromosome. One function at least that can be ascribed to the h_2 species is to make the protein coded for by cistron Q.

(5) Transcription of species l_3 is maximum by 2 minutes after induction, then the rate of transcription drops (see Fig. 6) (Sheldrick, 1969); this decline is most probably related to the known shutting off of the λ exonuclease, which is one of the enzymes coded for by l_3 (Radding, 1964a, b; Luzzati, 1969).

(6) Up to 5 minutes, transcription occurs only on the right half of the λ DNA (Fig. 5) (Naono, personal communication).

(7) λ DNA synthesis begins at about 5 minutes.

(8) The synthesis of lysozyme, a late enzyme, the expression of which

Table II
SYNTHESIS OF VARIOUS mRNA SPECIES AS A FUNCTION OF TIME AFTER HEAT INDUCTION

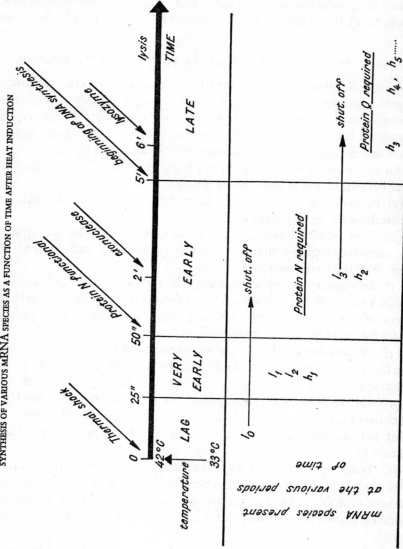

requires the Q function (Dambly, Couturier and Thomas, 1968), begins at around the sixth minute after heating (Kourilsky, unpublished results).

(9) Transcription proceeds on both halves and mainly on the H strand, and this pattern is maintained until lysis occurs at about 20 to 30 minutes.

DISCUSSION AND SUMMARY

The transcription pattern which summarizes our results is probably oversimplified. It is not unlikely that a number of early mRNA species have escaped our investigations. It should be emphasized as well that the location of the various mRNA species is not yet precise enough to allow exact mapping. The present results very clearly demonstrate, however, that λ transcription consequent to the induction of a prophage is not a random process but obeys an orderly chronological scheme. Four distinct, albeit complementary, mechanisms play a major role in the temporal control of gene transcription:

(1) A negative control of key genetic units by the C_I product.
(2) A positive control of early (and possibly late) genes by the N product.
(3) A triggering of late gene transcription by the Q product.
(4) The shutting off of synthesis of the repressor.

The regulation by the immunity substance, namely the repression phenomenon, is rather well understood. The nature of the three other mechanisms is unknown. It is of interest that the synthesis of one regulatory gene, C_I, which was thought to be constitutive, is actually subjected to some kind of regulation. These features of the λ system may be of fundamental importance if they apply to differentiating organisms as well as to this specific bacteriophage.

Acknowledgements

We wish to thank Dr. D. Luzzati, Dr. S. Naono and Dr. P. Sheldrick for communicating their unpublished results and Dr. A. Weissbach for correcting the manuscript.

This work was supported by grants from the Fonds de Développement de la Recherche Scientifique et Technique, the Commissariat à l'Energie Atomique, the Centre National de la Recherche Scientifique, the Ligue Nationale Française contre le Cancer, and the Fondation pour la Recherche Médicale Française.

REFERENCES

BØVRE, K., IYER, V. N., and SZYBALSKI, W. (1968). *Bact. Proc.*, 159.
CAMPBELL, A. (1962). *Adv. Genet.*, **11**, 101.
COHEN, S. N., and HURWITZ, J. (1967). *Proc. natn. Acad. Sci. U.S.A.*, **57**, 1759.
COHEN, S. N., and HURWITZ, J. (1968). *J. molec. Biol.*, **37**, 387.
DAMBLY, C., COUTURIER, M., and THOMAS, R. (1968). *J. molec. Biol.*, **32**, 67.

DAVIS, R. W., and DAVIDSON, N. (1968). *Proc. natn. Acad. Sci. U.S.A.*, **60**, 243.
DOVE, W. (1966). *J. molec. Biol.*, **19**, 187.
EISEN, H. A., PEREIRA DA SILVA, L., and JACOB, F. (1968). *C.r. hebd. Séanc. Acad. Sci., Paris*, **266**, 1176.
HOGNESS, D. S. (1966). *J. gen. Physiol.*, **49**, 29.
HOGNESS, D. S., and SIMMONS, J. R. (1964). *J. molec. Biol.*, **9**, 411.
HRADECNA, Z., and SZYBALSKI, W. (1967). *Virology*, **32**, 633.
JOYNER, A., ISAAC, L. N., ECHOLS, H., and SLY, W. S. (1966). *J. molec. Biol.*, **19**, 174.
KOURILSKY, P., and LUZZATI, D. (1967). *J. molec. Biol.*, **25**, 357.
KOURILSKY, P., LUZZATI, D., and GROS, F. (1967). *C.r. hebd. Séanc Acad. Sci., Paris*, **265**, 89.
KOURILSKY, P., MARCAUD, L., SHELDRICK, P., LUZZATI, D., and GROS, F. (1968). *Proc. natn. Acad. Sci. U.S.A.*, **61**, 1013.
LUZZATI, D. (1969). In preparation.
NANDI, V. S., WANG, J. C., and DAVIDSON, N. (1965). *Biochemistry, N.Y.*, **4**, 1687.
NAONO, S., and GROS, F. (1967). *J. molec. Biol.*, **25**, 517.
NAONO, S., and GROS, F. (1969). In preparation.
PTASHNE, M. (1967a). *Proc. natn. Acad. Sci. U.S.A.*, **57**, 306.
PTASHNE, M. (1967b). *Nature, Lond.*, **214**, 232.
PTASHNE, M., and HOPKINS, N. (1968). *Proc. natn. Acad. Sci. U.S.A.*, **60**, 1282.
RADDING, C. M. (1964a). *Biochem. biophys. Res. Commun.*, **15**, 8.
RADDING, C. M. (1964b). *Proc. natn. Acad. Sci. U.S.A.*, **52**, 965.
SKALKA, A. (1966). *Proc. natn. Acad. Sci. U.S.A.*, **55**, 1190.
SKALKA, A., BUTLER, B., and ECHOLS, H. (1967). *Proc. natn. Acad. Sci. U.S.A.*, **58**, 576.
SHELDRICK, P. (1969). In preparation.
TAYLOR, K., HRADECNA, Z., and SZYBALSKI, W. (1967). *Proc. natn. Acad. Sci. U.S.A.*, **57**, 1618.
THOMAS, R. (1966). *J. molec. Biol.*, **22**, 79.
WESTMORELAND, B. C., SZYBALSKI, W., and RIS, H. (1969). *Science*, **163**, 1343.

DISCUSSION

Wolpert: Dr Gros, could you speculate on how the positive control elements might work?

Gros: Elements which exert a direct positive control upon genes have already been described for certain bacterial systems. For instance in the arabinose system (Sheppard and Englesberg, 1966), gene activation, instead of being due to a release of repression, appears to be mediated by an inducing factor which is a protein synthesized under the control of a regulator gene (gene D). When arabinose is absent the protein does not behave as an inducer, but if arabinose is present it changes this protein into an active inducer. Such positive control mechanisms are known in a variety of situations in bacteria; the arabinose system is one, the same seems to hold for the phosphatase system, and for the rhamnose and maltose systems as well. In fact it is not unlikely that in bacteria the systems known to involve positive regulatory mechanisms are as numerous as those involving repression mechanisms and their release by inducers.

DISCUSSION

Wolpert: In this system gene D is continuously being transcribed, but in *your* (phage) system you implied that there was no transcription going on in the region which then became activated.

Gros: That is right. That is the basis for temporal control of course, which distinguishes the phage and bacterial systems. In the bacterial system, if you remove arabinose the system comes back to the original stage. The particular virtue of the model proposed in the λ system is that it imposes a strict and irreversible sequence of gene transcription.

Burke: This outward growth of messenger RNA in both directions presumably means that the genes are on different strands of the phage DNA molecule.

Gros: The promoters for these operons are situated on different strands.

Burke: So both strands can effectively be used at once, and one can get twice the normal product from a single initiation site. This is a very versatile situation.

Gros: This mostly creates a system which is more flexible in terms of controls. Since the promoter for C_I gene transcription and the promoter for the C_{II} OP operon are located opposite each other on two complementary strands, one possibility is that once derepression activates region x as a promoter for C_{II} OP, a conformational change is created in this region of the chromosome so that the C_I promoter is no longer efficient. This would explain the step of "commitment" during prophage induction leading to the cessation of C_I gene transcription.

Yet such reasoning should not be pushed too far since there are DNA-containing bacteriophages in which bidirectional transcription for distinct regions of the chromosome has not been found (such as T_7; Kubinski, Opara-Kubinska and Szybalski, 1966; Szybalski, 1969).

Stoker: How does it start again? How is the repressor triggered off again in a fresh infection leading to lysogenization? The transcription of the C_I region is turned off after induction and then turned on again when the phage DNA is released into a new uninfected K12.

Gros: How the C_I region is turned on during lysogenization is still a mystery. Another regulator gene has to be invoked, unless it is the repressor itself which sits on the new incoming λ DNA phage and acts as an inducer for its own synthesis.

Allison: But a repressor has to be formed in order to sit on the phage DNA. Could this be a position effect? Then when the phage DNA is incorporated into the bacterial system some kind of position effect would switch on the repressor.

Gros: This may very well be so. Another hypothesis is that genes C_{II} or

C_{III} could act as inducers of the C_I region in this particular case and no longer be active once lysogenization has been established, for it is known that genes C_{II} and C_{III} are acting during the phase of lysogenization and become turned off afterwards.

Allison: But what turns them off during lysogenization?

Gros: I don't know.

Bergel: May I ask a rather general question in order to link up this admirable and complex work with our rather tentative approaches to the problem of homeostatic regulators? This work seems to illustrate the possible complexity which Professor Stoker mentioned earlier and for which various explanations were forthcoming, partly negative and partly positive. Does it suggest that our present knowledge is too small and that perhaps our ideas about agents like chalones or more specific factors like NGF are too simplified, even inside mammalian systems?

Gros: Professor Iversen pointed out the possibility that chalones could represent some sort of general regulators acting at the level of DNA transcription. My feeling is that as long as people do not have in hand a system which is genetically simple and in which mechanisms can be analysed in terms of transcriptional and translational phenomena our approach to the study of differentiating factors will be difficult. This certainly doesn't mean that the problem is not very interesting in itself.

Möller: What do you know about your substances in biochemical terms?

Gros: Neither N nor Q has been purified yet. These are the main problems being pursued now. From 20 to 25 different protein peaks representing λ-specific products have been seen on acrylamide gels. If you irradiate the *E. coli* host with ultraviolet light and then superinfect it with phage, that is, under conditions where the host is not itself making specific proteins, it becomes possible to distinguish virus-coded proteins from the bacterial ones. One can establish their genetic origin quite accurately by infecting with appropriate deletion mutants. This was done by Ptashne (1967) when he first characterized the λ specific repressor. A few other early gene products could also be identified, but this approach has been used much more extensively by M. Schwartz (personal communication, 1968). These various proteins are now under study and will be purified eventually.

Möller: What is the current view on the mechanism of fixation of these proteins to the DNA?

Gros: I believe the assembly step is in great part based upon some self-recognition mechanism—in other words, on the structural proteins being able to interact by means of specific sites; at some stage they also have to recognize DNA, of course. The basis for this recognition is not very well known but it is interesting that one can reconstitute active phage particles if one complements *in vitro* phage structures lacking their tails with extracts containing tail proteins, or does the same thing with head-defective phages.

Bergel: Is there any connexion between this recognition problem and the discovery of restriction enzymes which must recognize certain viral DNA's as foreign products? Meselson and Yuan (1968) described them in *E. coli*.

Gros: You are right in raising the general problem of recognition systems, which are in fact quite numerous in nature. Enzymes taking part in the phenomenon of "restriction", which follows the host-induced modification phenomenon, belong to this category, and there are certainly many others; after all, methylating systems themselves recognize specific sequences in the gene, and glucosylating ones also. And you were right to say that a DNA recognition mechanism might also be important.

REFERENCES

KUBINSKI, H., OPARA-KUBINSKA, Z., and SZYBALSKI, W. (1966). *J. molec. Biol.*, **20**, 313.
MESELSON, M., and YUAN, R. (1968). *Nature, Lond.*, **217**, 1110.
PTASHNE, M. (1967). *Proc. natn. Acad. Sci. U.S.A.*, **57**, 306.
SHEPPARD, D., and ENGLESBERG, E. (1966). *Cold Spring Harb. Symp. quant. Biol.*, **31**, 345.
SZYBALSKI, W. (1969). *Proc. Can. Cancer Conf.*, **8**, in press.

THE HISTONES, THEIR INTERACTIONS WITH DNA, AND SOME ASPECTS OF GENE CONTROL

E. W. JOHNS

Chester Beatty Research Institute, Institute of Cancer Research: Royal Cancer Hospital, London

FRACTIONATION AND CHARACTERIZATION OF HISTONES

THE histones—the basic proteins associated with DNA in most somatic cell nuclei—were first isolated over 80 years ago by Kossel in 1884. The first suggestion that histone was not a homogeneous protein came from Stedman and Stedman (1950) who showed that by ethanol precipitation of histone sulphate two fractions could be obtained, one rich in arginine and the other rich in lysine. Since then many workers have described lysine-rich and arginine-rich histones and various intermediate types (see reviews by Phillips, 1962 and Murray, 1965). However when attempts were made to compare the fractions obtained in the various laboratories, many differences were found, and it became increasingly obvious that the histones were in fact a family of closely related proteins, and that the various workers were merely extracting different combinations of these proteins.

In our laboratory we have concentrated on the fractionation and characterization of these proteins and have developed methods which enable five main histone fractions to be isolated in a reasonably pure form. These fractions have been named according to their method, and order, of isolation, and it must be admitted that the nomenclature, like all histone nomenclatures, has become confusing. Therefore before the fractions are considered in more detail, the procedures which have resulted in their isolation will be briefly reviewed. For comparisons between nomenclatures and fractions described here, and those of other workers, see recent reviews by Hnilica (1967) and Butler, Johns and Phillips (1968).

Whole histone, as extracted from calf thymus deoxyribonucleoprotein with dilute hydrochloric acid, was chromatographed on carboxymethyl-cellulose using a step-wise elution and three main fractions were obtained (Johns et al., 1960). These were designated F_1, F_2 and F_3 in order of elution and were lysine-rich, intermediate, and arginine-rich respectively. Histone F_2 was subsequently separated into two components by a combination of

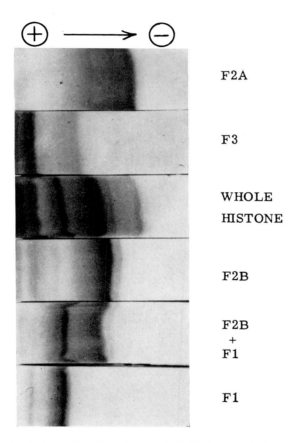

Fig. 1. Starch gel electrophoresis at pH 2·3 of whole histone and histone fractions F1, F2A, F2B, and F3 (from Johns et al., 1961).

chromatographic and selective extraction procedures (Johns and Butler, 1962) and the subfractions were designated F2A and F2B. Two methods were then developed for the large-scale preparation of these four fractions by the selective extraction of deoxyribonucleoprotein with various solvents, followed by acetone precipitation (Johns, 1964a). These fractions were all characterized by total and N-terminal amino acid analyses, and by starch gel electrophoresis at pH 2·3 (Johns et al., 1961). Fig. 1 shows the fractions run in starch gel under these conditions.

More recently, histone fraction F2A has been resolved into two components designated F2A1 and F2A2 (Phillips and Johns, 1965; Hnilica and Bess, 1965; Johns, 1967b). This now gives us the five fractions F1, F2A1, F2A2, F2B and F3, each one accounting for about 20 per cent of the total histone. A schematic diagram showing how all five fractions may be obtained during one preparation is given in Fig. 2.

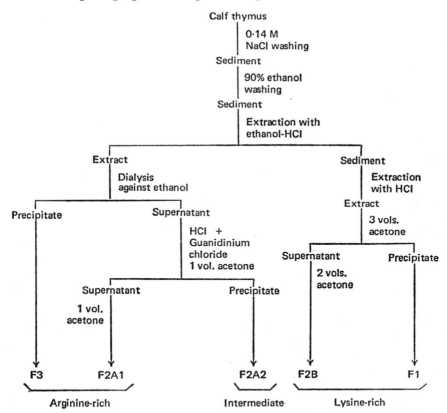

FIG. 2. A procedure for the isolation of the five main histone fractions from calf thymus (Johns, 1964a; Johns, 1967b).

The total and N-terminal amino acid analyses of these five fractions, prepared in the manner described, are given in Table I. It can be seen that they

Table I

THE AMINO ACID COMPOSITION (IN MOLES PER CENT) OF THE FIVE MAIN HISTONE FRACTIONS

		F1	F2B	F2A2	F2A1	F3
Aspartic acid		2·5	5·0	6·6	5·2	4·2
Threonine		5·6	6·4	3·9	6·3	6·8
Serine		5·6	10·4	3·4	2·2	3·6
Glutamic acid		3·7	8·7	9·8	6·9	11·5
Proline		9·2	4·9	4·1	1·5	4·6
Glycine		7·2	5·9	10·8	14·9	5·4
Alanine		24·3	10·8	12·9	7·7	13·3
Valine		5·4	7·5	6·3	8·2	4·4
Cystine/2		0·0	trace	trace	trace	1·0
Methionine		0·0	1·5	trace	1·0	1·1
Isoleucine		1·5	5·1	3·9	5·7	5·3
Leucine		4·5	4·9	12·4	8·2	9·1
Tyrosine		0·9	4·0	2·2	3·8	2·2
Phenylalanine		0·9	1·6	0·9	2·1	3·1
Lysine	Basic amino acids	26·8	14·1	10·2	10·2	9·0
ε-N-Methyl lysine		0·0	0·0	0·0	1·2	1·0
Histidine		trace	2·3	3·1	2·2	1·7
Arginine		1·8	6·9	9·4	12·8	13·0
N-terminal group		acetyl	proline	acetyl	acetyl	alanine

all have about 25 per cent basic amino acids. F1 and F2B are lysine-rich histones, F2A2 is an intermediate type, having a molar ratio of lysine to arginine of about one, and F2A1 and F3 are arginine-rich histones. There is a wide variation in the content of many amino acids, and there are also unique characteristics of each fraction. For example, F1, apart from its high lysine content, has twice as much proline as any other fraction. This is important when considering the secondary structure of these proteins, since proline tends to inhibit the formation of an α helix. F3 is the only fraction to contain cysteine (Deakin, Ord and Stocken, 1963) and has an alanine N-terminal group. F2B has a serine content twice that of any other fraction and has a proline N-terminal group. F2A1 has a very low proline content and a very high glycine content, and F2A2 is characterized by a very high leucine content. These characteristics can be very important since the values for certain amino acids can be used as a guide to the purity of a fraction and to the degree of cross contamination by other fractions.

As a whole the histones are only exceeded by the protamines in their arginine content and basicity, and the lysine-rich histone F1 is unique among known proteins in having over 25 per cent lysine. The molecular weights of the histones range from 12000 to about 20000; they are all quite small molecules.

Contrary to what was believed a few years ago, we now know that there are a very limited number of major histone fractions and it is possible that few further subdivisions of these fractions will occur. Hnilica (1967) has described the separation of F2B into two subgroups designated F2B1 and F2B2, and the interesting work of Kinkade and Cole (1966), Stellwagen and Cole (1968) and Bustin and Cole (1968) has shown that a type of microheterogeneity exists in the lysine-rich histone F1.

The evidence for a limited heterogeneity in the main histone fractions (with the possible exception of F1) comes from the carboxy and amino terminal analyses (Phillips, 1958; Phillips, 1963; Phillips and Simson, 1969) and from the amino acid sequence work which is being carried out now in a number of laboratories. Fambrough and Bonner (1966) have also shown that a similar but limited heterogeneity exists in the pea plant histones. That F2A1 is a homogeneous protein has been proved recently by the publication of the complete amino acid sequence of this fraction (histone IV in the nomenclature used) by Delange and his co-workers (1968a).

The amount of whole histone in a cell is approximately equal to the amount of DNA and as each fraction is about 20 per cent of this, the fractions are present in relatively large amounts. The possibility does therefore exist that minor components may be present which would not show up in the analyses, and would be too small to detect in the sequence work. It has been pointed out by Cruft (1966) that minor components like this may be extremely important, but if they exist then their functions must be quite different from those of the major fractions.

These five major histone fractions probably occur in all mammalian somatic cell nuclei. A wide variety of mammalian tissues of varying activity in protein synthesis and cell division, both normal and malignant, have been examined and in all cases all five fractions were present. It was interesting to find them so non-specific, but it is now apparent that the same five fractions exist in birds, reptiles, fish and some plants, and as far as it is possible to judge at the present time they are all identical.

The work of Delange and his co-workers (1968b) on the histone fraction IV (F2A1) from calf thymus and pea embryo suggests that the sequences of these histones will be very similar. The authors state that if they are as similar as they expect when both complete sequences are known, then such a finding will be unique for two homologous proteins so widely separated on the evolutionary scale. This similarity almost certainly accounts for the fact that the histones are not antigenic and indicates at least for fraction F2A1 that its function must be very dependent on the exact sequence or shape of the whole molecule, not just an active centre or part, and that any major variation due to mutation has been lethal. If this is the case then it indicates some important structural function in the chromatin where the shape of the whole molecule is important.

Since these histone fractions appear in all mammalian cells regardless of their differing functions, the next obvious question was, do the various fractions differ quantitatively in any way in different tissues? To enable us to investigate this problem we developed a method for the separation of the histone fractions by electrophoresis in polyacrylamide gel (Johns, 1967a). Fig. 3 shows the five fractions and whole histone separated in this way and stained with naphthalene black. F3 and F2B run together under these conditions but the other fractions are well enough separated to be cut out as separate bands from the gel. The slow-running band at the top of the gel is bovine plasma albumin added as a marker protein for determining relative mobilities. The colour can be eluted from the sliced bands using dimethyl sulphoxide, and standard curves of colour against amounts of the various fractions can be obtained. With this method it is now possible to measure quantitatively the amounts of the various histone fractions in a whole histone extract of about 100 μg. We have applied this technique so far to a number of mammalian tissues, both normal and malignant, but have found no significant differences. Fambrough, Fujimura and Bonner (1968) have however claimed that in young pea cotyledons the content of the lysine-rich histone F1 is significantly lower than in the other pea organs.

INTERACTIONS BETWEEN HISTONES AND DNA

I should like to turn now to the question of the interactions between these histone fractions and DNA. If specific interactions do occur, and by this I mean that if there are parts of the DNA molecule where one histone is allowed and the others are not, then the addition of DNA to a large excess of whole unfractionated histone should result in the precipitation of a nucleoprotein in which the various histone fractions are in the same proportions as they exist in the native nucleoprotein. To examine this possibility DNA was added to an 8-fold excess of whole histone in 2M-sodium chloride solution in which the complex would normally be dissociated, and dialysed down to 0·14M-sodium chloride over a period of 6 hours. Under these conditions the DNA and histone will recombine and precipitate. The precipitated nucleoprotein was removed and the remaining solution, which contained the histone which had not combined with the DNA, was readjusted

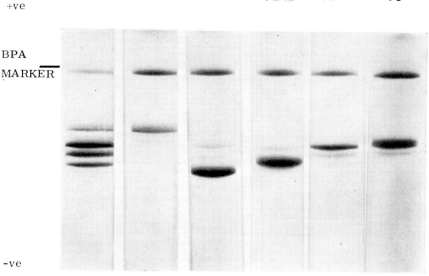

FIG. 3. Polyacrylamide gel electrophoresis at pH 2·4 of whole histone and histone fractions F1, F2A1, F2A2, F2B, and F3 (Johns, 1967a). BPA=bovine plasma albumin. (Reprinted with permission from Pergamon Press Ltd. Fig. 2 from Butler, Johns and Phillips, 1968.)

FIG. 4. Starch gel electrophoresis patterns of histones extracted from reconstituted nucleoproteins made by the successive addition of DNA to an excess of whole histone (Johns and Butler, 1964).

To face page 132

to 2M-sodium chloride and another addition of DNA made. This process was repeated twice more and so four nucleoproteins were obtained by the successive additions of DNA. The histones were extracted from these nucleoproteins and the DNA was also recovered.

Fig. 4 shows the starch gel electrophoresis patterns of the extracted histones. It can be seen that the first two additions of DNA do not combine with any lysine-rich histone, F1, and that the last addition produces a nucleoprotein containing lysine-rich histone only. The DNA has in fact fractionated the histone and thus there appears to be no specificity of interaction between histone fractions and parts of the DNA molecule, but simply the type of selective interaction one might expect between a polyanion and polycations of varying charges. However, the interesting point to note is that as the composition of the histone in the nucleoprotein changes there is also a marked change in the ratio of histone to DNA. We therefore decided to investigate the amounts of the various histone fractions required to combine with and precipitate a given amount of DNA. This was carried out simply by adding a given quantity of DNA to a 4-fold excess of each of the histone fractions in 2M-sodium chloride solution and dialysing down to 0·14M-sodium chloride. The precipitated nucleoproteins were then recovered and the ratios of histone to DNA were determined. In all cases all the DNA had been precipitated. The results of this experiment (Johns and Butler, 1964) showed that much less of F1 lysine-rich histone was required to precipitate the DNA (histone/DNA ratio 0·75 to 1) than of the other fractions and that the F2A fractions were the least efficient in this direction (histone/DNA ratio 2·3 to 1). Although F1, the lysine-rich histone, has the least affinity for DNA, since it is the last fraction to be taken up from a mixture of histones, it is nevertheless the most efficient in precipitation.

I would now like to discuss these results with special reference to the structure of nucleoprotein, but before I do so I would like to make two points.

The first is that a structural gene required to code for a protein containing approximately 500 amino acids, that is, with a molecular weight of about 55 000 (a fairly average size protein), would require about 1 500 nucleotides, assuming three nucleotides are required to code for one amino acid. This would be a portion of one strand of DNA with a molecular weight of about half a million, or a portion of the double helix of approximately one million molecular weight. Now the average molecular weight for the histones is about 16 000, and since there is an approximate one-to-one relationship between the number of basic amino acid side chains and the phosphate groups of the DNA, and an approximate one-to-one ratio of histone to

DNA in all cells, it will require about 60 molecules of histone to combine with such a gene. Therefore on this basis alone any specificity of interaction relevant to the problem of gene control is unlikely unless the order or sequence of histone fractions along the DNA is important. Another way of looking at this is that each average histone will cover a portion of the DNA which will ultimately code for about 10 amino acids. The second point is that one molecule of DNA of molecular weight about 6 million would be combined with about 300 to 400 histone molecules and since there are about 20 phosphate groups per complete turn of the DNA diad helix, each average molecule of histone would only occupy two to three turns.

With these points in mind it appears unlikely that the addition of individual histone fractions to DNA would throw any light on the specificity of interactions appropriate to the problem of gene control, as it seems probable that all genes will be covered by many different histones. Indeed it has been shown that the inhibition of the *in vitro* DNA-primed RNA synthesis by histones is non-specific, since in sufficient amount any fraction can cause the total inhibition of RNA synthesis (Barr and Butler, 1963).

The experiments I have described also indicate that there is no specificity of interaction between histones and specific parts of the DNA molecule, as the DNA is capable of fractionating the whole histone or of combining completely with any of the fractions. However, there is a marked difference in the manner in which the different histones combine with the whole of the DNA. Much less of F1, the lysine-rich histone, than of the other fractions is required to precipitate a given weight of DNA, and F2A is the least efficient in this direction. It is unlikely that these differences can be accounted for simply by the net positive charges on the histone molecules. F1 has a higher content of basic amino acids and fewer acidic amino acids than the other fractions but this is not enough to account for it being three times more efficient in precipitating the DNA. The most likely explanation is that the ability to satisfy and precipitate the DNA is proportional to the number of basic amino acids sterically available for combination with the phosphate groups of the DNA, as a result of the shape of the histone molecule. This would of course be determined by the primary structure.

Phillips and Simson (1962), Johns (1964*b*) and Butler, Johns and Phillips (1968) have shown that the primary structure of the histone molecules is very irregular in the spacing of the basic amino acids, and that there are regions in some fractions with ten or more non-basic amino acids together. The regular structure proposed for nucleoprotein in which every phosphate group is attached to a basic amino acid side chain, separated by three non-basic residues, is therefore no longer an acceptable theory.

It is more likely that the nucleoprotein molecule consists of DNA with the histones attached in the following manner. The F2A1 and F2A2 fractions which combine with the DNA in the ratio of 2·3 to 1 would have loops or bulges of the polypeptide chain (probably the more hydrophobic portions not containing the basic side chains) not combined with the DNA. F1, the lysine-rich histone, which combines with DNA in the ratio of 0·75 to 1, would be stretched round, along, or even across strands of DNA but covering much more DNA per weight of protein than the other fractions, and probably having very little α helical structure. The high content of proline found in F1, about 10 per cent, would certainly tend to prevent this fraction from having an α helical structure. Fractions F2B and F3 would be intermediate in this respect. This would be consistent with the results of Bradbury and his co-workers (1965) who have shown that F1 has the lowest α helix content of all the fractions, and also with the work of Laurence (1966) who has shown that the F2A fractions have the greatest ability to bind the anionic fluorescent dye 8-anilino naphthalene-1-sulphonic acid, which appears to combine mainly with the hydrophobic portions of the protein molecule.

With this situation in mind, with hydrophobic portions of the histone molecules not in intimate contact with the phosphate groups of the DNA, we need to turn to some recent X-ray evidence on the structure of deoxyribonucleoprotein. Wilkins, Zubay and Wilson (1959) first suggested that in native deoxyribonucleoprotein the DNA diad helix is coiled again to form what is now known as a super-helix or super-coil. Pardon (1966) and Pardon, Wilkins and Richards (1967) gave the dimensions of such a super-helix as 100 Å in diameter with a pitch of 120 Å. It appears from more recent studies by Richards and Pardon (personal communication) and Garrett (1968) that the supercoiling is entirely dependent on the histones. There is no supercoiling with DNA alone but when the deoxyribonucleoprotein is reconstituted using whole histone the super-coil returns. Basic polypeptides and protamine will not cause the DNA to assume this characteristic native conformation. It is possible then that the hydrophobic loops of the histones not in close contact with the phosphates of the diad helix may form hydrophobic bonds with corresponding loops further down the chain, or in some way cause an uneven tension in the diad helix which would result in the formation of a super-helix.

The model in Fig. 5 shows a length of the DNA diad helix, of about 34 turns, constructed from rubber tubing and pins, with part of it in the form of a super-coil. These rough models (scale 1 Å = 1 mm with each pin 7 mm apart, representing one base pair) can be very useful in the study of possible types of interactions between histones and DNA.

These 100 Å coils are almost certainly the basic units of the chromosomes and threads of approximately 100 Å can be seen clearly in many electron micrographs of chromatin ultrastructure (Mazia, 1966; Solari, 1965). The macrostructure of the chromosomes could then be built up from these coils by cross-linking the acidic amino acid side chains of the histones with di-valent cations such as magnesium or calcium. The whole structure, being held together by hydrophobic and ionic bonds, would probably be very susceptible to small changes in the ionic environment, and such changes may be part of the mechanism of the condensation and subsequent diffusion of the chromatin during the cell cycle. It is known that EDTA is necessary to break down the condensed structure of native deoxyribonucleoprotein *in vitro* (Zubay and Doty, 1959) and that the synthetic complex between histone fraction F1 and DNA requires the presence of EDTA to render it soluble (Johns and Forrester, 1969).

HISTONES AND GENE CONTROL

It is now well established that the histones are capable of inhibiting *in vitro* the synthesis of RNA formed on a DNA template in the presence of RNA polymerase (Huang and Bonner, 1962; Bonner and Huang, 1963; Allfrey, Littau and Mirsky, 1963; Barr and Butler, 1963; Hindley, 1963; Huang, Bonner and Murray, 1964). These authors have also noted a variation in the extent to which the various histone fractions can inhibit this reaction. Allfrey, Littau and Mirsky (1963) and Hindley (1963) concluded that the arginine-rich histones were the best suppressors of DNA template activity, whereas Barr and Butler (1963) and Huang, Bonner and Murray (1964) reported that the lysine-rich histone was the most efficient. Since these authors have carried out their experiments under very different conditions, using different ratios of histone to DNA, different ionic environments and histones prepared by different methods, it is not surprising that the results are at variance. Also, since DNA–histone complexes are invariably precipitated in the conditions used in most of these experiments it is not surprising that the DNA is unable to act as a template for RNA synthesis. Much more knowledge is required about the correct histone–DNA ratios to use and the precise ionic environments required to enable the nucleoprotein to remain soluble, before experiments of this kind will be meaningful (Johns and Forrester, 1969).

In my opinion there is no real evidence at the present time to show that *in vivo* the histones need to be removed from the DNA to enable it to be used as a template for RNA synthesis. In the active regions of *Drosophila* salivary

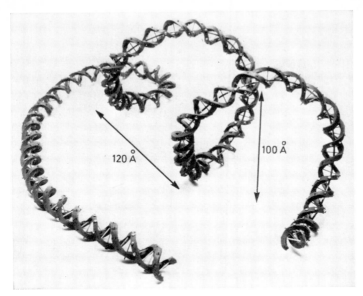

FIG. 5. A model of DNA (scale 1 mm = 1 Å) showing the approximate dimensions of the super-helix.

FIG. 6. Models of DNA, and RNA polymerase to the same scale (1 mm = 1 Å), showing the relative dimensions of these molecules and the possible steric effect of a super-helix on the ability of the RNA polymerase to move along the diad helix.

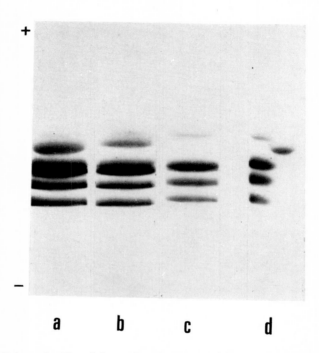

FIG. 7. Polyacrylamide gel electrophoresis patterns of duck erythroid cell histones and calf thymus histones. (a) Whole histone from duck erythrocytes, (b) whole histone from duck immature erythroid cells, (c) whole histone from calf thymus, (d) double gel: left-hand side, whole histone from calf thymus; right-hand side, histone FV or F2C from avian erythrocytes (from Dick and Johns, 1969).

glands (Swift, 1964) and in the active diffuse chromatin of interphase lymphocytes (Frenster, 1966) it is well established that there is an increase in the amount of acidic proteins and RNA, but never a significant decrease in the content of histone relative to DNA. Indeed, Sonnenberg and Zubay (1965) have shown that it is only necessary to open up the structure of deoxyribonucleoprotein to obtain an increase in template activity. It is likely then that the structure of the chromatin is critically important. When completely condensed, during mitosis, the chromatin is obviously all inactive, but during interphase when it is diffuse and has a more open structure, parts must be copied and parts rendered inactive. This then would be a good function for the super-coil. The X-ray evidence mentioned previously, which shows extensive supercoiling in native nucleoprotein, nevertheless still shows some diffraction patterns similar to isolated stretched DNA, so presumably some parts of the nucleoprotein molecule are present as straight stretches of diad helix. Unfortunately, at present there is no way of estimating the relative amounts of straight chain and super-coil. Nevertheless it is attractive to postulate that the histones are non-specific gene repressors, acting by causing supercoiling, and that other macromolecules, which must have the necessary specificity to recognize a particular gene, cause a derepression by breaking the very weak bonds of the super-coil and thus making that region available for transcription by RNA polymerase.

Here again the scale models are useful in showing that a large molecule like RNA polymerase would almost certainly be sterically hindered by a super-coil. Fig. 6 shows such a model with RNA polymerase made to the same scale, according to Fuchs and co-workers (1964). Although this is probably not the correct shape of the RNA polymerase molecule, it is undoubtedly very large, with a molecular weight between 500000 and 1000000 (Lubin, 1969), and would have great difficulty in negotiating a super-coil.

The question remains: what type of molecule is capable of recognizing a specific base sequence, and modifying the existing structure in some way so as to render it capable of transcription? The best candidates for this function at the present time would appear to be the acidic or non-histone nuclear proteins (Paul and Gilmour, 1968), or RNA (Huang and Bonner, 1965; Frenster, 1966).

THE AVIAN ERYTHROCYTE SPECIFIC HISTONE

There is one other aspect of our work I would like to mention as it has a specific bearing on the problem of the control of RNA and protein synthesis.

The mammalian erythrocyte does not synthesize RNA or protein and does not contain a nucleus. The avian erythrocyte also does not synthesize RNA or protein but it does contain a nucleus which is apparently completely inactive. It also contains a unique histone, not found in any other bird tissues, which has been designated FV or F2C (for references to the discovery, isolation and characterization of this fraction, see the reviews mentioned on p. 128). It would appear that during the evolutionary process, mammals have rejected the nucleus as being of no further use in the erythrocyte, whereas birds have simply developed a new histone to render it completely inactive. We have been examining this situation quantitatively with the polyacrylamide gel electrophoretic method (Johns, 1967a) and it appears that this unique fraction F2C is replacing the very lysine-rich histone F1. In Fig. 7, gel c is the normal calf thymus histone pattern, the double gel d shows where the unique fraction F2C runs relative to the other histone fractions, and gels a and b are total histones from duck erythrocytes and reticulocytes respectively. The quantitative measurements show that when F2C is present the content of the lysine-rich histone F1 drops to about 4 per cent. We think in fact that there is a complete loss of F1 and that this trace is due to the non-erythroid cells in the blood used (Dick and Johns, 1969).

We do not understand the significance of this change but feel that this final blanking off of the nucleus is probably concerned with the chromatin which must be active in all cells regardless of their functions—possibly the chromatin which codes for the enzymes and proteins and RNA which are needed for the normal living and functioning of any somatic cell.

It would be interesting to study a completely differentiated cell like this under the conditions described by Harris (1965, 1967), where the erythrocyte is persuaded by a new cytoplasm to become active again, and to see if the unique fraction disappeared and the lysine-rich histone returned.

SUMMARY

The histones of mammalian somatic cell nuclei have been separated into five well-characterized fractions designated F1, F2A1, F2A2, F2B and F3. The primary structure of these fractions and their interactions with DNA have been studied and it is concluded that the histones are combined with the DNA in a very irregular manner. The way in which the histones combine with the DNA is thought to give rise to the super-helical structure of deoxyribonucleoprotein, which in turn probably prevents the DNA from being used as a template for RNA synthesis. Other molecules, such as RNA or non-histone nuclear proteins, may change the structure in such a way as to allow the DNA template to be transcribed.

Quantitative studies on the avian erythrocyte specific histone are also described.

Acknowledgements

This work has been supported by grants to the Chester Beatty Research Institute (Institute of Cancer Research: Royal Cancer Hospital) from the Medical Research Council and the British Empire Cancer Campaign for Research.

REFERENCES

ALLFREY, V. G., LITTAU, V. C., and MIRSKY, A. E. (1963). *Proc. natn. Acad. Sci. U.S.A.*, **49**, 414–421.
BARR, G. C., and BUTLER, J. A. V. (1963). *Nature, Lond.*, **199**, 1170–1172.
BONNER, J., and HUANG, R. C. C. (1963). *J. molec. Biol.*, **6**, 169–174.
BRADBURY, E. M., CRANE-ROBINSON, C., PHILLIPS, D. M. P., JOHNS, E. W., and MURRAY, K. (1965). *Nature, Lond.*, **205**, 1315–1316.
BUSTIN, M., and COLE, R. D. (1968). *J. biol. Chem.*, **243**, 4500–4505.
BUTLER, J. A. V., JOHNS, E. W., and PHILLIPS, D. M. P. (1968). *Prog. Biophys. molec. Biol.*, **18**, 209–244.
CRUFT, H. J. (1966). *Ciba Fdn Study Grp Histones: their role in the transfer of genetic information*, p. 15. London: Churchill.
DEAKIN, H., ORD, M. G., and STOCKEN, L. A. (1963). *Biochem. J.*, **89**, 296–304.
DELANGE, R. J., FAMBROUGH, D. M., SMITH, E. L., and BONNER, J. (1968a). *Proc. natn. Acad. Sci. U.S.A.*, **61**, 1145–1146.
DELANGE, R. J., FAMBROUGH, D. M., SMITH, E. L., and BONNER, J. (1968b). *J. biol. Chem.*, **243**, 5906–5913.
DICK, C., and JOHNS, E. W. (1969). *Biochim. biophys. Acta*, **175**, 414–418.
FAMBROUGH, D. M., and BONNER, J. (1966). *Biochemistry, N.Y.*, **5**, 2563–2570.
FAMBROUGH, D. M., FUJIMURA, F., and BONNER, J. (1968). *Biochemistry, N.Y.*, **7**, 575–584.
FRENSTER, J. H. (1966). In *The Cell Nucleus: Metabolism and Radiosensitivity*, pp. 27–46. London: Taylor and Francis.
FUCHS, E., ZILLIG, W., HOFSCHNEIDER, P. H., and PREUSS, A. (1964). *J. molec. Biol.*, **10**, 546–550.
GARRETT, R. A. (1968). *J. molec. Biol.*, **38**, 249–250.
HARRIS, H. (1965). *Nature, Lond.*, **206**, 583–588.
HARRIS, H. (1967). *J. Cell Sci.*, **2**, 23–32.
HINDLEY, J. (1963). *Biochem. biophys. Res. Commun.*, **12**, 175–179.
HNILICA, L. S. (1967). *Prog. nucleic Acid Res. molec. Biol.*, **7**, 25–106.
HNILICA, L. S., and BESS, L. G. (1965). *Analyt. Biochem.*, **12**, 421–436.
HUANG, R. C. C., and BONNER, J. (1962). *Proc. natn. Acad. Sci., U.S.A.*, **48**, 1216–1222.
HUANG, R. C. C., and BONNER, J. (1965). *Proc. natn. Acad. Sci. U.S.A.*, **54**, 960–967.
HUANG, R. C. C., BONNER, J., and MURRAY, K. (1964). *J. molec. Biol.*, **8**, 54–64,
JOHNS, E. W. (1964a). *Biochem. J.*, **92**, 55–59.
JOHNS, E. W. (1964b). *Biochem. J.*, **93**, 161–163.
JOHNS, E. W. (1967a). *Biochem. J.*, **104**, 78–82.
JOHNS, E. W. (1967b). *Biochem. J.*, **105**, 611–614.
JOHNS, E. W., and BUTLER, J. A. V. (1962). *Biochem. J.*, **82**, 15–18.
JOHNS, E. W., and BUTLER, J. A. V. (1964). *Nature, Lond.*, **204**, 853–855.

JOHNS, E. W., and FORRESTER, S. (1969). *Biochem. J.*, **111**, 371–374.
JOHNS, E. W., PHILLIPS, D. M. P., SIMSON, P., and BUTLER, J. A. V. (1960). *Biochem. J.*, **77**, 631–636.
JOHNS, E. W., PHILLIPS, D. M. P., SIMSON, P., and BUTLER, J. A. V. (1961). *Biochem. J.*, **80**, 189–193.
KINKADE, J. M., and COLE, D. R. (1966). *J. biol. Chem.*, **241**, 5790–5805.
KOSSEL, A. (1884). *Z. Hoppe-Seyler's physiol. Chem.*, **8**, 511.
LAURENCE, D. J. R. (1966). *Biochem. J.*, **99**, 419–426.
LUBIN, M. (1969). *J. molec. Biol.*, **39**, 219–233.
MAZIA, D. (1966). In *The Cell Nucleus: Metabolism and Radiosensitivity*, pp. 15–25. London: Taylor and Francis.
MURRAY, K. (1965). *A. Rev. Biochem.*, **34**, 209–246.
PARDON, J. F. (1966). Thesis, University of London.
PARDON, J. F., WILKINS, M. H. F., and RICHARDS, B. M. (1967). *Nature, Lond.*, **215**, 508–509.
PAUL, J., and GILMOUR, R. S. (1968). *J. molec. Biol.*, **34**, 305–316.
PHILLIPS, D. M. P. (1958). *Biochem. J.*, **68**, 35–40.
PHILLIPS, D. M. P. (1962). *Prog. Biophys. biophys. Chem.*, **12**, 211–280.
PHILLIPS, D. M. P. (1963). *Biochem. J.*, **87**, 258–263.
PHILLIPS, D. M. P., and JOHNS, E. W. (1965). *Biochem. J.*, **94**, 127–130.
PHILLIPS, D. M. P., and SIMSON, P. (1962). *Biochem. J.*, **82**, 236–241.
PHILLIPS, D. M. P., and SIMSON, P. (1969). *Biochem. biophys. Acta*, **181**, 154–158.
SOLARI, A. J. (1965). *Proc. natn. Acad. Sci. U.S.A.*, **53**, 503–511.
SONNENBERG, B. P., and ZUBAY, G. (1965). *Proc. natn. Acad. Sci. U.S.A.*, **54**, 415–420.
STEDMAN, E., and STEDMAN, E. (1950). *Nature, Lond.*, **166**, 780–781.
STELLWAGEN, R. H., and COLE, D. R. (1968). *J. biol. Chem.*, **243**, 4456–4462.
SWIFT, H. (1964). In *The Nucleohistones*, pp. 169–183, ed. Bonner, J., and Ts'o, P. San Francisco: Holden-Day.
WILKINS, M. H. F., ZUBAY, G., and WILSON, H. R. (1959). *J. molec. Biol.*, **1**, 179–185.
ZUBAY, G., and DOTY, P. (1959). *J. molec. Biol.*, **1**, 1–20.

DISCUSSION

Möller: I should like to mention some work done at the Department of Cell Research in Stockholm (Killander and Rigler, 1965) concerned with the binding of acridine orange to DNA in lymphocytes. This dye binds to phosphate groups in DNA and the degree of binding can be quantified. Binding occurs only when acridine orange binding sites in the DNA are liberated by weakened interaction between DNA and nucleoprotein. In highly differentiated cells, like chicken erythrocytes or spermatozoa, there is virtually no binding of acridine orange.

The interesting point is that acridine orange binding can be triggered by a variety of inducing processes. If lymphocytes are treated with phytohaemagglutinin, the acridine orange binding ability increases markedly within 5 minutes (Killander and Rigler, 1965).

We have made mixed cultures of human lymphocytes. When lymphocytes from two different individuals are mixed a series of synthetic events starts; first RNA and protein synthesis starts and after 24 hours DNA synthesis begins. The reaction is considered to represent specific immunological recognition of transplantation antigens. Binding of acridine orange starts to increase within 45 minutes of mixing the cells. The binding intensity increases during the first 3 to 6 hours, then it declines to background level after 24 hours at the same time as DNA synthesis starts (Andersson *et al.*, 1969). This reaction occurs only when there is a genetic difference between the lymphocytes. This pattern of binding appears to represent an extremely early recognition phenomenon. The structural changes in the DNA nucleoprotein complex revealed by this technique appear to be necessary for cell division, since division never takes place unless these changes have occurred, but they are by themselves insufficient for the initiation of DNA synthesis. Thus, *all* cells in a mixed lymphocyte culture show increased acridine orange binding, but only 1 per cent start to synthesize DNA and divide. Additional recognition mechanisms are therefore necessary for the actual induction of DNA synthesis.

Johns: Do you know what percentage of the phosphate groups are bound to the acridine orange? What is the evidence that it is an actual link to the phosphate groups?

Möller: Between 50 per cent and 90 per cent of the phosphate groups in DNA became bound in stimulated cells. The evidence for linkage to phosphates has been reported by Rigler (1964).

Gros: Is this evidence that the dye is linked to the phosphates of DNA rather than intercalated among the superposed bases?

Johns: There can be many types of linkage with dye binding.

Allison: The binding of acridine orange to DNA is certainly complex (Steiner and Beers, 1961) and does not always involve intercalation of the type described by Lerman (1963). Phosphate groups are important, but they are not the only groups involved in the interaction, which is in any case not confined to nucleic acids. I have found that phosphoproteins, for instance, bind acridine orange.

I would just add to what Dr Möller has said: the ability of nuclei in lymphocytes to act as templates for RNA synthesis with added polymerase is increased at the same time as he finds his increase in acridine binding after treatment with phytohaemagglutinin (Weissmann *et al.*, 1967). A parallel increase can be obtained by treating these nuclei with trypsin, which of course breaks down basic proteins.

Johns: If you strip the histones from DNA you will certainly get an increase in template activity, but the point is, is this how it happens *in vivo*? Can DNA be transcribed whether the histones are there or not?

Allison: I agree, and this leads to my second point. In the transforming lymphocyte you have a physiological derepression which can be produced with a protease in fairly low concentrations. This is much more likely to be a specific effect on histones than some kind of ionic effect of the kind which Dr Möller mentioned.

Wolpert: It is interesting here that when John Gurdon transfers nuclei from say a tadpole gut epithelial cell into a frog egg which has been enucleated, there is an enormous swelling of the nucleus. And H. Harris also reported this when he fused different mammalian cells.

Johns: I believe Professor Harris feels that electrolytes are responsible for the reactivation of the chicken erythrocyte nucleus after fusion with a mammalian cell (see Harris, 1969).

Gros: I fully agree that there are a variety of ways of increasing *in vitro* the RNA-forming capacity of deoxynucleoproteins extracted from chromatin. One of the best documented is the use of ammonium sulphate.

I wanted to ask Dr Johns if there are nuclei that are known *not* to contain histones. I believe that certain algae have been claimed to have particularly well-organized nuclear structures without histone proteins?

Johns: The literature is very confusing on this point. Professor Iwai claims that there are histones in algae (Iwai, 1964) and in some protozoa (Iwai, 1966). The problem however is really one of the definition of a histone and this is extremely difficult. The first world conference on histone chemistry attempted to do this, and produced the following definition: histones are basic proteins that at some time are associated with DNA (see Bonner and Ts'o, 1964). One tends to say that if a basic nuclear protein is not like a mammalian histone it is not a histone, but unicellular organisms may have their own types of histones quite different from the mammalian type.

Bergel: The problem is that one cannot distinguish whether the two are married or only living together!

Johns: There are basic proteins associated with nuclear ribosomes, so basic nuclear protein is not a sufficient definition.

Gros: Have you any comment on the histone-bound RNA that has been reported by Professor J. Bonner?

Johns: I have not tried to repeat this work myself and so would not like to comment.

Allison: It is interesting that the only repressor we know anything about (the Lac repressor) turns out to be an acidic protein.

Gros: That is a good point. We know most about two repressors—the Lac and λ-specific ones. The "lactose repressor" has been extensively purified but the λ repressor has not, at least to my knowledge. It has been "characterized" by double-labelling experiments (Ptashne, 1967). The Lac repressor has been purified to the extent that its chemistry begins to be reasonably well established. It has a low isoelectric point.

Dr Johns, you gave a model of RNA polymerase consisting of six subunits; I am not sure that this is any longer accepted by the experts but the possibility that the polymerase dissociates into subunits before it becomes attached to the promoters is still very much supported by facts and it may be that you have a hole or cavity between the big and the small subunits, which would fit in with your model.

REFERENCES

ANDERSSON, J., KILLANDER, D., MÖLLER, E., and MÖLLER, G. (1969). *J. exp. Med.*, in press.
BONNER, J., and TS'O, P. O. P. (eds.) (1964). *The Nucleohistones*, p. 15, ed. Bonner, J., and Ts'o, P. O. P. San Francisco: Holden-Day.
HARRIS, H. (1969). *Ciba fndn. Symp. Control Processes in Multicellular Organisms.* London: Churchill. In Press,
IWAI, K. (1964). In *The Nucleohistones*, pp. 59–65, ed. Bonner, J., and Ts'o, P. O. P. San Francisco: Holden-Day.
IWAI, K. (1966). *Expl Cell Res.*, **43**, 696–699.
KILLANDER, D., and RIGLER, R. (1965). *Expl Cell Res.*, **39**, 701.
LERMAN, L. S. (1963). *Proc. natn. Acad. Sci. U.S.A.*, **49**, 94.
PTASHNE, M. (1967). *Proc. natn. Acad. Sci. U.S.A.*, **57**, 306.
RIGLER, R., Jr. (1964). In *Proc. II Int. Congr. Histochem. Cytochem.*, p. 223, ed. Schiebler, T. H., Pearse, A. G. E., and Wolff, H. H. Heidelberg: Springer.
STEINER, R. F., and BEERS, R. F., Jnr. (1961). *Polynucleotides, Natural and Synthetic Nucleic Acids.* Amsterdam: Elsevier.
WEISSMANN, G., TROLL, W., BRITTINGER, G., and HIRSCHHORN, R. (1967). *J. Cell Biol.*, **35**, 140a. (Abstract of paper presented at 7th annual meeting of American Society for Cell Biology.)

ENZYMES AND ISOENZYMES

Cecil L. Leese

Chester Beatty Research Institute, London

HOMEOSTASIS AND ENZYME ADAPTATION

At each level of biological organization, homeostatic control mechanisms, mediated through a variety of regulatory substances, maintain the organism in dynamic equilibrium within the relatively narrow confines of conditions consistent with healthy viability. An essential feature of these mechanisms is that they enable the cell, the tissue or the whole organism to adapt to changes in both internal and external environmental conditions. Modulations in the rates of antagonistic processes are frequently recognizable as the basis of the adaptability necessary for conserving the appropriate *milieu intérieur* essential for maintaining an efficient functional state in the organism. The interrelationships within metazoan endocrine systems, not discussed at this symposium, are classical examples of homeostatic control operating through opposing feedback processes between hormone-secreting tissues and their target cells. The corollary is that failure in adaptive homeostatic control underlies the development of many functional disorders and of irreversible pathological states.

Homeostatic control at the metabolic level can be identified with modulation in the activities of adaptive enzyme systems, as opposed to constitutive enzymes. In this manner the direction and intensity of metabolite flux within the complex circuitry of interdependent metabolic pathways may be altered in response to changed environmental factors. Adaptive changes in enzyme activities occur at two main levels of molecular organization. The first is the genetic level, at which the total cellular population of enzyme molecules is determined by the rate of information transfer from genome to polysome, through which the rate of enzyme synthesis is regulated at the levels of transcription and translation. The second is the level of the preformed enzyme molecule, the activity of which is modulated by interaction with a variety of small molecular reactants.

At the genetic level the details of enzyme induction and repression in microorganisms, so fruitfully collated in the operon model of Jacob and Monod (1961), have demonstrated the way in which cell metabolism may be controlled through the regulation of enzyme synthesis by alterations in the environmental concentrations of small effector molecules. An important characteristic of certain of these adaptive systems is the operation of polycistronic gene units through which effector molecules regulate the coordinated synthesis of the several enzymes catalysing individual reactions of multi-stage metabolic pathways.

In metazoan organisms the situation is more complex because of the greater amount of genetic information contained within cells and the physiological interdependence of cells and tissues through humoral and neural interaction. Nevertheless, examples of induction and repression of enzyme synthesis in mammalian systems dependent on nutritional, hormonal and neural factors have been increasingly reported (Pitot, 1967). Evidence has also accrued to indicate that gene-linked coordinated enzyme synthesis, analogous to the polycistronic systems of microorganisms, operates in the homeostatic control of mammalian cell metabolism. The opposing metabolic pathways of glycolysis and gluconeogenesis in rodent liver, have, for example, been shown by Weber and his co-workers (1966) to be controlled by a small number of rate-limiting enzyme-catalysed steps in each of the two pathways. These rate-limiting stages are those which are effectively thermodynamically irreversible, and the enzymes catalysing these reactions appear to be synthesized in a synchronized manner in response to hormonal and other factors. The direction of carbohydrate metabolism in hepatic tissue is determined by the amounts of the key rate-controlling enzymes of the two pathways. Other enzymes catalysing reversible reactions in these pathways are largely unaffected by humoral conditions and are analogous to the constitutive enzymes of microbial metabolism.

At the level of enzymes, the catalytic properties of enzyme molecules have been studied for many years, though in the idealized circumstances of *in vitro* conditions which may bear only a tenuous relationship to their function *in vivo*. The catalytic properties of enzymes are well described in terms of enzyme–substrate dissociation constants, turnover numbers, reactant activation and inhibition, co-factor and activator requirements, competition for common metabolites and obligatory coupling between metabolic pathways. These, and the more recent concepts of effects on activity of conformational changes in enzyme structure induced by allosteric reactions at sites other than the active centre (Monod, Changeux and Jacob, 1963), exemplify the way in which the activity of preformed enzyme molecules may be modulated in response to changes in the concentrations of various low molecular weight substances.

ISOENZYMES

A subject making an increasing impact on the study of cellular metabolism and its control is the demonstration that many enzymes exist in multiple forms, each form catalysing the same biochemical interconversion. This is a general biological phenomenon, in that enzyme polymorphism of this type has been discovered in cells and tissues of species of most phyla. Enzyme polymorphism was foreshadowed by the early report of Meister (1950) that crystalline bovine heart lactate dehydrogenase was electrophoretically heterogeneous, and since that time more than 150 enzymes from plant and animal sources have been shown to exist in isoenzymic forms. Interest in the field has been intensive in the past decade (see Wróblewski, 1961; Vesell, 1968a).

Much of the growth in this field is due to refinements in electrophoretic techniques coupled with specific histochemical staining procedures for detecting enzyme activities. These have simplified the separation and detection of isoenzymic forms of particular catalytic proteins. Many reports of isoenzymic systems have been based on these techniques, but although they demonstrate differences in charge and possible differences in molecular size they sometimes afford little evidence for the true molecular basis of the variant forms of enzymes and much less information about their possible physiological significance. Before reliable conclusions of this kind can be reached much more information is required about the amino acid composition and primary sequence, the secondary, tertiary and quaternary structures, the catalytic properties, the antigenicity, and the tissue distribution patterns in relation to physiological function. Extensive information of this type is available for only a few of the isoenzymic systems at present known, from which conclusions can be made about the physiological role of these polymorphic forms in metabolic control mechanisms. At present the term isoenzyme is applied widely and indiscriminately to all systems exhibiting polymorphism, irrespective of molecular basis, and additional criteria are needed to distinguish between the various types of isoenzymic systems as they become apparent.

The molecular basis of isoenzymes

Genetic and non-genetic factors contribute to the formation of isoenzymes. The simplest situation is that in which separate genes encode separate monomeric enzymes composed of single polypeptide chains differing in amino acid sequence and physical, catalytic and immunological properties. The mammalian erythrocyte carbonic anhydrase isoenzymes B and C, composed of single polypeptide chains, appear to belong to this class. The third isoenzyme of this series, carbonic anhydrase A, closely resembles the B variant, differing possibly by a minor sequence change which may be artifactual (Edsall, 1968; Nyman and Lindskog, 1964).

In other systems isoenzymes arise through modifications introduced into single catalytically active gene products after translation. In one mode, single polypeptide chains may aggregate to yield a series of isoenzymes of different size. This occurs with glutamate dehydrogenase, where aggregation is accompanied by changes in substrate specificity (Bitensky, Yielding and Tomkins, 1965). Modification of single gene products may also occur through alteration of the original polypeptide chains by decarboxylation of glutamate and aspartate residues, deamidation of glutamine and asparagine residues, or deletion of terminal amino acid residues with retention of catalytic activity. The latter phenomenon is responsible for the formation of certain yeast hexokinase variants arising through the action of proteolytic enzymes (Kaji, 1965; Gazith et al., 1968).

Conjugation of proteins with small molecules may also be responsible for the formation of isoenzymes. Butterworth and Moss (1966) showed that several human kidney alkaline phosphatase isoenzymes, while functionally similar, contained varying numbers of sialic acid residues removable by preincubation with neuraminidase, which thereby simplified the isoenzyme distribution pattern. Association of protein molecules with substrates, co-factors and other low molecular weight substances may be the explanation for isoenzymes separated according to molecular charge. Ursprung and Carlin (1968) have demonstrated that the distribution pattern of alcohol dehydrogenase from *Drosophila* is markedly dependent on the presence of nicotinamide adenine dinucleotide (NAD) in the electrophoretic support medium.

These examples, together with chemical modifications made *in vitro* which may alter distribution patterns of isoenzymes, illustrate the care required to distinguish true isoenzymic variation existing *in vivo* and artifacts introduced by analytical techniques. This is of critical importance in determining the physiological significance of particular isoenzyme systems.

A major group of isoenzymes consists of those in which two or more genes encode the sequence of inactive polypeptide subunits which aggregate to give catalytically active multimeric enzymes. The multimers may be homopolymers composed of identical polypeptides or heteropolymers consisting of aggregates of different polypeptide subunits. The polymeric enzymes frequently show marked differences in physical, catalytic and immunological properties. The polypeptide aggregates exhibit, with varying degrees of facility, the property of reversible dissociation permitting the formation of hybrid molecules *in vivo* or, in appropriate conditions, *in vitro*. Among this group of multimeric enzymes are the well-known examples of lactate dehydrogenase, creatine kinase and aldolase.

A further basis for isoenzyme formation exists in the ability of protein molecules of identical amino acid composition and sequence to assume an equilibrium mixture of more than one thermodynamically stable configuration. If the "conformers" are such that exposed charge differences are significant and the rate of interconversion between the conformers is slow, then they should be demonstrable by electrophoretic or ion-exchange techniques (Epstein and Schecter, 1968). Conformational isoenzymes of this type have been demonstrated in several systems, including the mitochondrial malate dehydrogenase (Kitto, Wassarman and Kaplan, 1966) and the brain-type creatine kinase isoenzymes of certain avian species (Dawson, Eppenberger and Eppenberger, 1968).

Physiological significance of isoenzymes

Enzyme polymorphism is a widespread phenomenon, generated through the operation of several genetic and epigenetic molecular mechanisms. The physiological significance of most of these isoenzymic systems is at present enigmatic. Yet more detailed analysis of certain examples has demonstrated how enzyme variants, differing in catalytic properties, function in distinct metabolic pathways, at different locations in cells and in different tissues of the same organism. In these situations isoenzymes may subserve various metabolic requirements and operate in cellular control mechanisms. Several examples of these are discussed below.

Aspartokinase of E. coli

The control of branched metabolic pathways in microorganisms, recently reviewed by Stadtman (1968), demonstrates effectively how the elaboration of isoenzymic variants, coupled with enzyme repression and end-product feedback inhibition, permits a tight control of the intermediary metabolites of divergent pathways. One of the clearest examples of this type of metabolic control is the aspartokinase isoenzyme system of *E. coli*. In this organism lysine, methionine and threonine are each derived from aspartate by a branched biosynthetic pathway in which the first common step is the phosphorylation of aspartic acid. This pathway is represented in simplified form in Fig. 1.

The regulation of intermediates in this sequence is effected by a complex series of enzyme repression and feedback controls ensuring the efficient flow of metabolites through the subsidiary branched pathways, according to the metabolic demands made on the organism at a given time (Cohen, 1965; Stadtman, 1966). Three aspartokinases are elaborated in this organism, catalysing the initial phosphorylation step. Each enzyme is sensitive to

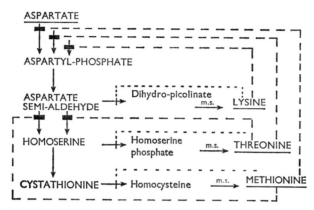

Fig. 1. Pathway of aspartate metabolism in *E. coli*. M.s., multi-stage pathway.

one of the individual amino acid end-products. The lysine-sensitive enzyme is susceptible to both feedback inhibition and repression by lysine. The threonine-dependent enzyme is subject to feedback inhibition by threonine and to multivalent repression through the combined effects of threonine and isoleucine (Freundlich, 1963). The synthesis of the third variant is strongly repressed by methionine but shows no feedback inhibition (Patte, Truffa-Bachi and Cohen, 1966). These authors also demonstrated that the threonine- and methionine-specific aspartokinases both carry specific homoserine dehydrogenase activities controlling the later branch in the pathway leading to the two amino acids, and that these dehydrogenase activities were controlled by the same repressive and feedback mechanisms as regulate the kinase activities.

Superimposed on these primary control factors are a series of fine control mechanisms operating in each of the individual branch pathways. Thus lysine not only regulates the synthesis and activity of the lysine-specific aspartokinase but through feedback inhibition and repression of enzymes catalysing the stages immediately following the branch point ensures that intermediates required by the methionine and threonine pathways are effectively utilized.

Lactate dehydrogenase (LDH)

The existence of five electrophoretically different isoenzymes of LDH, distributed in ontogenetic, tissue and species-specific patterns in the cytoplasm of mammalian and avian cells, has been confirmed in many reports since their original discovery by Markert and Møller (1959). These isoenzymes have subunit structure and are composed of random tetrameric aggregates of two different polypeptide subunits, the M and H types, which themselves are inactive and may be obtained from the tetramers by mild denaturing procedures (Appella and Markert, 1961; Markert, 1963*a*). Analysis of amino acid composition and tryptic peptide degradation products (Markert, 1963*b*; Wachsmuth, Pfleiderer and Wieland, 1964) confirmed that the M and H polypeptide subunits were different in both

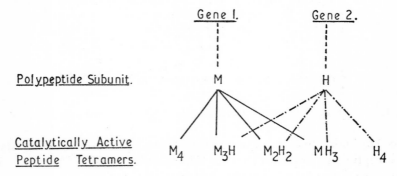

Fig. 2. The constitution of LDH isoenzymes.

amino acid composition and sequence, suggesting that they were encoded by separate gene loci. Each polypeptide subunit (molecular weight= 36 000) bears a cysteine residue essential for catalytic activity in the tetramers, and Fondy and co-workers (1965) have shown that in the chicken heart (H) and muscle (M) LDH isoenzymes minor sequence differences occur in the dodecapeptide degradation products containing this cysteine residue.

The concept that separate gene loci controlled the synthesis of LDH-M and LDH-H subunits was supported by the observation of a mutant allele of the H locus inherited as a single autosomal codominant gene in mice (Shaw and Barto, 1963), and of similar mutants in the H locus (Boyer, Fainer and Watson-Williams, 1963) and in the M locus (Nance, Claflin and Smithies, 1963) in man. In such cases 15 electrophoretically distinct LDH activity bands may be distinguished in place of the usual five. The constitution of LDH isoenzymes is summarized in Fig. 2, which shows that the catalytically active tetramers consist of two homopolymers, M_4 and H_4, and three hybrid heteropolymers containing both M and H subunits. Random hybridization of M and H subunits appears to occur *in vivo*, generally affording a binomial distribution of the five variant enzyme activities. Markert (1963b) demonstrated that a similar association pattern could be accomplished *in vitro* from dissociated M and H subunits. Moreover hybridization could be accomplished *in vitro* between M and H subunits derived from a wide variety of vertebrate species (Markert, 1964), demonstrating species similarity between the subunit types.

The discovery of a lactate dehydrogenase isoenzyme system (LDH-X) in mature testes and sperm of certain avian and mammalian species, and the fact that this unique enzyme also existed in allelic forms, disclosed a third gene encoding the subunits of which this variant is composed (Zinkham, 1968; Hawtrey and Goldberg, 1968). In structural characteristics LDH-X appears to be closely related to the common LDH isoenzymes. Hybrids between LDH-X and the other isoenzymes can seldom be demonstrated *in vivo* but catalytically active tetramers can be formed *in vitro* from the polypeptide products of all three genes. The unique tissue distribution and ontogenetic behaviour of LDH-X suggests that the cistron responsible for its synthesis is active only during testicular maturation and spermatogenesis and that it must be permanently repressed in cells of all other tissues. Conversely the appearance of LDH-X in the course of spermatogenesis is accompanied by a repression of the M and H loci (Hawtrey and Goldberg, 1968), presenting an interesting picture of gene activation and differential synthesis of gene products.

Differential synthesis of a similar nature occurs in the course of embryonic and postnatal development (Fine, Kaplan and Kuftinec, 1963; Lindsay, 1963; Gerhardt *et al.*, 1964; Auerbach and Brinster, 1967). Foetal tissues of a given species frequently resemble each other in their LDH isoenzyme distribution patterns, but these diverge in the course of organogenesis, with the unique adult tissue LDH distribution patterns developing in late embryonic and postnatal growth. Human foetal heart and liver tissue, for example, both show LDH isoenzyme patterns in which the M_2H_2 heteropolymer predominates. During embryogenesis these patterns are modified until soon after birth the heart and liver tissues exhibit the adult LDH patterns in which the H_4 and M_4 homopolymers, respectively, are the predominant forms (Gerhardt *et al.*, 1964). Each species exhibits individual developmental patterns in isoenzyme tissue differentiation. The isoenzymes are, therefore, interesting and useful markers of tissue and cell differentiation.

In any consideration of the role of LDH isoenzymes two aspects are usually stressed. The first is the highly characteristic species-specific tissue distribution patterns, generally regarded as a genetic mechanism affording cells and tissue with particular metabolic advantages closely related to physiological function. The second is the characteristic differences in chemical, physical and, more particularly, catalytic properties of the individual isoenzymes. The marked differences in substrate inhibition of LDH-M_4 and LDH-H_4 have for some time been regarded as the key to their physiological significance. In the standard conditions of *in vitro* analysis, usually performed at 25°C, purified LDH-M_4 shows little inhibition with increasing concentrations of pyruvate or lactate. In contrast LDH-H_4 is strongly inhibited at high concentrations of both substrates. This is usually interpreted to mean that the M_4 isoenzyme could operate favourably in tissues with low oxygen consumption and a glycolytic metabolism which may be exposed to transient anaerobiosis where elevated levels of glycolytic end products may be encountered. In contrast, in tissues with a high aerobic metabolism LDH isoenzymes predominantly composed of the H-subunit would favour the channelling of pyruvate into the tricarboxylic acid cycle and oxidative pathways, because high pyruvate concentrations would limit the conversion to lactate.

With a few important exceptions a broad relationship can be detected in avian and mammalian tissues between the LDH-isoenzyme pattern and the metabolic and physiological characteristics of the tissue. Heart and brain have a sustained oxidative metabolism and therefore a predominance of LDH-H subunits in their isoenzymic composition, whereas voluntary muscle, which has sudden energy requirements, is characterized by an LDH-M type pattern. A similar relationship has been demonstrated in various types of avian muscle (Wilson, Cahn and Kaplan, 1963) and in various zones of the rat kidney where there is a close correlation between LDH-isoenzyme distribution and oxidative metabolism (Kaplan, 1965). A further example is illustrated in Fig. 3, in which a strong orientation towards the LDH-H_4 variant is observed in the region of the gastric mucosa occupied by oxyntic cells in which a highly oxidative metabolism is associated with the physiological function of acid secretion (Leese, 1965; Yasin and Bergel, 1965).

A recent revaluation (Vesell, 1965, 1968*b*) of the aerobic-anaerobic concept of LDH isoenzyme function has raised certain controversial issues. First, concentrations of substrates inhibiting the isoenzymes *in vitro* are not encountered *in vivo*. Secondly, the substrate inhibition characteristics of LDH-H_4 and LDH-M_4 are not as distinct when measured at higher temperatures closer to physiological conditions. Kaplan, Everse and Admiraal

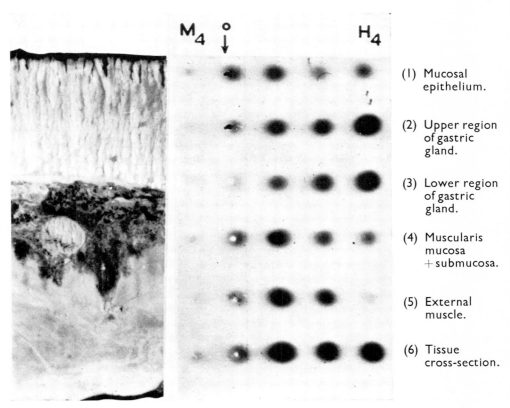

FIG. 3. Distribution of LDH isoenzymes in the body region of gastric mucosa. O: origin.

(1968) have, however, emphasized that an "abortive" ternary complex formed between stoicheiometric amounts of pyruvate, oxidized nicotinamide adenine dinucleotide (NAD^+) and LDH (preferentially with the $LDH-H_4$ enzyme) competes with the reduced coenzyme (NADH) in the reduction of pyruvate. They suggest that such complexes may determine catalytic differences between LDH isoenzymes under *in vivo* conditions of enzyme, substrate and co-factor concentrations. Cellular NAD^+–NADH ratios would be of critical interest in this connexion.

Ontogenetic changes in tissue LDH isoenzyme distribution patterns and the differential appearance of LDH-X in spermatogenesis suggest that the rates of M and H subunit synthesis might be independently modified by a variety of factors. Goodfriend and Kaplan (1964) found, with two exceptions, that hormones alter the level of LDH activity in target tissues without modifying the isoenzyme distribution pattern. They demonstrated, however, an increase in the differential rate of synthesis of M subunits in the seminal vesicles of young testosterone-treated rats and in the uteri of immature rats and rabbits, where oestradiol induced isoenzyme patterns similar to those of the mature organ. The oestradiol effect on the selective synthesis of M subunits was blocked by actinomycin D whereas puromycin caused a proportionate fall in the levels of both enzymes. Section of motor nerves supplying muscles in the rabbit also modified the LDH isoenzyme distribution, but although there was a decrease in M subunits in the gastrocnemius muscle the opposite effect was observed in the soleus muscle (Dawson and Kaplan, 1965). Experiments of this nature must be carefully controlled histologically to ensure that modifications of isoenzyme patterns are not due solely to alterations in the relative types of cells present in the tissues.

The possibility that oxygen might have an effect on the differential synthesis of LDH subunits was realized from the observations of Dawson, Goodfriend and Kaplan (1964) that cultured tissue explants generally showed an increased relative rate of M subunit synthesis which was suppressed when the oxygen tension of the gas phase was raised. A similar effect was noted in a number of established cell lines in culture, where low oxygen tensions permitted a relative increased rate of M subunit synthesis without greatly modifying the levels of H subunit production (Goodfriend, Sokol and Kaplan, 1966).

Unpublished observations (C. L. Leese, D. A. Gilbert, R. Yasin and J. Biffen) indicate that the isoenzyme composition of cells may oscillate during their generation time. This phenomenon is illustrated in Fig. 4, which shows the behaviour of LDH and aldolase isoenzymes in monolayer cultures of BHK21-PYY cells over the course of a generation time (about 12 hours) after synchronization by thermal shock—in this case a 2-hour reduction of temperature to 20°C. The B/P ratio is a measure of the LDH isoenzyme distribution based on reactivity towards α-ketobutyrate (B) and pyruvate (P) (Plummer et al., 1963). The FDP/FMP ratio, measured by the reaction rates towards fructose-1,6-diphosphate and fructose-1-phosphate respectively, indicates the relative proportions of the major aldolase isoenzyme activities A and B (Rutter et al., 1963).

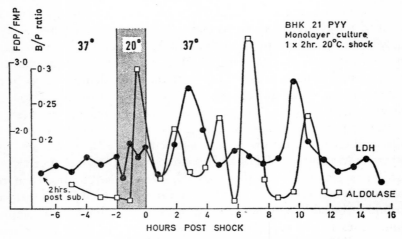

FIG. 4. Periodic variations in LDH and aldolase isoenzyme distribution in synchronized cell culture.

After the shock period both isoenzyme systems show marked periodic changes in composition at frequencies considerably less than the generation time. Of additional interest are the very rapid changes occurring during the shock period at 20° or at lower temperatures. The frequencies observed in these early experiments are within the time scale required for polypeptide synthesis and it is tempting to speculate that selective control of subunit synthesis regulates isoenzymic composition in response to varying metabolic demands at different times during the generation time of the cell. Whether this is a valid interpretation of the data or whether there are alternative explanations must await experimental refinement to define the true nature of the frequencies, amplitudes and phasing of the isoenzymic variations.

One of us (Gilbert, 1968) has suggested that oscillatory behaviour in isoenzymic composition of this type might be a further example of periodic changes in metabolic parameters in which the properties of cells, e.g. in the course of fertilization, division, differentiation, senescence and oncogenesis, are determined by patterns of rhythmic variation in levels of cell constituents arising through the operation of numerous metabolic control circuits.

Aldolase

Mammalian tissues contain three principal forms of fructose-1,6-diphosphate (FDP) aldolase activities, designated A, B and C, that are the principal forms of the enzyme present in skeletal muscle, liver and brain respectively. Each variant reversibly catalyses the cleavage of FDP and of fructose-1-phosphate (FMP) to triose derivatives (Fig. 5.) Adult tissues

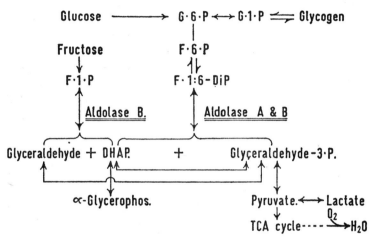

Fig. 5. Role of aldolase A and B in carbohydrate metabolism. DHAP: dihydroxyacetone phosphate.

exhibit characteristic aldolase isoenzyme distribution patterns with a marked similarity in organ-specific patterns over a wide range of vertebrate species (Penhoet, Rajkumar and Rutter, 1966; Blostein and Rutter, 1963). Aldolase A is present in all tissues in variable amounts whereas the B and C variants are more restricted in distribution.

Although similar in their general structural features, aldolases A, B and C are markedly different in catalytic effects on the substrates FDP and FMP. Aldolase A cleaves FDP at 50 times the rate that it cleaves FMP, whereas the FDP/FMP activity ratio for aldolase B is close to unity (Rutter et al., 1963) and for aldolase C it is ten (Rutter et al., 1968). The kinetic data for both the cleavage reaction (hexose to triose) and the reverse aldol condensation of triose derivatives indicate that aldolase A is physiologically adapted to a glycolytic function, whereas aldolase B is relatively more efficient in FDP syntheses. Aldolase B thus shows a greater metabolic diversification in that it readily permits gluconeogenesis and, through its ability to cleave FMP, plays a role in fructose metabolism in tissues such as liver. These concepts are further supported by the inhibitory effects of the various adenosine phosphates on the two aldolase isoenzymes (Spolter *et al.*, 1965) and by the tissue distribution patterns.

Other glycolytic enzymes, including glycogen phosphorylase (Krebs and Fisher, 1962), α-glycerophosphate dehydrogenase (Rouslin, 1967), glyceraldehydephosphate dehydrogenase (Papadopoulos and Velick, 1967) and phosphoglucomutase (Joshi *et al.*, 1967), have been reported to exist in multiple forms with differences in catalytic properties such that those forms existing in muscle preferentially favour a glycolytic metabolism whereas those in liver, where a greater metabolic diversity is required, permit the operation of the reverse gluconeogenic pathway.

Developmental changes in aldolase isoenzyme distribution have been observed in rodent liver, kidney and intestine, where aldolase A is the principal embryonic activity. This slowly changes to the characteristic adult aldolase B activity of these organs in the course of embryogenesis (Rutter and Weber, 1965). A similar sequence is reported to occur in rat brain, where the foetal aldolase A activity changes to the specific aldolase C pattern of the adult brain (Rutter *et al.*, 1968). These developmental changes indicate that the synthesis of the three isoenzymes is independently regulated by three gene loci. At present, however, it is difficult to define the genetic basis of the aldolase variants because of conflicting evidence concerning their structure (see Rutter *et al.*, 1968).

Physicochemical evidence indicates that native muscle and liver aldolase variants are trimeric aggregates of two different polypeptide subunits. They possess considerable differences in amino acid composition and sequence, even in the region of the active centre of the molecules close to a lysine residue forming a Schiff base with the reactant dihydroxyacetone phosphate (Horecker, 1965; Lai, Hoffee and Horecker, 1965). Carboxypeptidase releases three C-terminal tyrosine residues from each native enzyme molecule, a feature of particular interest in that the loss of these terminal residues markedly reduces the catalytic activity of the enzymes and modifies the FDP/FMP activity ratios.

Recently, however, the proposed trimeric subunit structure for these molecules has been queried because of the finding that reversible dissociation of mixtures of two of the parent enzymes (AB, AC, or BC) gave five electrophoretically distinct variants compatible with a tetrameric subunit structure, comparable to the LDH isoenzyme system. Similar five-banded electrophoretic patterns have also been demonstrated in extracts of tissues containing more than one of the parent isoenzymes (Penhoet, Rajkumar and Rutter, 1966). Support for the tetrameric structure of aldolase molecules has, interestingly, been obtained electron microscopically (Rutter *et al.*, 1968). Thus there is a structural dilemma regarding the aldolase molecules which must be resolved before conclusions can be reached about the genetic basis of these isoenzymes.

Hexokinase

The recent discovery of multiple forms of hexokinase (glucose-ATP-6-phosphotransferase) activities in mammalian cells and tissues has contributed greatly to the understanding of the mechanisms of glucose transport and phosphorylation in the initial stages of carbohydrate metabolism (Gonzalez *et al.*, 1964; Katzen and Schimke, 1965; Katzen, 1966). A striking aspect of this work is the explanation afforded for the insulin-dependent glucose utilization in certain tissues, e.g. heart and skeletal muscle and certain adipose tissue, and the insulin independence of these processes in many other organs.

Electrophoretic and ion-exchange techniques have demonstrated the presence of three hexokinase isoenzymes (HK1, HK2 and HK3) in the supernatant fraction of mammalian tissue extracts. These enzymes have a high affinity for glucose (K_m: 10^{-4} M to 10^{-6} M) and a broad substrate specificity embracing a range of hexose derivatives, and they constitute the so-called "low K_m" hexokinase group. A unique hexokinase, also with a low K_m, has been detected in mature testes and is thought to be a form of the enzyme specifically associated with sperm (Katzen, 1967).

These enzymes are a marked contrast to the "high K_m" hexokinase (HK4) found in mammalian liver ($K_m = 1 \cdot 6 \times 10^{-2}$ M). This enzyme, the hepatic glucokinase, has a narrow substrate specificity confined effectively to glucose. It is inducible by insulin and high carbohydrate intake and is suppressed by starvation, high protein intake and in diabetes, the effect in the last being relieved by insulin treatment (Katzen, 1966). The induction of this enzyme involves *de novo* protein synthesis, and the high K_m value for glucose, together with the inefficient product inhibition by glucose-6-phosphate, is consistent with the physiological function of hepatic carbohydrate storage under conditions of high dietary carbohydrate intake and the maintenance of physiological blood glucose levels (Sharma, Manjeshwar and Weinhouse, 1964). Hepatic glucokinase has a molecular weight of 48 000 (Pilkis and Krahl, 1966), half that of the "low K_m," hexokinase group, each with a molecular weight of 96 000 (Grossbard and Schimke, 1966a). Whether these and other factors reflect a subunit relationship between the various hexokinase isoenzymes is not at present known.

In the low K_m group of hexokinase isoenzymes, HK1, HK2 and HK3 are generally distributed throughout mammalian tissues in varying proportions in tissue-specific patterns. The rat enzymes have been most extensively investigated but other mammalian species, including man, have counterparts to these three isoenzymes, although minor species differences occur in electrophoretic properties. All members of the low K_m group show marked differences in chemical, physical and catalytic properties (Schimke and Grossbard, 1968) and in their stability towards trypsin degradation, guanidine denaturation, reactivity towards thiol reagents and inactivation by extreme pH conditions. The HK2 variant is the least stable of the three members of the group. The relationship between these soluble hexokinase variants and bound hexokinase in cells is not entirely clear, although Grossbard and Schimke (1966b) have shown that hexokinase activity released from particulate, bound material exhibits similar isoenzyme patterns with the same catalytic properties. The three forms do not appear to be artifacts due to tissue autolysis, as observed, for example, in some of the yeast hexokinase variants (Gazith et al., 1968), although trypsin treatment differentially alters the migratory properties of the enzymes. At present there is no evidence suggesting that these isoenzymes are interconvertible.

Katzen and Schimke (1965) showed that the integrity of the HK2 isoenzyme was dependent on the presence of thiol in vitro. The absence of thiol caused the appearance of two electrophoretically distinct sub-bands of this variant (HK2a and HK2b). Both of these sub-bands were also observed in fasted rats, and in animals with diabetes induced with streptozotocin (Katzen, 1966) from which the HK2a activity rapidly disappears. Treatment of diabetic animals with insulin rapidly restores HK2a activity within 30 to 40 minutes, but at levels which never exceed those seen in normal animals. A similar insulin effect was observed in intact heart muscle from diabetic animals in vitro. Thus the insulin effect on HK2a differs markedly, in both time course and activity levels, from the insulin-induced synthesis of hepatic glucokinase.

HK1 and HK3 show no comparable insulin or nutritional dependence. Katzen (1967) has shown that the insulin-dependent HK2 isoenzyme is widely distributed in tissues but that it is only where it is the predominant form of the low K_m group that the tissue exhibits an insulin-dependent glucose uptake, e.g. in skeletal and heart muscle and the epididymal fat pad of young animals. In those tissues where HK1 and HK3 activities are comparable to that of HK2, glucose utilization becomes independent of the hormone. Katzen (1967) suggests that the multiple hexokinase activities form an integral part of the glucose transmembrane carrier system, actively transporting glucose into the cell and catalysing its phosphorylation at the same time. The varying properties of the three variant hexokinase activities in the translocase system indicate that insulin maintains the integrity of the transport mechanism where the HK2 variant predominates.

SUMMARY

Increasing reports of enzymes existing in multiple forms, each catalysing the same biochemical reaction, have indicated that this is a general biological phenomenon. Although the significance of most of these isoenzymic systems is at present unknown, those which have been investigated in greater detail indicate that enzyme multiplicity may arise through several genetic and non-genetic molecular mechanisms. The properties of several of these isoenzymic systems have been discussed to show how they may play a significant role, through the interaction of both genetic and humoral factors, in biochemical control mechanisms and in homeostatic regulation at the metabolic level.

Acknowledgements

This investigation has been supported by grants to the Chester Beatty Research Institute, Institute of Cancer Research: Royal Cancer Hospital, from the Medical Research Council and the British Empire Cancer Campaign for Research.

REFERENCES

APPELLA, E., and MARKERT, C. L. (1961). *Biochem. biophys. Res. Commun.*, **6**, 171–176.
AUERBACH, S., and BRINSTER, R. L. (1967). *Expl Cell Res.*, **46**, 89–92.
BITENSKY, M. W., YIELDING, K. L., and TOMKINS, G. M. (1965). *J. biol. Chem.*, **240**, 663–667.
BLOSTEIN, R., and RUTTER, W. J. (1963). *J. biol. Chem.*, **238**, 3280–3285.
BOYER, S. H., FAINER, D. C., and WATSON-WILLIAMS, E. J. (1963). *Science*, **141**, 642–643.
BUTTERWORTH, P. J., and MOSS, D. W. (1966). *Nature, Lond.*, **209**, 805–806.
COHEN, G. N. (1965). *A. Rev. Microbiol.*, **19**, 105–126.
DAWSON, D. M., EPPENBERGER, H. M., and EPPENBERGER, M. E. (1968). *Ann. N.Y. Acad. Sci.*, **151**, 616–626.
DAWSON, D. M., GOODFRIEND, T. L., and KAPLAN, N. O. (1964). *Science*, **143**, 929–933.
DAWSON, D. M., and KAPLAN, N. O. (1965). *J. biol. Chem.*, **240**, 3215–3221.
EDSALL, J. T. (1968). *Ann. N.Y. Acad. Sci.*, **151**, 41–63.
EPSTEIN, C. J., and SCHECTER, A. N. (1968). *Ann. N.Y. Acad. Sci.*, **151**, 85–101.

FINE, I. H., KAPLAN, N. O., and KUFTINEC, D. (1963). *Biochemistry, N.Y.*, **2**, 116–121.
FONDY, T. P., EVERSE, J., DRISCOLL, G. A., CASTILLO, F., STOLZENBACH, F. E., and KAPLAN, N. O. (1965). *J. biol. Chem.*, **240**, 4219–4234.
FREUNDLICH, M. (1963). *Biochem. biophys. Res. Commun.*, **10**, 277–282.
GAZITH, J., SCHULZE, I. T., GOODING, R. H., WOMACK, F. C., and COLOWICK, S. P. (1968). *Ann. N.Y. Acad. Sci.*, **151**, 307–331.
GERHARDT, W., ØVLISEN, B., CLAUSEN, J., and ANDERSON, H. (1964). In *Protides of the Biological Fluids*, pp. 203–206, ed. Peeters, H. Amsterdam: Elsevier.
GILBERT, D. A. (1968). *J. theoret. Biol.*, **21**, 113–122.
GONZALEZ, C., URETA, T., SANCHEZ, R., and NIEMEYER, H. (1964). *Biochem. biophys. Res. Commun.*, **16**, 347–352.
GOODFRIEND, T. L., and KAPLAN, N. O. (1964). *J. biol. Chem.*, **239**, 130–135.
GOODFRIEND, T. L., SOKOL, D. M., and KAPLAN, N. O. (1966). *J. molec. Biol.*, **15**, 18–31.
GROSSBARD, L., and SCHIMKE, R. T. (1966a). *Fedn Proc. Fedn Am. Socs exp. Biol.*, **25**, 220.
GROSSBARD, L., and SCHIMKE, R. T. (1966b). *J. biol. Chem.*, **241**, 3546–3560.
HAWTREY, C., and GOLDBERG, E. (1968). *Ann. N.Y. Acad. Sci.*, **151**, 611–615.
HORECKER, B. L. (1965). *Israel J. med. Sci.*, **1**, 1148–1161.
JACOB, F., and MONOD, J. (1961). *J. molec. Biol.*, **3**, 318–356.
JOSHI, J. G., HOOPER, J., KUWAKI, T., SAKURADA, T., SWANSON, J. R., and HANDLER, P. (1967). *Fedn Proc. Fedn Am. Socs. exp. Biol.*, **26**, 557.
KAJI, A. (1965). *Archs Biochem. Biophys.*, **112**, 54–64.
KAPLAN, N. O. (1965). *J. cell. comp. Physiol.*, **66**, suppl. 1, 1–10.
KAPLAN, N. O., EVERSE, J., and ADMIRAAL, J. (1968). *Ann. N.Y. Acad. Sci.*, **151**, 400–412.
KATZEN, H. M. (1966). *Biochem. biophys. Res. Commun.*, **24**, 531–536.
KATZEN, H. M. (1967). In *Advances in Enzyme Regulation*, vol. 5, pp. 335–356, ed. Weber, G. Oxford: Pergamon Press.
KATZEN, H. M., and SCHIMKE, R. T. (1965). *Proc. natn. Acad. Sci. U.S.A.*, **54**, 1218–1225.
KITTO, G. B., WASSARMAN, P. M., and KAPLAN, N. O. (1966). *Proc. natn. Acad. Sci. U.S.A.*, **56**, 578–585.
KREBS, E. G., and FISHER, E. H. (1962). *Adv. Enzymol.*, **24**, 263–290.
LAI, C. Y., HOFFEE, P., and HORECKER, B. L. (1965). *Archs Biochem. Biophys.*, **112**, 567–579.
LEESE, C. L. (1965). *Eur. J. Cancer*, **1**, 211–216.
LINDSAY, D. T. (1963). *J. exp. Zool.*, **152**, 75–89.
MARKERT, C. L. (1963a). *Science*, **140**, 1329–1330.
MARKERT, C. L. (1963b). In *Cytodifferentiation and Macromolecular Synthesis*, pp. 65–84, ed. Locke, M. New York: Academic Press.
MARKERT, C. L. (1964). *VI Int. Congr. Biochem.*, New York, Abstracts, IV–104, 320.
MARKERT, C. L., and MØLLER, F. (1959). *Proc. natn. Acad. Sci. U.S.A.*, **45**, 753–763.
MEISTER, A. (1950). *J. biol. Chem.*, **184**, 117–129.
MONOD, J., CHANGEUX, J. P., and JACOB, F. (1963). *J. molec. Biol.*, **6**, 306–329.
NANCE, W. E., CLAFLIN, A., and SMITHIES, O. (1963). *Science*, **142**, 1075–1077.
NYMAN, P. -O., and LINDSKOG, S. (1964). *Biochim. biophys. Acta*, **85**, 141–151.
PAPADOPOULOS, C. S., and VELICK, S. F. (1967). *Fedn Proc. Fedn Am. Socs exp. Biol.*, **26**, 557.
PATTE, J. C., TRUFFA-BACHI, P., and COHEN, G. N. (1966). *Biochim. biophys. Acta*, **128**, 426–439.
PENHOET, E. E., RAJKUMAR, T. V., and RUTTER, W. J. (1966). *Proc. natn. Acad. Sci. U.S.A.*, **56**, 1275–1282.
PILKIS, S. J., and KRAHL, M. E. (1966). *Fedn Proc. Fedn Am. Socs exp. Biol.*, **25**, 523.
PITOT, H. C. (1967). In *Molecular Biology*, Pt. II, pp. 383–423, ed. Taylor, J. H. London: Academic Press.
PLUMMER, D. T., ELLIOT, B. A., COOKE, K. B., and WILKINSON, F. H. (1963). *Biochem. J.*, **87**, 416–422.

Rouslin, W. (1967). *Fedn Proc. Fedn Am. Socs exp. Biol.*, **26**, 557.
Rutter, W. J., Blostein, R. E., Woodfin, D. M., and Weber, C. S. (1963). In *Advances in Enzyme Regulation*, vol. 1, pp. 39–56, ed. Weber, G. Oxford: Pergamon Press.
Rutter, W. J., Rajkumar, T., Penhoet, E., Kochman, M., and Valentine, R. (1968). *Ann. N.Y. Acad. Sci.*, **151**, 102–117.
Rutter, W. J., and Weber, C. S. (1965). In *Developmental and Metabolic Control Mechanisms and Neoplasia*, pp. 195–218. Baltimore: Williams and Wilkins.
Schimke, R. T., and Grossbard, L. (1968). *Ann. N.Y. Acad. Sci.*, **151**, 332–350.
Sharma, C., Manjeshwar, R., and Weinhouse, S. (1964). In *Advances in Enzyme Regulation*, vol. 2, pp. 189–200, ed. Weber, G. Oxford: Pergamon Press.
Shaw, C. R., and Barto, E. (1963). *Proc. natn. Acad. Sci. U.S.A.*, **50**, 211–214.
Spolter, P. D., Adelman, R. C., DiPietro, D. L., and Weinhouse, S. (1965). In *Advances in Enzyme Regulation*, vol. 3, pp. 79–89, ed. Weber, G. Oxford: Pergamon Press.
Stadtman, E. R. (1966). *Adv. Enzymol.*, **28**, 41–154.
Stadtman, E. R. (1968). *Ann. N.Y. Acad. Sci.*, **151**, 516–530.
Ursprung, H., and Carlin, L. (1968). *Ann. N.Y. Acad. Sci.*, **151**, 456–475.
Vesell, E. S. (1965). *Science*, **150**, 1590–1593.
Vesell, E. S. (ed.) (1968a). *Ann. N.Y. Acad. Sci.*, **151**, 1–689.
Vesell, E. S. (1968b). *Ann. N.Y. Acad. Sci.*, **151**, 5–13.
Wachsmuth, E. D., Pfleiderer, G., and Wieland, T. (1964). *Biochem. Z.*, **340**, 80–94.
Weber, G., Singhal, R. L., Stamm, N. B., Lea, M. A., and Fisher, E. A. (1966). In *Advances in Enzyme Regulation*, vol. 4, pp. 59–81, ed. Weber, G. Oxford: Pergamon Press.
Wilson, A. C., Cahn, R. D., and Kaplan, N. O. (1963). *Nature, Lond.*, **197**, 331–334.
Wróblewski, F. (ed.) (1961). *Ann. N.Y. Acad. Sci.*, **94**, 655–1030.
Yasin, R., and Bergel, F. (1965). *Eur. J. Cancer*, **1**, 203–209.
Zinkham, W. H. (1968). *Ann. N.Y. Acad. Sci.*, **151**, 589–610.

DISCUSSION

Gros: The distribution pattern of isoenzymes provides the cell with a very elegant way of modulating its response to hormonal and allosteric factors. You say that the H and M polypeptide chains are different, but in what respect? Do they differ immunologically, for example?

Leese: Dissociation of the tetramers, or mixtures of them, produces two polypeptides separable electrophoretically. Amino acid analysis and peptide mapping of the homopolymers M_4 and H_4 show marked differences. The hybrid molecules in general show the expected intermediate analytical data. Indeed the properties of the hybrid molecules generally fall between the values for the homopolymers in a wide range of physical and biochemical parameters. They do differ immunologically. Antibodies produced to either of the homopolymers are specific for that homopolymer in precipitin and enzyme inactivation effects and show no reactivity with the other homopolymer. The hybrid molecules, however, are inactivated by either anti-H_4 or anti-M_4, as would be expected.

Gros: Is it certain that these distribution patterns are regulated simply by the rate at which the subunits are provided by the cell, or could you visualize other types of control mechanisms that would perhaps switch from one form to another?

Leese: LDH isoenzymes in tissue extracts usually approximate to a binomial distribution and this is generally assumed to be equated with a process of random association of available subunits. This is not necessarily the same as the rate of synthesis, and other types of mechanisms determining the availability of subunits for aggregation may operate. For example, because of different charges on the subunits, association with other subcellular components might play a role in determining the rates and stoicheiometry of subunit association.

Vernon: May I sound a slightly sceptical note about the biological significance of isoenzymes? I fancy that many of them are characterized by finding multiple bands on starch gel electrophoresis. There is a very interesting case with aspartate amino transferase from heart muscle in which you can also find multiple bands which can be partially separated, and they also have different kinetic properties in the sense that certain inhibition constants are different. But they do not differ in their polypeptide chains at all; the primary sequence is precisely the same and they are in fact conformational isomers.

Leese: This is quite true. The mitochondrial malate dehydrogenase isoenzymes are conformational isomers having closely similar kinetic properties and yet are separable on the basis of charge differences.

Vernon: Yes, and this is a general phenomenon. I certainly wouldn't be convinced by the amino acid analysis that the M and H polypeptide chains were different.

Leese: Perhaps the most convincing evidence in the case of LDH comes from the immunological differences of the two extreme forms.

Vernon: No, that won't do, because if a molecule is unfolding it can also behave differently immunologically. The various subforms of transaminase will do this.

Leese: This may be true and will depend on the stability of the conformers and their ease of interconvertibility. I think this serves to re-emphasize the importance of investigating isoenzymic systems from as many different aspects as possible. For LDH I think there is sufficient diverse evidence to support the subunit structure and the unique nature of the subunits themselves.

Vernon: I think this is true, but when you said that 150 enzymes have been shown to have isoenzymes, I fancy that the number that are well authenticated is very much smaller. For example, I don't necessarily think that hexokinases 1, 2 and 3 are really different. They also may be conformational isomers.

Leese: They could be, but as yet there is so little evidence concerning their structure that no definite conclusions can be made about their molecular constitution.

Gros: How do you explain those oscillating patterns which you observe after synchronization by low temperature? Is it governed by variation in rates of synthesis or some sort of cyclic feedback?

Leese: This we cannot tell at present. The frequency of oscillation in these early experiments is, as I mentioned, within a time scale compatible with polypeptide synthesis, but this is purely speculative. The sampling time, for example, was arbitrarily chosen at 30 minutes, which may completely mask the real frequency of the changes. There may be several alternative explanations for the phenomenon. One of our problems is obtaining reproducibility in these systems, so any conclusion about the origin of oscillating behaviour is at present impossible.

Elsdale: It is interesting that the aldolase level went up into a big peak actually during your 2-hour cold-shock treatment.

Leese: The measurements are FDP/FMP activity ratios and a large change in this can be brought about by a relatively small change in the absolute activity of one or other of the isoenzymes.

Allison: Surely you wouldn't have very much synthesis at the low temperature?

Leese: The changes observed within the shock period occur within a matter of minutes and cannot be associated with protein synthesis. Again we cannot make any sound conclusion about the origin of these changes.

Forrester: What degree of synchrony do you get by this type of cold-shock procedure? If you want to examine changes in these enzymes through an undisturbed cell cycle it might be better to obtain a synchronized population by harvesting mitotic cells using the method of Robbins and Marcus (1964).

Leese: We have no accurate measure of the degree of synchrony in the monolayer cultures at present, but with suspension cultures we have obtained synchrony by cold shock of the order of 80 per cent, based on cell counts over the intermitotic period. The Robbins and Marcus technique of synchronization is an alternative we hope to employ in this work.

Stoker: The mean intermitotic time of these cells is about 12 hours, and your cycling went on for about 12 hours and then you had four more hours, and it looked as though it had stopped. What happens if you go on after 16 hours?

Leese: We do not know if this is a system in which the oscillating behaviour is damped. We aim to establish cultures for at least one generation time before the shock period to enable the cells to adapt to the culture conditions and to establish the pre-shock isoenzyme patterns. As you have implied it would be necessary to continue observations for longer periods to see if and when damping occurs.

REFERENCE

ROBBINS, E., and MARCUS, P. I. (1964). *Science*, **144**, 1152–1153.

INTERFERONS AS POSSIBLE REGULATORS

N. B. Finter

Virus Research Department, Pharmaceuticals Division, Imperial Chemical Industries Limited, Macclesfield

Interferons are certainly regulators in the sense that they can influence the growth of viruses in cells. Cells infected with a virus usually form interferon, and it may well be that viruses are the most important and frequent natural stimulus for interferon formation. In turn, the interferon formed can render other cells partially or completely resistant to virus infection.

Since interferons may be of value for medical purposes it is important to know how they are formed and how they act, and these topics are also of great interest from the point of view of fundamental biochemistry. They have therefore been studied in detail, and these studies are considered in the accompanying paper (Burke, 1969). Here, interferons will be considered in relation to two other topics—their effects on the production of interferon by cells, and their possible effects in cells not infected with viruses.

First, however, some of the properties of interferons will be reviewed briefly. Details will be found in the monograph edited by Finter (1966), and in reviews by Baron and Levy (1966) and Finter and Bucknall (1968).

The interferons form a family of proteins which are produced by the cells of mammals, amphibians, fish and birds. Similar substances appear to be formed in plants. Interferons have molecular weights within a range from about 25 000 to 160 000, and those species which have been studied in detail, namely rabbits, mice and chickens, seem to form several interferons in response to different stimuli, or sometimes even to the same stimulus. These interferons differ in molecular weight, but have apparently identical biological properties, except that, for example, less high molecular weight interferon than low molecular weight interferon is excreted in the urine of rabbits, probably simply because of the different sizes of the molecules.

The most obvious property of interferons is their antiviral activity, and indeed they were discovered by Isaacs and Lindenmann (1957) because of this property, and even today can be assayed only by making use of it. Available evidence suggests that in cells treated with interferon the ribosomes are altered so that they do not respond normally to the information in viral RNA, but still respond normally to cell messenger RNA. This alteration may be due to the presence of a new protein added to existing ribosomes, or to the incorporation of a protein into newly formed ribosomes. In any event the consequence is that new viral proteins are not formed and hence no new virus particles are made. The remarkable feature of the antiviral action of interferons is that it is not directed solely against one class of virus. On the contrary, interferons are active against the great majority of viruses, whether these are large or small, or contain RNA or DNA.

Interferons are very potent substances. Thus for example, chick interferon has been purified 20000 times, and contained 1·6 million units of biological activity per mg of protein, but such preparations are probably still far from pure (Fantes, 1968). Thus interferons rank with the most potent biologically active substances known.

Interferons are generally much more active when tested in cells of their own species than in cells of other species and they have frequently been termed "species" specific. The specificity however appears to lie at the level of the phylogenetic family rather than at the level of the species (Bucknall, 1967), and recently certain preparations of human interferon have been shown to be highly active in rabbit cells (Desmyter, Rawls and Melnick, 1968).

Interferons are proteins, sensitive to trypsin, but rather unusually stable to extremes of pH, particularly pH 2. They are poor antigens and probably non-antigenic in the species in which they are formed. They are also remarkably non-toxic, both to cells in tissue culture and to animals.

Apart from their antiviral effects, very few other alterations are known to be produced in cells by interferons. However, treatment of cells with interferon preparations can influence the quantity of interferon subsequently formed by those cells in response to an inducer. The literature contains contradictory reports, some saying that interferon treatment of cells increases the subsequent yield of interferon, and others that interferon treatment depresses the yield. It now appears that interferon preparations can produce both effects: treatment of mouse L cells with interferon for 2 hours primed them so that, when they were subsequently exposed to an inducer, ultraviolet-irradiated Newcastle disease virus (NDV_{UV}), interferon was formed in the same quantity as in control cells, but more rapidly. In contrast, if the cells were treated for 24 hours with relatively large amounts of interferon, much less interferon was subsequently formed in response to NDV_{UV}, or none at all (Paucker and Boxaca, 1967).

At this point, a general comment must be made about effects obtained with interferon preparations. It must be remembered that in almost all studies with interferons so far, only relatively crude preparations have been available for use. There is therefore always the question whether effects obtained with such a preparation are due to the interferon or to other substances also present. Since these hypothetical substances could result from treatment of cells with the very stimulus used to provoke interferon formation, comparable preparations from non-stimulated cells do not provide an adequate control.

There are two partial solutions to this dilemma. First, the effects of different preparations of, for example, mouse interferon, made from different cells and using different inducers of interferon formation, can be compared in the hope that the only or principal common factor will be interferon. Second, the consequences of progressive purification of an interferon preparation can be observed: if the relative amounts of antiviral activity per mg protein and of some other activity increase strictly in parallel, this suggests that the two activities may reside in the same molecule.

From studies with chick cells and relatively crude preparations of chick interferon, Friedman (1966) suggested that the observed priming and depressing effects on interferon production were in fact due to interferon itself. In confirmation Paucker and Golgher (1969) found that when mouse interferon was purified, its priming and depressing effects on subsequent formation of interferon in response to an inducer, NDV_{UV}, could not be separated from the antiviral activity. In contrast, a "blocker" of interferon production found in the medium of cells infected with an arbovirus was separated from interferon during purification (Isaacs, Rotem and Fantes, 1966).

The same problem of specificity applies also to those other effects which have been seen with interferon preparations. In early studies, it was reported that interferon preparations increased aerobic glycolysis (Isaacs, Klemperer and Hitchcock, 1961), produced changes in the morphology of human amnion cells (Gresser, 1961), reduced cellular RNA metabolism (Levy, Snellbaker and Baron, 1963; Sonnabend, 1964), and slowed cell division (Paucker, Cantell and Henle, 1962). However, in subsequent studies with more highly purified interferon preparations, effects on aerobic glycolysis (Lampson et al., 1963), on RNA metabolism (Levy and Merigan, 1966) and on the rate of cell division (Baron, Merigan and McKerlie, 1966) were not found. Nevertheless, Paucker and Golgher (1969) find that relatively highly purified preparations of mouse interferon still reduce the rate of growth of mouse L cells, and to an extent proportional to the concentration added. Although their results must be open to the criticism that the interferon was still not sufficiently pure their evidence does suggest that the observed effects were due to interferon itself.

Do interferons have any other effects in cells in the absence of virus infection? At present, it appears not, though this may merely reflect our ignorance. However, the following argument may be valid. If a culture of actively growing cells is treated with interferon, the cells become resistant to virus infection because of changes produced in the ribosomes, and they also divide more slowly. If the interferon is removed, the rate of cell division returns quite quickly to normal, and after a few cell divisions the cells again become fully sensitive to infection with viruses. Presumably there is loss of the postulated Antiviral Protein, responsible for the antiviral state, from existing ribosomes, or dilution of the ribosomes produced in consequence of interferon treatment by normal ribosomes. This surely means that for normal cellular function, "normal" ribosomes are preferable to "interferon" ribosomes: otherwise, ribosomes of the "interferon" type must have been selected during the course of evolution to become the normal type, because of the obvious survival value of permanent resistance to virus infection. It thus seems probable that some cellular messages are transcribed inefficiently, or perhaps not at all, in interferon-treated cells, and this may be because they resemble viral messenger RNA more closely than most other cell messages. One consequence could be the reported decrease in the rate of division of cells treated with interferon. However it seems very improbable that any overall damage to an animal results from exposure of its cells to interferon for periods of several days at a time, since this must happen not infrequently during natural virus infections.

In this connexion two points seem relevant. First, during normal cell mitosis and in cells during metaphase arrest, multiplication of viruses is inhibited (Marcus and Robbins, 1963), as it is in interferon-treated cells. Second, it is well known that during mitosis there is a marked decrease in the rate of protein synthesis. Ribosomes obtained from HeLa cells arrested in metaphase incorporated labelled amino acids into acid-insoluble material at a rate about three times less than ribosomes from control cells. However, when metaphase ribosomes were treated with a suitable concentration of trypsin their ability to incorporate amino acids increased to 85 per cent of the control level (Salb and Marcus, 1965). This effect was interpreted as being due to removal of a Translational Inhibitory Protein from the ribosomes. These same workers (Marcus and Salb, 1966) also found that if ribosomes harvested from interferon-treated cells were also treated with trypsin (a much smaller amount had to be used), they responded to Sindbis virus RNA in the same way as control ribosomes. They suggest that a Translational Inhibitory Protein is also present on the ribosomes of interferon-treated cells. Thus there seem to be analogies between normal cells during mitosis and cells treated with interferon, but whether this is more than coincidental remains for further study.

There is much evidence (reviewed by Baron, 1966) that the formation and action of interferon plays an important role in the body as an antiviral defence mechanism. Isaacs, Cox and Rotem (1963), considering the nature of the fundamental stimulus for the formation of interferon, suggested that this was a response to the entry of a "foreign" nucleic acid into cells, and obtained suggestive experimental evidence in favour of this hypothesis. Such a response would obviously provide a homeostatic mechanism whereby the expression of unwanted messages, whether viral in origin or derived fortuitously from other cells in an animal, would be suppressed. This hypothesis however has not been fully confirmed in later experiments. As far as viruses are concerned, the nucleic acid has definitely been implicated as the stimulus to the formation of interferon, though there is at present controversy over whether single-stranded or double-stranded material is involved in this. As far as other interferon inducers are concerned, it is difficult to see what for example there is in common between such inducers as viruses, pyran copolymer (Merigan, 1968), *Brucella abortus* (Youngner and Stinebring, 1964) and malaria parasites (Huang, Schultz and Gordon, 1968). Equally, the formation of interferon in response to some of these stimuli may possibly be fortuitous.

In summary, there is at present no evidence that interferons have a regulatory function in cells not exposed to virus infection, but the possibility that such a function may be found in the future cannot be completely excluded.

SUMMARY

Interferons are certainly regulators in the sense that they render cells resistant to infection with viruses. They belong to a family of antiviral proteins, with molecular weights in the range 25000–160000, which are formed by cells of fish, amphibians, birds and mammals in response to certain stimuli, including viruses. They can only be measured in terms of their antiviral properties, and are very potent substances: thus chick interferon contains more than 1·6 million units of antiviral activity per mg of protein.

It is believed that interferons produce their antiviral effects by altering the way in which the ribosomes respond to viral messenger RNA molecules. Otherwise, treatment of cells with interferon has very few known effects. According to circumstances, it may increase or decrease the production of interferon by the cells in response to a stimulus. Also, the rate of cell division may be slowed. In all such studies, highly purified interferon must be used if meaningful results are to be obtained.

The possibility that interferons have a regulatory function in normal cells cannot be completely excluded, but there is at present no evidence for such a function.

REFERENCES

BARON, S. (1966). In *Interferons*, pp. 268–292, ed. Finter, N. B. Amsterdam: North Holland.
BARON, S., and LEVY, H. B. (1966). *A. Rev. Microbiol.*, **20**, 291–318.
BARON, S., MERIGAN, T. C., and McKERLIE, M. L. (1966). *Proc. Soc. exp. Biol. Med.*, **121**, 50–52.
BUCKNALL, R. A. (1967). *Nature, Lond.*, **216**, 1022–1023.
BURKE, D. C. (1969). This volume, p. 171.
DESMYTER, J., RAWLS, W. E., and MELNICK, J. L. (1968). *Proc. natn. Acad. Sci. U.S.A.*, **69**, 69–76.
FANTES, K. H. (1968). *Ciba Fdn Symp. Interferon*, pp. 78–90. London: Churchill.
FINTER, N. B. (ed.) (1966). *Interferons*. Amsterdam: North Holland.
FINTER, N. B., and BUCKNALL, R. A. (1968). In *Recent Advances in Pathology* (4th edn), pp. 429–447, ed. Robson, J. M., and Stacey, R. S. London: Churchill.
FRIEDMAN, R. M. (1966). *J. Immun.*, **96**, 872–877.
GRESSER, I. (1961). *Proc. natn. Acad. Sci. U.S.A.*, **47**, 1817–1822.
HUANG, K., SCHULTZ, W. W., and GORDON, F. B. (1968). *Science*, **162**, 123–124.
ISAACS, A., COX, R. A., and ROTEM, Z. (1963). *Lancet*, **2**, 113–116.
ISAACS, A., KLEMPERER, H. G., and HITCHCOCK, G. (1961). *Virology*, **13**, 191–199.
ISAACS, A., and LINDENMANN, J. (1957). *Proc. R. Soc. B*, **147**, 258–267.
ISAACS, A., ROTEM, Z., and FANTES, K. H. (1966). *Virology*, **29**, 248–254.
LAMPSON, G. P., TYTELL, A. A., NEMES, M. M., and HILLEMAN, M. R. (1963). *Proc. Soc. exp. Biol. Med.*, **112**, 468–478.
LEVY, H. B., and MERIGAN, T. C. (1966). *Proc. Soc. exp. Biol. Med.*, **121**, 53–55.
LEVY, H. B., SNELLBAKER, L. F., and BARON, S. (1963). *Virology*, **21**, 48–55.
MARCUS, P. I., and ROBBINS, E. (1963). *Proc. natn. Acad. Sci. U.S.A.*, **50**, 1156–1164.
MARCUS, P. I., and SALB, J. M. (1966). *Virology*, **30**, 502–516.
MERIGAN, T. C. (1968). *Ciba Fdn Symp. Interferon*, pp. 50–60. London: Churchill.
PAUCKER, K., and BOXACA, M. (1967). *Bact. Rev.*, **31**, 145–156.
PAUCKER, K., CANTELL, K., and HENLE, W. (1962). *Virology*, **17**, 324–334.
PAUCKER, K., and GOLGHER, R. R. (1969). Personal communication.
SALB, J. M., and MARCUS, P. I. (1965). *Proc. natn. Acad. Sci. U.S.A.*, **54**, 1353–1358.
SONNABEND, J. A. (1964). *Nature, Lond.*, **203**, 496–498.
YOUNGNER, J. S., and STINEBRING, W. R. (1964). *Science*, **144**, 1022–1023.

[For discussion of this paper, see pp. 176–179.]

INTERFERONS AS POSSIBLE REGULATORS—BIOCHEMICAL ASPECTS

D. C. BURKE

Department of Biological Chemistry, Marischal College, University of Aberdeen

THE wide variety of substances that induce the formation of interferon can be divided into two broad groups. First there are the viruses, and this group include those substances that induce interferon formation because they contain viruses, for example statolon, which is a preparation containing a mould virus. Virus-induced interferon formation requires both cellular RNA synthesis and protein synthesis, and interferon is synthesized *de novo*. Second, there are the non-viral substances such as endotoxin which are only effective in animals and which appear to act by releasing preformed interferon. Recently it has been shown that double-stranded RNA will also induce interferon formation. The double-stranded RNA can be a synthetic molecule of one strand of polyinosinic acid and another of polycytidylic acid (poly IC), or it can be the naturally occurring double-stranded molecules formed in virus replication or found in reovirus.

Virus-induced interferon formation has been widely formulated as a derepression mechanism in which the inducer interacts with the host genome, resulting in the formation of a new messenger RNA and the protein interferon. There is good evidence that the inducer is a functional nucleic acid, and it has been suggested that the first stage in virus-induced interferon formation is the formation of double-stranded RNA (Skehel and Burke, 1968*a*, *b*).

However other interpretations of these data are possible, and results obtained with ultraviolet-inactivated myxoviruses as inducers suggest that single-stranded RNA molecules can act as inducers (S. S. Gandhi and Burke, unpublished observations). Since there is now also good evidence that single-stranded RNA molecules can induce low titres of interferon (Baron *et al.*, 1969) it seems likely that there must be some cellular mechanism that is activated by these nucleic acid molecules. The nature of this mechanism is obscure. It is also well established that interferon formation can be induced by double-stranded RNA molecules. It is not known whether these double-stranded molecules, which are much better inducers than single-stranded RNA, are effective because they are an essential trigger for the formation of interferon, or whether this is the most effective way of getting foreign RNA into cells. Nor is anything known of the intracellular fate of double-stranded RNA. Interferon induction by double-stranded RNA molecules was thought at first to proceed by a similar mechanism to virus-induced interferon formation (Field et al., 1968), but some recent evidence suggests that the process, which is less sensitive than the virus-induced process to metabolic inhibitors, may involve the release of preformed interferon (Finkelstein, Bausek and Merigan, 1968).

Whatever the outcome, it is clear that viruses and probably double-stranded RNA molecules initiate interferon production by a process involving RNA and protein synthesis that resembles derepression, and it is important to discover what cellular control mechanisms are concerned in this process. Very little is known about the events whereby nucleic acids initiate interferon formation. If the derepression mechanism is correct then the inducer molecule must interact specifically with the host genome, either directly or through a cytoplasmic intermediate, resulting in the activation of a normally dormant nuclear gene. Interferon is strikingly species-specific, a phenomenon readily explicable if interferon is a cellular protein, whose synthesis is normally repressed. If this were so, then it might be expected that uninfected cells would produce low levels of interferon "constitutively", and there have been reports that this is so (Wagner and Smith, 1967), but it is difficult to exclude the possibility that interferon formation in this case was due to contamination with virus or endotoxin.

One way in which information about the control of nuclear events can be obtained is through the use of cell fusion, either spontaneous or virus-induced (cf. Harris, 1968), and there has been one report of the use of this technique to study interferon formation (Guggenheim, Friedman and Rabson, 1968). They found that chick red blood cells, which do not normally produce interferon, produced low titres of both chick and human interferon after fusion with the AH-1 line of human cells—cells which themselves produce interferon. The cell fusion agent was ultraviolet-inactivated Sendai virus which also acted as the interferon inducer. The authors suggested that activation of the dormant chick genes for interferon was a consequence of fusion with the human cells, presumably through the mediation of some cytoplasmic agent.

In the Jacob-Monod model for derepression, control is exerted at the transcriptional level but there is increasing evidence for control at the translational level, and indeed it has been questioned whether the Jacob-Monod model, for which there is such strong evidence in bacterial cells, is applicable to animal cells at all (Harris, 1968). It is therefore relevant to ask whether the production of interferon is controlled at the transcriptional or translational stage.

Production of virus-induced interferon is inhibited by low doses of several inhibitors of RNA synthesis, including actinomycin, at doses where the inhibitors principally act on DNA-directed RNA synthesis (Walters, Burke and Skehel, 1967). This suggested that interferon production required new RNA synthesis and that control was at the transcriptional level. A direct attempt to look for messenger RNA synthesis in interferon-producing cells was unsuccessful (Burke and Walters, 1966), but this is not surprising since interferon messenger RNA represents such a small percentage of total RNA synthesis. Production is also inhibited by several inhibitors of protein synthesis (Burke, 1966), showing that protein synthesis is required, although this observation does not discriminate between transcriptional and translational control.

To summarize, the available evidence suggests that viral nucleic acids, possibly through the formation of a double-stranded RNA, cause the unmasking of a latent property of the cellular DNA by an unknown mechanism. This leads to the transcription of an interferon messenger RNA and formation of interferon.

What control mechanisms are involved in the action of interferon? After cells are treated with interferon, cellular RNA and protein synthesis must occur before the cells are protected against virus challenge, and this has usually been interpreted as a requirement for the synthesis, by a derepression mechanism, of an antiviral protein, which is itself the active agent. This sequence of events would provide a means of producing a large number of antiviral protein molecules after stimulation with a few molecules or even a single molecule of interferon. There are however other ways of interpreting the requirement for RNA and protein synthesis; for example, it may be that new ribosomes have to be synthesized before interferon is effective. Whatever is the explanation, the evidence available to date suggests that, in interferon-treated cells, cellular ribosomes can attach to and translate cellular messenger RNA molecules normally but cannot attach to viral messenger RNA molecules (Joklik and Merigan, 1966; Levy and Carter, 1968). Alternatively it has been suggested (Marcus and Salb, 1967) that ribosomes can attach to viral messenger RNA with somewhat reduced efficiency but are unable to translate the messenger RNA. The former theory now seems to be more likely but both involve action at the translational level, and suggest that interferon acts by bringing about a change in the ribosome so that it cannot function normally on viral messengers but translates cellular messengers normally. However, an intensive search for changes in the composition of ribosomes from interferon-treated cells has been unsuccessful and it is possible that treatment with interferon inhibits other intracellular events unique to the virus-infected cell, such as the functioning of a particular transfer RNA molecule. If interferon acted by inhibiting an event unique to the virus-infected cell then the lack of toxicity of interferon preparations would be readily explained. There are, however, persistent reports of the slowing of protein synthesis (Johnson, Lerner and Lancz, 1968) and of cell division (Paucker and Golgher, 1969) by partially purified preparations of interferon, and it is possible that interferon action is selective rather than specific. Action at the ribosome would account for interferon's effectiveness against both DNA and RNA viruses, for both make messenger RNA. It also implies that all viral messenger RNA molecules differ in some way from cellular messenger RNA molecules, and such a difference could be an important control factor. The basis for this difference is not known although one obvious possibility could be a difference in the initiation triplets, about which little is known in mammalian cells. The continued synthesis of viral protein in cells whose rate of protein synthesis has been drastically inhibited as a consequence of virus infection (Martin and Kerr, 1968) also indicates that the translation of viral messenger RNA may differ from that of cellular messenger RNA.

It is of interest to note that although interferon inhibits the formation of T antigen in cells productively infected with SV 40 virus, where the viral genome and messenger RNA exist as free entities in the infected cell, it does not inhibit formation of T antigen in cells which have been transformed by the virus, suggesting that in this latter case control of the viral antigen operates by the same mechanism as the control of cellular protein synthesis (Oxman et al., 1967).

In conclusion, it appears that interferon exerts its antiviral effect at the level of the ribosome, and indicates a control mechanism which may be of importance, at least in the virus-infected cell. The mechanism by which interferon acts is still unknown, despite intensive work. Since virus infections are widespread in nature, the availability of a normally repressed mechanism for inhibiting their growth could offer considerable evolutionary advantages.

SUMMARY

There is strong evidence that double-stranded RNA molecules will induce interferon formation, although it is possible that single-stranded RNA molecules are also effective. It is not known whether double-stranded RNA molecules are effective because they are the trigger for the events leading to interferon formation or because this is the most effective way of getting foreign RNA into cells. Neither is it known what events lie between the entry of the double-stranded RNA molecule and the release of interferon. Results obtained with metabolic inhibitors strongly suggest that DNA-directed RNA synthesis is essential for interferon formation, and control is therefore at the transcriptional level. This interpretation is also suggested by recent cell-fusion experiments.

Before interferon can exert its effect new protein and RNA synthesis are required. This has usually been interpreted as reflecting the formation of an antiviral protein which is itself the active agent. However other explanations are possible. The evidence suggests that in interferon-treated cells the ribosomes can attach normally to cellular messenger RNA molecules but not to viral messenger RNA molecules. The signals responsible for ribosome attachment to mammalian messenger RNA molecules are not understood, and it is not possible to pursue this interpretation further than to say that it suggests a difference between all viral messenger RNA molecules and those of the host, which could be a control mechanism of importance.

REFERENCES

BARON, S., BOGOMOLOVA, N. N., BILLIAU, A., LEVY, H. B., and BUCKLER, C. E. (1969). Personal communication.
BURKE, D. C. (1966). In *Interferons*, pp. 55–86, ed. Finter, N. B. Amsterdam: North Holland.
BURKE, D. C., and WALTERS, S. (1966). *Biochem. J.*, **101**, 25–26P.
FIELD, A. K., TYTELL, A. A., LAMPSON, G. P., and HILLEMAN, M. R. (1968). *Proc. natn. Acad. Sci. U.S.A.*, **61**, 340–346.
FINKELSTEIN, M. S., BAUSEK, G. H., and MERIGAN, T. C. (1968). *Science*, **161**, 465–468.
GUGGENHEIM, M. A., FRIEDMAN, R. M., and RABSON, A. (1968). *Science*, **159**, 542–543.
HARRIS, H. (1968). *Nucleus and Cytoplasm*. Oxford: Clarendon Press.
JOHNSON, T. C., LERNER, M. P., and LANCZ, G. J. (1968). *J. Cell Biol.*, **36**, 617–624.
JOKLIK, W. K., and MERIGAN, T. C. (1966). *Proc. natn. Acad. Sci. U.S.A.*, **56**, 558–565.
LEVY, H. B., and CARTER, W. A. (1968). *J. molec. Biol.*, **31**, 561–577.
MARCUS, P. I., and SALB, J. M. (1967). *Virology*, **30**, 502–516.
MARTIN, E. C., and KERR, I. M. (1968). In *The Molecular Biology of Viruses*, pp. 15–46, ed. Crawford, L. V., and Stoker, M. G. P. London: Cambridge University Press.
OXMAN, M. N., BARON, S., BLACK, P. H., TAKEMOTO, K. K., HABEL, K., and ROWE, W. P. (1967). *Virology*, **32**, 122–127.
PAUCKER, K., and GOLGHER, R. R. (1969). Personal communication.
SKEHEL, J. J., and BURKE, D. C. (1968a). *J. gen. Virol.*, **3**, 35–42.
SKEHEL, J. J., and BURKE, D. C. (1968b). *J. gen. Virol.*, **3**, 191–199.
WAGNER, R. R., and SMITH, T. J. (1967). *Ciba Fdn Symp. Interferon*, pp. 95–106. London: Churchill.
WALTERS, S., BURKE, D. C., and SKEHEL, J. J. (1967). *J. gen. Virol.*, **1**, 349–362.

DISCUSSION

Gros: Is it established that protection against viral infection by interferon is a reversible process in all cases?

Finter: It was not reversed during an 11-day period in chick cells kept in culture with interferon and serum-free medium, so that they did not divide (Isaacs and Westwood, 1959), although I do not think this particular type of experiment has ever been repeated. If interferon-treated cells are allowed to grow and divide in the absence of interferon, they again become sensitive to viral infection (Paucker and Cantell, 1963).

Gros: If this is so in all situations it would suggest that interferon has to be attached (stoicheiometrically) to some newly formed piece of the translation machinery, and once the inducer is diluted out the inhibition can resume.

Möller: How does interferon, which is a large molecule, penetrate the cell?

Finter: It is not known how interferon enters a cell. It appears that small amounts are needed to render a cell resistant to virus infection, possibly only a few molecules per cell (Buckler, Baron and Levy, 1966). This could imply an effect entirely at the cell surface, but the evidence increasingly suggests that interferon does penetrate into the cell by some mechanism which is not yet understood (Levine, 1966).

Burke: We have no positive evidence that interferon gets inside because we cannot label it radioactively or by fluorescent antibody, and we can't recover it from treated cells. We could label the impure preparation with radioactive iodine and follow the fate of all the proteins that are there; this has been tried. But to follow the interferon one needs to have pure material.

Möller: If it happened to be fixed on to ribosomes, which an ordinary albumin solution wouldn't be, this would be significant.

Gros: In this respect, can you recover interferon activity from interferon-producing cells by treating ribosomes with detergents or salts?

Finter: In general, only extracellular interferons can be detected in reasonable quantities. However, all methods for the detection and measurement of interferons involve antiviral tests, usually in tissue cultures, and such tests are probably very insensitive in absolute terms.

Gros: You could possibly break up the cells and try to recover the ribosomal fraction.

Burke: In the chick cell system intracellular titres are very low. The material appears to be released very rapidly. In the L cell system material is found intracellularly and this has never been looked at; it would be interesting to do so.

Möller: Is it possible that interferon acts directly on free messenger RNA of viral origin, or on messenger RNA fixed to ribosomes?

Burke: The problem is that its actions appear to be mediated by an antiviral protein, so there is no point in taking a cell-free system and adding interferon to it. You have to take the components of treated cells and look at *their* behaviour in the cell-free system. What is holding up progress in this field is that no one has a good system for looking at protein synthesis from mammalian cells directed by viral RNA. Marcus and Salb (1966) made the first claim and it does not appear to be substantiated.

Bergel: Dr Jacques, is it possible that phagosomes take up the proteins and import them into the cell by endocytosis as the first stage of the lysosomal process, and then combination with the lysosome bag takes place? The question is whether, prior to the action of the lysosomal proteases—that is, before the secondary lysosomes are formed—the proteins might exert their effect?

Allison: Some proteins are known to become associated with membranes directly. This is true of phospholipase enzymes, streptolysins and some other bacterial toxins which can penetrate lipid layers. So it is conceivable that interferons could penetrate the cell membrane directly.

Jacques: If the interferon molecules have to enter the cell in order to exert their antiviral activity, they might indeed be taken in by endocytosis. In that case, they would probably not be able to reach extravacuolar target receptors without crossing the phagosomal or the lysosomal membrane.

Allison: Does interferon not have to enter the *nucleus* to initiate the new synthesis of antiviral protein? Interferon has to induce antiviral protein before it can prevent virus replication.

Burke: It could do this through interaction at the cytoplasmic level.

Subak-Sharpe: What is the evidence that nuclear genes are involved? We know there is DNA in mitochondria.

Burke: The evidence is not really very good. Species specificity of course argues for a nuclear gene. You would have to say that the mitochondrial genes differ from species to species, which is possible but does not seem very likely. In work using hybridized cells, human cells which produced interferon were fused with chick erythrocytes which are nucleated but do not produce interferon. Ultraviolet-irradiated Sendai virus was the fusing agent and also acted as an inducing agent for interferon; there was some evidence of the formation of chick interferon by this system. This was suggestive evidence for the activation of the nuclear genes (Guggenheim, Friedman and Rabson, 1968); and chick erythrocytes don't contain mitochondria.

Ormerod: Mitochondrial DNA codes for only about ten proteins. Some of the structural components of mitochondria are coded by its own DNA, so there would not be much DNA left over to code for other proteins.

Subak-Sharpe: However, we do not know that all the mitochondrial DNA's are identical, which you are assuming. And as I understand it from Dr D. Wilkie of University College London the number of DNA molecules per mitochondrion may still be in the order of 20–30 to one, and not one to one.

Stoker: On the problem of recognition, how do the interferon-treated ribosomes recognize the difference between viral and host messenger RNA? The normal ribosomes and the interferon-treated ribosomes presumably don't distinguish host message, and treat it the same—whether it is the initiation site or some other recognition point—and only recognize the viral message. So is it a positive recognition on the virus message, a distinct recognition site, or is it the absence of a recognition site on the host message? In an earlier symposium the question came up of whether interferon-treated ribosomes can distinguish synthetic messenger plant virus RNA and so on (Wolstenholme and O'Connor, 1968); is there any further work on this, or is it still so difficult to look at ribosome function and protein synthesis in the animal cell system?

Burke: There is no clear-cut answer. One observation with a bearing on this is that in many virus-infected cells one of the actions of the virus is to depress the rate of host cell protein synthesis, and yet viral protein synthesis continues (Martin and Kerr, 1968). We don't know how this works but it is almost like a reversed interferon effect. That might argue for a difference between host and viral messenger RNA.

Stoker: There are many interferon-sensitive viruses which don't inhibit the host protein synthesis.

Burke: Another drawback is that this is an all-or-none theory—you either recognize viral messenger or you don't—whereas viruses vary very widely in their sensitivity to interferon, by a factor of about 10000. So you must postulate something which can accommodate this variability.

Finter: Equally, cell messenger RNA molecules may possibly vary in their sensitivity to the effects of interferon treatment, since for example the growth of treated cells is slowed. Perhaps these RNA molecules differ qualitatively.

Bergel: I was very interested in Dr Finter's remarks about a slowing-down effect of some interferons, at least, on cell growth. Presumably many different viruses are present in the body, and if a substance is present which has a slowing effect on cells, this in itself must regulate growing tissue of any person not living in a virus-free environment.

Finter: I presume you are thinking here not of overt virus infections but of sub-clinical infections, for example of the nasopharynx with adenoviruses, myxoviruses and so on. I imagine that here the amounts of interferon formed may be too small to have any effects on cell growth. There are of course also latent virus infections, for example of the skin with herpes simplex virus. Here there must be even less stimulus to interferon formation, or indeed perhaps none at all if the viral DNA becomes integrated into the cell genome in the same way as prophage, which is a possibility.

Burke: One other aspect of interferon that bears on the general problems we are discussing is the question of recognition and species specificity, even though it isn't absolute for interferon. It does show that in some way the molecule is recognized or is received by the target cell. We know that the loss of interferon from solution by uptake by cells is the same for interferon from any species, but what discriminates between heterologous and homologous interferons is quite unknown; there must be some recognition signal.

REFERENCES

BUCKLER, C. E., BARON, S., and LEVY, H. (1966). *Science*, **152**, 80.
GUGGENHEIM, M. A., FRIEDMAN, R. M., and RABSON, A. S. (1968). *Science*, **159**, 542–543.
ISAACS, A., and WESTWOOD, A. (1959). *Nature, Lond.*, **184**, 1232.
LEVINE, S. (1966). *Proc. Soc. exp. Biol. Med.*, **121**, 1041.
MARCUS, P. I., and SALB, J. M. (1966). *Virology*, **30**, 502–516.
MARTIN, E. M., and KERR, I. M. (1968). In *The Molecular Biology of Viruses*, pp. 15–46, ed. Crawford, L. V., and Stoker, M. G. P. London: Cambridge University Press.
PAUCKER, K., and CANTELL, K. (1963). *Virology*, **21**, 22.
WOLSTENHOLME, G. E. W., and O'CONNOR, M. (eds.) (1968). *Ciba Fdn Symp. Interferon*. London: Churchill.

LYSOSOMES AND HOMEOSTATIC REGULATION

PIERRE J. JACQUES*

Laboratoire de Chimie Physiologique, Université Catholique de Louvain, Belgium

LYSOSOMES proper, but above all the family of vacuoles which compose the lysosomal apparatus, play a decisive role in various aspects of homeostasis, whether one maintains the original meaning of this useful concept as the constant composition of the *milieu intérieur* despite the permanent flow of matter which characterizes life, or extends the definition of homeostasis to the dynamic regulated equilibrium qualifying the steady-state concentration, not only of individual molecules, but also of the various types of cells and non-protoplasmic supporting structures occurring in a normal complex organism.

STRUCTURE AND FUNCTIONS OF THE LYSOSOMAL APPARATUS

As described by de Duve and Wattiaux (1966), the lysosomal apparatus is composed of a variety of cytoplasmic vacuoles (lysosomes, phagosomes and postlysosomes) which can recognize each other in that they selectively fuse when they meet in suitable conditions, thus bringing together their respective contents without allowing them the opportunity to escape into the cytoplasmic matrix.

Lysosomes are characterized by the presence of many acid hydrolases (Barrett, 1969), and, in certain cells, an acid peroxidase, bacteriostatic basic proteins, and incompletely defined substances capable of binding carcinogens (for references see Jacques, 1969a). Primary lysosomes have not yet taken part in digestive events; they are thought to be formed by the endoplasmic reticulum and the Golgi apparatus (Novikoff, Essner and Quintana, 1964) and they discharge their contents outside the cell or into secondary lysosomes (digestive vacuoles). The latter contain, in addition, the substrates of intracellular digestion brought mainly by the phagosomes; they are the sites of intracellular *digestion, resorption* of low molecular weight

*Chercheur Agrégé du Fonds de la Recherche Fondamentale Collective.

products of hydrolysis, *storage* of insoluble products and indigestible substrates, and of *parasitism* by Mycobacteria or some protozoa.

Phagosomes are either hetero- or autophagosomes. Heterophagosomes are formed at the cell surface through endocytosis (de Duve, 1963; Jacques, 1969b) and carry extracellular substances to the digestive vacuoles when they act as prelysosomes (de Duve and Wattiaux, 1966; Straus, 1967), or back to the extracellular compartment, in regurgitation (Gordon and Jacques, 1966) and in diacytosis (Jacques, 1966, 1968). Autophagosomes appear to arise through fusion, end to end, of fragments of the endoplasmic reticulum (Ericsson, 1969) around a zone of the cytoplasmic matrix, thus sequestrating various types of the cell's own structures or free molecules, which are later exposed to digestion within secondary lysosomes.

In privileged cells like amoebae, the remnants of digestion contained in telolysosomes (residual bodies) are regularly discharged into the medium by exocytosis, whereas in the majority of metazoan cells which have lost this capacity of defecation they are retained in the telolysosomes as a permanent residue which indicates or perhaps causes senescence. Telolysosomes may eventually become *postlysosomes*, by losing their enzymic equipment.

The functional relationships between the various members of the lysosomal apparatus are shown schematically in Fig. 1 in a way which emphasizes the main function of the apparatus in protozoa and reticuloendothelial cells: digestion and storage, primarily for nutritional or scavenging purposes.

However, as repeatedly stressed earlier (Jacques, 1966, 1968, 1969b), *vesicular transport* is another property and sometimes the main function of the lysosomal apparatus. This is already apparent in Fig. 1 although only its ancillary role in digestion is shown, by three converging lines for the supply of matter (lysosomal enzymes, and endocytosed or sequestered compounds) meeting in the digestive vacuoles, from which furthermore starts the vesicular pathway for the evacuation of indigestible products. In other cells, vesicular transport is the major function of the lysosomal apparatus, sometimes to the extent that lysosomes proper have little if any role to play in it. Such an extreme example of specialization is given by the endothelial cells lining blood capillaries where numerous phagosomes are formed all along the cell surface and seldom have the opportunity of meeting the rare scattered lysosomes, since a very short intracellular journey separates them from the other side of the cell (Majno, 1965).

To the cell physiologist, the lysosomal apparatus appears as only a part of a larger cellular organ (Jacques, 1968) which can be found equally in plant or animal cells and corresponds roughly to the vacuome (Accoyer, 1924;

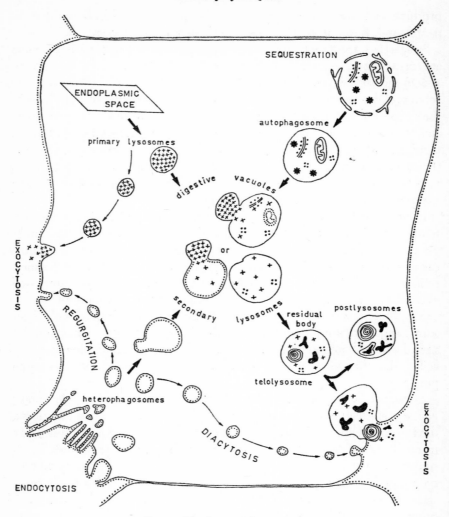

FIG. 1. The lysosomal apparatus.

Parat and Painleve, 1924; Parat, 1928) or the vacuolar apparatus (Bensley, 1911; Guillermond, 1934) of the early cytologists. Indeed, the expression "vacuolar system" has been revived since the electron microscope and techniques of biochemical cytology have permitted the inframicroscopic world to be investigated, and it has been used to designate the Golgi apparatus plus the endoplasmic reticulum and the nuclear membrane (De Robertis, Nowinski and Saez, 1960) or the lysosome–phagosome system (de Duve and Wattiaux, 1966).

These apparently divergent views can be reconciled by considering the *vacuolar apparatus*, resulting from an invagination of the plasma membrane, to be composed of two major parts (Fig. 2). The *endoplasmic apparatus* is

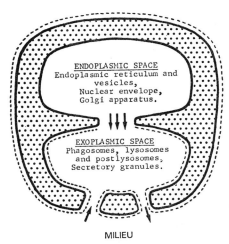

FIG. 2. The vacuolar apparatus.

primarily concerned with the biosynthesis and packaging of secretory products. The other part of the vacuome was called the *exoplasmic apparatus* by de Duve (1969) and comprises the lysosomal apparatus and non-lysosomal secretory granules (e.g. zymogens); it is thus involved in the vesicular transport, storage or catabolic breakdown of compounds which originate outside the cell, in the cytoplasmic matrix or in the endoplasmic space. Well developed in some protozoa, the vacuome seems to have appeared early in the evolution of intracellular organization. Its formation has analogies, at the cellular level, to the invagination of the entoblastic sheet which characterizes the gastrulation of so many embryos; in this case also the initial digestive pouch usually establishes a two-way traffic with the external medium and part of its endodermal wall further invaginates and gives rise to the parenchyma of the exocrine glands which discharge their secretions into the lumen of the digestive tract, without becoming accessible to its contents.

ROLE OF THE LYSOSOMAL APPARATUS IN HOMEOSTATIC REGULATION

Unicellular organisms

If the concept of homeostasis can be extended to unicellular organisms, it there includes the elementary mechanisms evolved by the vulnerable protoplasm for the purposes of survival. Here already the conspicuous lysosomal apparatus of protozoa plays an important role; advantage is taken of its heterophagic activity for nutrition, which also represents an effective means of tearing living potential enemies to pieces; furthermore, massive autophagy is the response of *Tetrahymena* cells to prolonged starvation (Elliott and Bak, 1964).

Multicellular organisms

The organization of multicellular organisms is characterized by several features: the specialization of groups of cells constituting the protoplasmic compartment of the body; the appearance of fluid compartments representing the *milieu intérieur* in which the cells are bathed, the fluids being involved in absorptive or secretory exchanges with the external world; and the building of solid extracellular matrices which give support to the enlarged organism. These features, which serve the new needs resulting from communal cell life, led many cells to utilize the lysosomal apparatus for sometimes very peculiar purposes.

Only the wandering or the sessile macrophages composing the reticuloendothelial (Aschoff, 1924a, 1924b) or histiocytic (Chevremont, 1948) system have kept the conspicuous and complete lysosomal apparatus of protozoa. They use it to clear the organism of old cells or even entire tissues, or of undesirable invaders ranging from protozoa to inanimate particles.

The other cells live in a *milieu intérieur* whose composition is kept constant (the genuine definition of homeostasis) and where they find a continuous supply of almost unvarying food which is furthermore entirely assimilable —that is, which leads to few if any waste products. This comfortable life has had two major consequences: (1) the almost general miniaturization of the lysosomal apparatus whose constituents are usually reduced to submicroscopic dimensions and thus make precious space available for the novel cytoplasmic structures required by the specialization of cells; and (2) the no less general loss of the faculty of defecation which threatens most metazoan cells with the dangerous overloading of the lysosomal apparatus, a process which invariably causes death when it lasts long enough as, for instance, in inborn lysosomal diseases (Hers, 1965; Hers and van Hoof, 1969).

The adaptation of the elementary functions of the lysosomal apparatus to special purposes is already seen in egg cells which store within endocytic vacuoles or in secretory granules the nutrients necessary for the development of the embryo. The delayed degradation of that yolk probably requires the lysosomal enzymes, since the latter are frequently observed in yolk platelets which usually show a multivesicular structure (Dalcq, 1963; de Duve and Wattiaux, 1966).

Later in the embryo or in the adult animal the lysosomal apparatus intervenes in homeostatic processes in two major ways. By its vesicular transport activity it contributes to the transfer of macromolecules between the body compartments. By its digestive activity, on the other hand, it plays a dominant role in the catabolic side of the turnover of soluble constituents of body fluids, of insoluble extracellular supporting structures and of the protoplasmic mass.

Catabolic aspects of turnover, detoxication and modelling of the body
(a) *Body fluids*

The concept of homeostasis was introduced by Claude Bernard to emphasize how constant the physical properties and the chemical composition of the *milieu intérieur* (primarily blood plasma and interstitial fluid) can remain despite the continuous renewal of its constituents and the instability of the environment.

As far as the physiological constituents of circulating fluids are concerned, their constant concentration results from an exact balance between the rate at which they are poured into the humoral compartment and the rate at which they are withdrawn from it. The lysosomal apparatus participates in this elementary type of homeostasis primarily by ensuring the intracellular breakdown of circulating cells and macromolecules or by transporting them to another compartment.

Thus, native homologous plasma proteins such as serum albumin and transferrin are taken up and digested within the lysosomes of liver cells (Gordon and Jacques, 1966); at least partially, the degradation of protein hormones like insulin may involve a similar process (Jacques and Wattiaux-de Coninck, 1969) which can, no doubt, be important to micromolecules as well.

The renewal of the cellular phase of blood also requires heterodigestion within lysosomes, after phagocytosis of red cells by cells of the reticuloendothelial system (Essner, 1960).

Reticuloendothelial cells also cooperate in keeping constant the *milieu intérieur* when they pick up and degrade (or simply store) foreign bodies introduced in the blood stream: numerous micro- and macromolecules, inert colloids, viruses and bacteria, many of which are antigenic to the host.

In all these instances, *endocytosis*, the gateway for extracellular compounds towards the exoplasmic space of cells, is implicated and by itself can account for the decrease in extracellular concentration which is the contribution of the lysosomal apparatus to this elementary type of homeostasis. The ultimate fate of the engulfed compounds (digestion, storage or discharge) is unimportant in this respect, although it determines various noteworthy processes such as the blockade of the reticuloendothelial system, the overloading of the lysosomal apparatus, the parasitism of the exoplasmic space by Mycobacteria (e.g. *Mycobacterium tuberculosis* and *M. leprae*) or protozoa, the activation of certain viruses (lethal heterodigestion), and the processing of antigens.

It may be noted that for body fluids such as lymph or the interstitial fluid whose composition at least partly depends on that of blood plasma, the lysosomal apparatus intervenes in homeostasis by supplying (instead of removing) macromolecules transported by vesicles from the plasma into these other fluids.

Inasmuch as part of the plasma protein reabsorbed from primary urine by the nephrons escapes digestion within the secondary lysosomes of tubular cells and is discharged back into the vascular compartment, this homeostatic function also applies to blood plasma.

(b) *Extracellular supporting structures*

The catabolic component of the turnover of bone matrix (Vaes, 1965, 1969) and of cartilage (Fell and Dingle, 1963; Dingle, 1963) involves the heterodigestive capacity of the lysosomes of osteoclasts and chondrocytes respectively, in a very different manner. In this case, the voluminous insoluble substrates to be hydrolysed cannot, because of their size, be seized and engulfed by the cell. Instead, lysosomal enzymes are *secreted* into the medium so that the lytic reactions take place extracellularly. It is believed however, that after this obligatory extracellular stage, digestion may be completed inside the cell, after resorption of small fragments by endocytosis.

(c) *Protoplasmic compartment*

Several distinct cellular mechanisms play a part in the digestion of the cell's own substance by its own lysosomal enzymes: autophagy and crinophagy in which the lytic process occurs within secondary lysosomes, and autolysis in the cytoplasmic matrix after partial release of the lysosomal enzymes from the exoplasmic space.

In contrast, the breakdown of entire cells and even organs within the lysosomes of histiocytes is based, as already described for the phagocytosis of red cells, on the endocytic and digestive capacity of the macrophages.

In *autophagy*, zones of the cytoplasm, usually containing organelles such as mitochondria, endoplasmic reticulum, secretory granules, and less frequently lysosomes and fragments of the brush border, become sequestered by a continuous membrane which may originate from the endoplasmic reticulum itself or from endocytic vesicles (Ericsson, 1964, 1969; de Duve and Wattiaux, 1966). This usual, poorly specific type of autophagic prehension can be contrasted to that observed in the endocrine pancreas and named *granulolysis* by Rouiller and his co-workers (Orci et al., 1968) where smaller autophagic vacuoles most often contain a single type of organelle, for example secretory granules.

The direct fusion of individual secretory granules with lysosomes has been observed by Smith and Farquhar (1966) in the pituitary gland of lactating animals. To this process, which leads to the destruction of surplus secretory substances by the cell which biosynthesized them, de Duve (1969) has given the name of *crinophagy*.

Autophagy occurs in physiological conditions (Ericsson, 1964, 1969; Novikoff and Shin, 1964; Ericsson, Trump and Weibel, 1965) and is considerably stimulated in the liver by the hormone glucagon (Ashford and Porter, 1962; Deter and de Duve, 1967). The reality of the role of autophagy in cellular physiology is best shown by the fact that the inborn absence of a lysosomal enzyme usually results in the progressive overloading of the lysosomal apparatus of various tissues by substances—most often normal metabolites—which can act as substrates of the missing enzyme and are usually present in the cytoplasmic matrix (Hers, 1965; Hers and Van Hoof, 1969). In these cases, autodigestion in lysosomes participates in the degradative part of the turnover of protoplasmic constituents, and this may be considered as an extension to the intracellular environment of the role played by the lysosomes in the classic type of homeostasis described above (p. 185). This probably holds true also for cellular *autolysis*, since homogenates of many different organs currently contain a detectable percentage of extragranular lysosomal hydrolases.

It is most remarkable that complex organisms submitted to prolonged starvation seek survival in the same way as unicellular organisms (see p. 186), by triggering a wave of autophagic activity; this is especially conspicuous in liver cells (Novikoff, Essner and Quintana, 1964; Swift and Hruban, 1964).

As Professor Bergel has indicated, studies on normal or anarchical cell growth must play an important part in the development of the concept of homeostasis. Indeed, the dynamic constancy of the number of cells of a given type at a given place in the organism (at a given stage of its development) is a phenomenon equivalent, in my opinion, to the constancy of the steady-state concentration of given molecular species in body fluids, which impressed Claude Bernard and led him to introduce the concept of homeostasis. In this connexion also, the lysosomal apparatus plays an important part.

The *differentiation* of keratinizing cells of the epidermis involves autophagy (Farquhar and Palade, 1965).

The participation of lysosomes in the *resorption of temporary organs*, in the embryo, has been recognized (Weber, 1963; Scheib, 1963; Eeckhout, 1965); it affects the genital ducts of Muller and of Wolff in the male and the female respectively, the interdigital membranes, and the tadpole tail during metamorphosis. Three different mechanisms, sometimes linked, may take part in these phenomena: autodigestion of the cells required to disappear, either through (1) *autolysis*, in accordance with the earlier biochemical concept which emphasized lysosomes as "suicide" particles (de Duve, 1959; Allison, 1965), or through (2) *autophagy*, as is the case for the regression of Mullerian ducts (Scheib, 1965); or (3) *heterodigestion* of these cells within lysosomes of macrophages arising from the histiocytic transformation of local cells, or invading the degenerating tissue from other parts of the body.

Such phenomena have also been encountered at later stages of development: soon after birth, in the course of the profound restructuring of kidneys, lungs and intestine (Clark, 1957; Moe and Behnke, 1962; Balis and Conen, 1964); later, in the resorbing thymus (Sachs *et al.*, 1962) or the uterus after delivery. Autophagy predominates in these processes. It is also conspicuous in the course of ageing (Jamieson and Palade, 1964), in the prostatic atrophy following castration (Brandes, Gyorkey and Groth, 1962; Swift and Hruban, 1964) and, more generally, as a response to various stress conditions (reviews by Weissman, 1965; de Duve and Wattiaux, 1966) such as exposure to cold, irradiation, hypoxia, metabolic inhibitors and diets deficient in vitamin E or potassium.

Through the resulting cellular autolysis, the release of acid hydrolases into the cytoplasm can intervene not only in physiological necrobiosis (see above) or pathological effects on the cell as in silicosis (Allison, Harington and Birbeck, 1966), but provided it does not kill the cell, it could cause alterations of the genetic material leading to malignant transformation (Allison and Paton, 1965).

Vesicular transport

The concentration of macromolecules, and probably also of several micromolecules, in various body fluids largely depends on the exchange of substances between adjacent compartments of the body through vesicles belonging to the lysosomal apparatus.

Such vesicular transport activity has been recognized, primarily through morphological studies, in several organs, including the intestinal mucosa, especially in the newborn suckling animal, the placental membranes, the choroid plexus, the ciliary body, the endothelium of blood capillaries and the peritoneal mesothelium.

Macromolecules are not transferred indiscriminately. The binding to intralysosomal substances (Brambell, 1966), the selectivity of endocytosis and the substrate specificity of lysosomal enzymes (Jacques, 1966, 1968, 1969b) may be relevant to this property, but other factors affecting the "sieving" capacity probably play a role in those cases where the molecular weight affects the ability of macromolecules to be transferred from one compartment to the other.

REGULATION OF LYSOSOMAL ACTIVITIES

The importance of the intervention of lysosomes in homeostatic phenomena, of which I have analysed various aspects, is most apparent when one considers the numerous interactions between lysosomal activities and the pre-eminent homeostatic regulators, the hormones. Besides this, other physiological factors and many pharmacological agents are capable of modulating one or another of the principal processes through which lysosomal functions are exerted: endocytic or autophagic prehension, the level of lysosomal enzymes, the rate of their exocytic discharge, and so on.

We shall limit ourselves to the few most striking aspects of this already well-documented chapter of cytopharmacology, of which two paragraphs still remain to be written: those concerning the mechanism of acidification of the secondary lysosomes, and the intracellular movement of vacuoles.

While it appears that liver lysosomes take part in the catabolic inactivation of protein hormones like insulin (Jacques and Wattiaux-de Coninck, 1969), lysosomes of thyroid cells in contrast account for the necessary hydrolytic stage in the elaboration of thyroid hormones, after endocytosis of thyroglobulin has been stimulated by thyrotropic hormone (TSH) (Novikoff and Vorbrodt, 1963; Novikoff, Essner and Quintana, 1964; Bauer and Meyer, 1964; Wollman, Spicer and Burstone, 1964; Wetzell, Spicer and Wollman, 1965).

Endocytosis is also stimulated by insulin in adipose cells (Barnett and Ball, 1960); it is sensitive, in reticuloendothelial cells, to a variety of steroids, of which several increase it whereas others depress or inhibit it (Nicol and Bilbey, 1960).

Autophagy in liver cells is considerably stimulated by glucagon (see p. 187).

Exocytosis of lysosomal enzymes is strikingly accelerated in cultured osteoclasts after treatment with parathyroid hormone (Vaes, 1965).

Furthermore, it is most likely that the stabilizing effect on lysosomes displayed *in vivo* as well as *in vitro* by cortisone and other drugs (Beaufay, van Campenhout and de Duve, 1959; de Duve, Wattiaux and Wibo, 1963; Weissmann and Thomas, 1964; Weissmann, 1967) accounts for their antiinflammatory activity and may influence the exchanges between the exoplasmic space and the cytoplasmic matrix.

The *level of lysosomal enzymes* considerably increases in two distinct situations. An increase has been observed after the accumulation of various substances (iron, polyvinylpyrollidone, carbon, sucrose, egg white) in the lysosomes of various cells (liver, spleen, kidneys, bone, cartilage; for references see de Duve and Wattiaux, 1966). This response of cells is not constant: the concentration of lysosomal enzymes remains normal in the liver after uptake of Triton WR-1339 (Wattiaux, 1966), of horseradish peroxidase and yeast invertase (Jacques, 1968) or of iodine-labelled insulin; other compounds like suramin (Smeesters and Jacques, 1968) increase, depress or do not affect the concentration of different hydrolases in liver cells. Furthermore, the significance of this cellular response remains obscure; in particular, there is usually no relation between the substrate-specificity of the enzymes whose activity increases most, and the chemical nature of the substance whose uptake or storage initiated the response.

A considerable increase in the concentration of lysosomal enzymes, associated this time with that of the rate of endocytosis, has been observed as a major feature of the histiocytic transformation of monocytes (Cohn and Benson, 1965*a,b,c,d*) and in the transformation of small lymphocytes into lymphoblasts which is triggered by phytohaemagglutinin (Robineaux, 1968). It cannot be excluded that one might find an analogous change in the course of malignant transformation, since it is accepted that cancer cells are endowed with considerable endocytic activity and since an increase of catheptic activity in certain tumours has already been related to the acquisition of malignancy (Maver and Greco, 1951; Greenstein, 1954).

As the topic of cancer has been raised (and its presence is noticeable in the background of several contributions to this symposium) it should be mentioned that, in addition to the chromosomal breaks in which Allison (see p. 188) incriminates lysosomes, the latter have also been variously implicated in cellular multiplication (Dougherty, 1964; Robbins and Gonatas, 1964; Gahan and Maple, 1966).

Finally, it is worth noting that three of the factors which Professor Stoker tells us stimulate cellular growth (see pp. 264-271) are known to stimulate or even to induce endocytosis, an effect which is probably necessary for the nutrition of cells cultured in media devoid of blood plasma. Thus *insulin* stimulates pinocytosis in adipose tissue (see p. 189); the necessity of *spreading* on a suitable surface for the induction of phagocytosis by leucocytes has long been recognized (Fenn, 1921; North, 1968); and the factor responsible for the histiocytic transformation of monocytes, which is present in newborn calf *serum* (Cohn and Benson, 1965c) has been recently identified as a gamma globulin with antibody activity directed against cell membranes (Cohn and Parks, 1967).

REFERENCES

ACCOYER, L. (1924). *C.r. Séanc. Soc. Biol.*, **91,** 665.
ALLISON, A. C. (1965). *Discovery, Lond.*, **26,** 8.
ALLISON, A. C., HARINGTON, J. S., and BIRBECK, M. (1966). *J. exp. Med.*, **124,** 141.
ALLISON, A. C., and PATON, G. R. (1965). *Nature, Lond.*, **207,** 1170.
ASCHOFF, L. (1924a). *Ergebn. inn. Med. Kinderheilk.*, **26,** 1.
ASCHOFF, L. (1924b). *Lectures on Pathology.* New York: Hoeber.
ASHFORD, T. P., and PORTER, K. R. (1962). *J. Cell Biol.*, **12,** 198.
BALIS, J. U., and CONEN, P. E. (1964). *Lab. Invest.*, **13,** 1215.
BARNETT, R. J., and BALL, E. G. (1960). *J. biophys. biochem. Cytol.*, **8,** 83.
BARRETT, A. J. (1969). In *Lysosomes in Biology and Pathology,* vol. 2, ed. Dingle, J. T., and Fell, H. B. Amsterdam: North Holland. In press.
BAUER, W. C., and MEYER, J. C. (1964). *Science,* **145,** 1431.
BEAUFAY, H., VAN CAMPENHOUT, E., and DE DUVE, C. (1959). *Biochem. J.*, **73,** 617.
BENSLEY, R. (1911). *Am. J. Anat.*, **12,** 37.
BRAMBELL, F. W. R. (1966). *Lancet,* **2,** 1087.
BRANDES, D., GYORKEY, F., and GROTH, D. P. (1962). *Lab. Invest.*, **11,** 339.
CHEVREMONT, M. (1948). *Biol. Rev.*, **23,** 267.
CLARK, S. L. (1957). *J. biophys. biochem. Cytol.*, **3,** 349.
COHN, Z. A., and BENSON, B. (1965a). *J. exp. Med.*, **121,** 153.
COHN, Z. A., and BENSON, B. (1965b). *J. exp. Med.*, **121,** 279.
COHN, Z. A., and BENSON, B. (1965c). *J. exp. Med.*, **121,** 835.
COHN, Z. A., and BENSON, B. (1965d). *J. exp. Med.*, **122,** 455.
COHN, Z. A., and PARKS, E. (1967). *J. exp. Med.*, **125,** 1091.
DALCQ, A. M. (1963). *Ciba Fdn Symp. Lysosomes,* p. 226. London: Churchill.
DE DUVE, C. (1959). In *Subcellular Particles,* p. 128, ed. Hayashi, T. New York: Ronald Press.
DE DUVE, C. (1963). *Ciba Fdn Symp. Lysosomes,* p. 128. London: Churchill.
DE DUVE, C. (1969). In *Lysosomes in Biology and Pathology,* vol. 1, ed. Dingle, J. T., and Fell, H. B. Amsterdam: North Holland. In press.
DE DUVE, C., and WATTIAUX, R. (1966). *A. Rev. Physiol.*, **28,** 435.
DE DUVE, C., WATTIAUX, R., and WIBO, M. (1963). *II Int. Pharmacol. Meeting,* Prague, vol. 5, p. 97. Oxford: Pergamon Press.

DE ROBERTIS, E., NOWINSKI, W., and SAEZ, F. (1960). *General Cytology.* London: Saunders.
DETER, R. L., and DE DUVE, C. (1967). *J. Cell Biol.*, **33,** 437.
DINGLE, J. T. (1963). *Ciba Fdn Symp. Lysosomes*, p. 384. London: Churchill.
DOUGHERTY, W. J. (1964). *J. Cell Biol.*, **23,** 25A.
EECKHOUT, Y. (1965). La Métamorphose caudale des Amphibiens Anoures. Doctoral thesis, Université Catholique, Louvain, Belgium.
ELLIOTT, A., and BAK, I. J. (1964). *J. Cell Biol.*, **20,** 113.
ERICSSON, J. L. E. (1964). *Acta path. microbiol. scand.*, suppl. 168, 1.
ERICSSON, J. L. E. (1969). In *Lysosomes in Biology and Pathology*, vol. 2, ed. Dingle, J. T., and Fell, H. B. Amsterdam: North Holland. In press.
ERICSSON, J. L. E., TRUMP, B. F., and WEIBEL, J. (1965). *Lab. Invest.*, **14,** 1341.
ESSNER, E. (1960). *J. biophys. biochem. Cytol.*, **7,** 329.
FARQUHAR, M. G., and PALADE, G. E. (1965). *J. Cell Biol.*, **26,** 263.
FELL, H. B., and DINGLE, J. T. (1963). *Biochem. J.*, **87,** 403.
FENN, W. O. (1921). *J. gen. Physiol.*, **3,** 575.
GAHAN, P. B., and MAPLE, A. J. (1966). *J. exp. Bot.*, **17,** 151.
GORDON, A. H., and JACQUES, P. (1966). In *Labelled Proteins in Tracer Studies*, p. 127, ed. Donato, L., Milhaud, G., and Sirchis, J. Brussels; Euratom 2950 d, f, e.
GREENSTEIN, J. P. (1954). *Biochemistry of Cancer.* New York: Academic Press.
GUILLERMOND, A. (1934). *Actual. Scient. ind.*, **171,** 108.
HERS, H. G. (1965). *Gastroenterology*, **48,** 625.
HERS, H. G., and VAN HOOF, F. (1969). In *Lysosomes in Biology and Pathology*, vol. 2, ed. Dingle, J. T., and Fell, H. B. Amsterdam: North Holland. In press.
JACQUES, P. J. (1966). *Revue Quest. scient.*, **27,** 99.
JACQUES, P. J. (1968). Academic thesis. Louvain: Librairie Universitaire, Belgium.
JACQUES, P. J. (1969a). *Biochem. J.*, **111,** 25 P.
JACQUES, P. J. (1969b). In *Lysosomes in Biology and Pathology*, p. 326, vol. 2, ed. Dingle, J. T., and Fell, H. B. Amsterdam: North Holland. In press.
JACQUES, P. J., and WATTIAUX-DE CONINCK, S. (1969). *VI Meeting Fed. Europ. Biochem. Soc.*, Madrid. Abstract No. 897, p. 279.
JAMIESON, J. D., and PALADE, G. E. (1964). *J. Cell Biol.*, **23,** 151.
MAJNO, G. (1965). In *Handbook of Physiology*, sect. 2, vol. 3, p. 2293, ed. Hamilton, W. F., and Dow, P. Washington: American Physiological Society.
MAVER, M. E., and GRECO, A. E. (1951). *J. natn. Cancer Inst.*, **12,** 37.
MOE, H., and BEHNKE, O. (1962). *J. Cell Biol.*, **13,** 168.
NICOL, T., and BILBEY, D. L. J. (1960). In *Reticuloendothelial Structure and Function*, p. 301, ed. Heller, J. H. New York: Ronald Press.
NORTH, R. J. (1968). *J. Reticuloendothelial Soc.*, **5,** 203.
NOVIKOFF, A. B., ESSNER, E., and QUINTANA, N. (1964). *Fedn Proc. Fedn Am. Socs exp. Biol.*, **23,** 1010.
NOVIKOFF, A. B., and SHIN, W. Y. (1964). *J. Microscopie*, **3,** 187.
NOVIKOFF, A. B., and VORBRODT, A. (1963). *J. Cell Biol.* **19,** 53A.
ORCI, L., JUNOD, A., PICTET, R., RENOLD, A. E., and ROUILLER, C. (1968). *J. Cell Biol.*, **38,** 462.
PARAT, M. (1928). *Archs Anat. microsc.*, **24,** 73.
PARAT, M., and PAINLEVE, J. (1924). *C.r. hebd. Séanc. Acad. Sci., Paris*, **179,** 612.
ROBBINS, E., and GONATAS, W. H. (1964). *J. Cell Biol.*, **21,** 429.
ROBINEAUX, R. (1968). *XII Int. Congr. Cell Biol.*, Brussels. Amsterdam: Excerpta Medica Foundation, International Congress Series, **166,** 6.
SACHS, G., DE DUVE, C., DVORKIN, B. S., and WHITE, A. (1962). *Expl Cell Res.*, **28,** 597.
SCHEIB, D. (1963). *Ciba Fdn Symp. Lysosomes*, p. 264. London: Churchill.

SCHEIB, D. (1965). *C.r. hebd. Séanc. Acad. Sci., Paris*, **260**, 1252.
SMEESTERS, C., and JACQUES, P. J. (1968). *XII Int. Congr. Cell Biol.*, Brussels. Amsterdam: Excerpta Medica Foundation, International Congress Series, **166**, 82.
SMITH, R. E., and FARQUHAR, M. G. (1966). *J. Cell Biol.*, **31**, 319.
STRAUS, W. (1967). In *Enzyme Cytology*, p. 239, ed. Roodyn, D. B. New York: Academic Press.
SWIFT, H., and HRUBAN, Z. (1964). *Fedn Proc. Fedn Am. Socs exp. Biol.*, **23**, 1026.
VAES, G. (1965). *Expl Cell Res.*, **39**, 470.
VAES, G. (1969). In *Lysosomes in Biology and Pathology*, vol. 1, ed. Dingle, J. T. and Fell, H. B. Amsterdam: North Holland. In press.
WATTIAUX, R. (1966). In *Etude expérimentale de la surcharge des lysosomes*. Gembloux: Duculot.
WEBER, R. (1963). *Ciba Fdn Symp. Lysosomes*, p. 282. London: Churchill.
WEISSMANN, G. (1965). *New Engl. J. Med.*, **273**, 1084.
WEISSMANN, G. (1967). *A. Rev. Med.*, **18**, 97.
WEISSMANN, G., and THOMAS, L. (1964). *Recent Prog. Horm. Res.*, **20**, 215.
WETZEL, B. K., SPICER, S. S., and WOLLMAN, S. H. (1965). *J. Cell Biol.*, **25**, 593.
WOLLMAN, S. H., SPICER, S. S., and BURSTONE, M. S. (1964). *J. Cell Biol.*, **21**, 191.

DISCUSSION

Möller: Is there a difference between pinocytosis and phagocytosis, or is it a matter of the size of the particle ingested?

Jacques: I recently analysed this question (Jacques, 1969) and I came to the view expressed by many other authors also that there is no difference between pinocytosis and phagocytosis other than the ability of cells to form large or small phagosomes. To my knowledge there is only one good piece of evidence against this view (Z. A. Cohn, personal communication): dilute cyanide, antimycin A or 2,4-dinitrophenol inhibited pinocytosis by macrophages in culture without affecting the phagocytic activity of the same cells.

Möller: Are the vacuoles that are formed invaginations of existing membrane or do they represent new synthesis of membrane?

Jacques: Although some experimental results have been held to indicate a *de novo* synthesis of membrane material during the endocytic invagination of the cell membrane (Karnovsky, 1962), this question cannot yet be answered.

Möller: Phagocytosis is not induced by all substances. In the case of antigenic material such as bacteria or foreign red cells, antibodies (opsonins) have to be present on the foreign cells in order for phagocytosis to occur and complement is also needed for this process. Apparently there must be some specific recognition signals which induce uptake. What is known about these inductive processes in phagocytosis?

Jacques: First I must mention that most soluble antigens and some viruses by the simple fact of their presence in body fluids can be taken up, in the absence of any induction, by the numerous types of cells which in a complex organism pinocytose in a continuous manner. Secondly, induction, as it occurs in pinocytosing amoebae (Holter, 1959; Chapman-Andresen, 1965) or macrophages (Cohn and Parks, 1967) and in phagocytosis of larger bodies (Fenn, 1921; Berry and Spies, 1949), is thought to be caused by the concentration of suitable prey at the cell–medium interface, whether they remain naked or become coated by immune or ordinary plasma molecules. As far as endocytosis is concerned, the recognition mechanism you are referring to may thus be based on purely physico-chemical phenomena with little specificity; however, some authors (Bona and Ghyka, 1969) put the recognition of the homologous or heterologous character of molecules at the level of the endocytic act.

Allison: Dr Jacques discussed two aspects of lysosomes and homeostasis; one is the role of lysosomes in the body in general and the other, which he touched on, and which I think is just as interesting, is the role at the cellular level. It is found that if lysosomes are activated either by uptake of foreign material or in some other way, for example by treatment with drugs that affect lysosomal membranes or with vitamin A or anti-cellular antibodies, this stimulates synthesis of lysosomal enzymes within these cells and often also their release into the extracellular medium (de Duve and Wattiaux, 1966; Dingle, Fell and Coombs, 1967). So here is a control system at the cellular level where the *use* of a particular set of enzymes leads to the synthesis of more enzymes of a similar kind. This is a striking adaptive phenomenon.

Möller: That leads to the question of why we have these processes; they are obviously useful in macrophages to remove various intruders and old red cells, and in tissue culture cells where they may serve nutritional purposes, but what is their function in the body as a whole? I would like Dr Allison to comment on the effect of carcinogens on lysosomes.

Allison: Most people accept that many carcinogenic materials are taken up into lysosomes in living cells and that the highest concentration of carcinogens that persists in cells is in lysosomes (Allison, 1969). Nobody really knows whether this is relevant to carcinogenesis. The second point, which Dr Jacques mentioned, is that lysosomal enzymes, in particular DNase, can produce chromosome damage (Allison and Paton, 1965). Gillian Paton and I have more recently found that if you take cultures of human diploid cells and add DNase in the appropriate conditions for getting it into the cells (with magnesium sulphate), many chromosome breaks are produced. We think that deletions or rearrangements initiated in this way represent one type of chromosomal mutation, rather like the Philadelphia chromosome, which might lead to carcinogenesis. This is an interesting but tenuous chain of evidence. So there are two reasonably solid facts, but whether they are relevant to carcinogenesis we do not know.

Möller: Does cortisone interfere with the effect of carcinogens?

Allison: In some circumstances it will prevent carcinogenesis in the skin.

Möller: Does vitamin A increase it?

Allison: Yes, in some circumstances (Polliack and Levij, 1967; Prutkin, 1968).

Möller: Suppose that agents which "labilize" the membranes, such as vitamin A, also tend to make the target cell in question more susceptible to surveillance mechanisms, such as cellular immunity; such agents would then have two effects which may antagonize each other with regard to the expression of neoplasia.

Allison: This is certainly true. As far as lysosomal membrane labilization is concerned, it is found that the so-called co-carcinogenic agents such as compound A, the croton oil factor, and some non-ionic detergents, are more efficient than the carcinogens themselves (Weissmann *et al.*, 1968).

Möller: What about the normal function of lysosomes in tissues? Dr Jacques suggested the scavenger function, but lysosomes are present in most cells in the body, although they appear to be rather inactive *in vivo*.

Jacques: Even in complex organisms lysosomes proper may be involved in the nutrition of most cells, by their ability to break down endocytosed circulating macromolecules (heterodigestion) or, in emergency situations, the cell's own cytoplasm (autodigestion). That same digestive activity may be applied to highly specialized purposes, for instance, by thyroid follicle cells for the production of thyroid hormone and by kidney tubule cells for the breakdown of valuable material recovered from urine. The scavenger activity of reticuloendothelial cells itself can acquire physiological significance when concerned with such processes as the destruction of senescing red blood cells.

Möller: You mentioned Brambell's finding on the passage of gamma globulins across the intestinal cells shortly after birth. In rats uptake of immune globulins from the milk occurs during the first 14 days of life and then suddenly stops. Is there any correlation between this and lysosomal activity in the intestinal cells?

Jacques: However attractive and plausible it may be, Brambell's theory (Brambell, 1966), which does provide an answer to your question, remains hypothetical. Do the immune proteins cross the intestinal mucosa or the placental membranes between cells or through them? If they are transferred through cells, is it after endocytosis or by simple permeation? If they are endocytosed, do they reach the lysosomes, and if they do, how can they be discharged by the cells without the other constituents of lysosomes? These are all open questions.

REFERENCES

ALLISON, A. C. (1969). In *Lysosomes in Biology and Pathology*, vol. 2, pp. 138–164, ed. Dingle, J. T., and Fell, H. B. Amsterdam: North Holland. In press.
ALLISON, A. C., and PATON, G. R. (1965). *Nature, Lond.*, **207**, 1170.
BERRY, L. J., and SPIES, T. D. (1949). *Medicine, Baltimore*, **28**, 239.
BONA, C., and GHYKA, G. (1969). *J. theoret. Biol.* In press.
BRAMBELL, F. W. R. (1966). *Lancet*, **2**, 1087.
DE DUVE, C., and WATTIAUX, R. (1966). *A. Rev. Physiol.*, **28**, 435.
CHAPMAN-ANDRESEN, C. (1965). *Archs Biol.*, **76**, 189.
COHN, Z. A., and PARKS, E. (1967). *J. exp. Med.*, **125**, 213.
DINGLE, J. T., FELL, H. B., and COOMBS, R. R. (1967). *Int. Archs Allergy appl. Immun.*, **31**, 283.
FENN, W. (1921). *J. gen. Physiol.*, **4**, 373.
HOLTER, H. (1959). *Int. Rev. Cytol.*, **8**, 481.
JACQUES, P. J. (1969). In *Lysosomes in Biology and Pathology*, vol. 2, p. 326, ed. Dingle, J. T., and Fell, H. B. Amsterdam: North Holland. In press.
KARNOVSKY, M. L. (1962). *Physiol. Rev.*, **42**, 143.
POLLIACK, A., and LEVIJ, I. S. (1967). *Nature, Lond.*, **216**, 187.
PRUTKIN, L. (1968). *Cancer Res.*, **28**, 1021.
WEISSMANN, G., TROLL, W., VAN DUUREN, B. L., and SESSA, G. (1968). *Biochem. Pharmac.*, **17**, 2421–2434.

REGULATORY MECHANISMS IN ANTIBODY SYNTHESIS

Göran Möller

Department of Bacteriology, Karolinska Institute, Stockholm

Various mechanisms operate to regulate the intensity and duration of the immune response. The sequential appearance of different immunoglobulin classes and the gradual increase of antibody affinity are other characteristics of antibody synthesis which are carefully controlled. An important regulating factor is antigen itself, which determines not only specificity, but also a variety of other properties of the antibodies produced (for review see Sela, 1967). Antigen may also turn off the immune response specifically by inducing immunological paralysis. The serum concentration of different immunoglobulin classes also plays a regulating role, since the concentration in the serum of each class determines the catabolic rate of that class.

One regulatory mechanism appears to be of particular importance and has been studied carefully in recent years (for review see Uhr and Möller, 1968). This is the ability of antibody to suppress its own synthesis. The phenomenon of antibody suppression was known at the end of the last century and was frequently used in vaccination against bacterial toxins, since it was found that a mixture of antibodies and toxin neutralized the toxic effects, suppressed a detectable immune response, but still primed the treated individuals so that they developed a normal secondary response to subsequent injections of toxins or to natural infection.

The basic characteristic of antibody suppression is its strict specificity. By itself this constitutes the strongest argument for the hypothesis that antibody suppresses by combining with the antigen, since this is the only specific property of antibody molecules. It was early discovered that the primary response was more amenable to suppression by antibody than the secondary and, as a rule, that antibody given before or during the first days after the antigen suppressed more efficiently than antibodies given later. Recent findings suggest that antibody suppression acts in a negative feedback system during the primary immune response to restrict the number of antibody-producing cells. However, the same phenomenon also has strong selective properties and changes the qualitative characteristics of the antibodies produced in the course of the immune response and appears to be largely responsible for the well-known increase in antibody affinity during antibody synthesis.

Suppression by antibody is the basic mechanism in the immunological enhancement of tumour homografts—the paradoxical phenomenon that humoral antibodies against certain tumours may facilitate their growth in histoincompatible recipients, where they are otherwise rejected. The principle has also been applied to clinical medicine: rhesus isoimmunization caused by incompatible pregnancies can be prevented by treating the mothers with anti-Rh antibodies shortly after delivery.

The present paper attempts to review some aspects of antibody suppression and to present data supporting the current concept of its mechanism.

QUANTITATIVE REGULATION OF CELLULAR ANTIBODY SYNTHESIS BY ANTIBODY

General aspects

It has been known for over half a century that mixing antigen with excess antibody can suppress the antibody response (for references see Uhr and Möller, 1968).

The possibility that actively formed serum antibody can act as a "feedback" mechanism was suggested by studies indicating that passively administered antibody injected as long as 5 days after immunization was still competent to inhibit the antibody response (Uhr and Baumann, 1961). Studies of the formation of antibodies to sheep red blood cells made by Rowley and Fitch (1964) and by Möller and Wigzell (1965) have added further evidence for a possible regulatory role by serum antibody, in particular as these latter reports were concerned with the effect of passively administered serum antibody on antibody-forming cells using the agar plaque technique (Jerne and Nordin, 1963).

The basic findings on antibody suppression at the serological level have shown that the phenomenon is immunologically specific. Thus, antibodies will suppress the immune response only to those antigenic determinants against which they are directed, whereas other determinants present on the same cell or inoculate will induce a normal immune response (Möller, 1963a; Brody, Walker and Siskind, 1967).

Characteristics of the cellular immune response

In the direct agar plaque technique, described by Jerne and Nordin (1963), the number of cells producing antibody to sheep red cells (plaque-forming cells, PFC) starts to increase after a latent period of approximately 24 hours and thereafter develops exponentially, reaching a peak four to five days after immunization. Subsequently, there is a rapid decay of the number of PFC. 19S antibody titres parallel the number of PFC (Jerne and Nordin, 1963; Wigzell, Möller and Andersson, 1966). Although extrasplenic antibody synthesis occurs in lymph nodes and peripheral blood it is quantitatively of minor importance and shows the same kinetics as splenic production. Serum 7S antibody is detected by 5–7 days after immunization as a rule and increases during the first 12–20 days.

The role of cell division in the development of PFC is not yet settled. It has been demonstrated by Jerne (1966) that about 25 per cent of the PFC cells from immune animals incorporate tritiated thymidine after incubation with the isotope for one hour *in vitro*. However, Tannenberg and Malaviya (1968) found that all PFC had incorporated thymidine given *in vivo*, which indicated that they all were derived by division.

It seems clearly established that the PFC do not belong to a single clone of antibody-producing cells but are derived from multiple clones which are recruited at different times after immunization. This conclusion is derived from demonstrations that the development of PFC occurred in colonies of different sizes in anatomically different sites in the spleen (Playfair, Papermaster and Cole, 1965; Kennedy *et al.*, 1966).

The plaque assay has been modified to detect PFC producing non-haemolytic antibodies (Dresser and Wortis, 1965; Sterzl and Riha, 1965; Weiler, Melletz and Brenninger-Peck, 1965). In the modified method a "developing" antiserum, produced in a foreign or the same species against immunoglobulin of the spleen cell donor species, is used to cause the appearance of PFC which are not revealed by the addition of complement alone (indirect PFC). With this technique the usual decrease of the PFC after 4–5 days does not occur, and the number of PFC is sustained at high levels for a longer period. It is assumed that the PFC detected in this way represent IgG-producing cells of various subclasses.

Suppression of the primary immune response

(a) *Antibody suppression of the 19S response.* Suppression of the primary immune response has been demonstrated at the cellular level with passively transferred antibody directed against sheep red cells and bacterial lipopolysaccharide antigens (Rowley and Fitch, 1964, 1965a, b, 1966; Möller and Wigzell, 1965; Britton and Möller, 1966). In analogy with previous results at the serological level it was found that large amounts of passively transferred antibody given before or at the same time as the antigen completely suppressed the development of 19S-producing PFC, as well as the morphological changes and the increase of spleen weight that accompany the primary response (Sahiar and Schwartz, 1966; Rowley and Fitch, 1964). Hyperimmune sera were efficient whether absorbed on to the antigen or injected by a separate route (Rowley and Fitch, 1964; Möller and Wigzell, 1965). Suppression has been shown to be immunologically specific.

Rowley and Fitch (1964) found 19S and 7S antibodies to be equally efficient in suppressing cellular antibody produced against red cells in rats. In this case, inhibition was compared with sera taken at different times after immunization and not with purified fractions from the same serum. In other experiments (Möller and Wigzell, 1965) purified (Sephadex G-200) fractions from the same serum were used and it was found that only 7S antibodies efficiently suppressed the immune response, whereas 19S was comparatively inefficient. However, 19S antibodies could also suppress the development of PFC if the ratio of antibody to antigen was increased 100–200 times, based on haemagglutination titres (Möller and Wigzell, 1965). Presumably the discrepancies in results are related to the difficulties inherent in comparisons of this type, as discussed previously.

In contrast to these results Henry and Jerne (1968) found that 19S antibodies always increased the number of antibody-producing cells after passive transfer, whereas 7S antibodies were suppressing. These findings are of great interest, since they suggest biologically different functions of different immunoglobulin classes.

In certain experimental systems, antibody-induced suppression of the immune response appears to require very small quantities of antibodies. For example, suppression of cellular 19S synthesis against lipopolysaccharide antigens of *Escherichia coli* can be achieved with the 7S fraction of hyperimmune sera by using amounts which contain no antibody detectable by haemolysis and haemagglutination using red cells sensitized with the antigen (Britton and Möller, 1968).

It was found that passively transferred serum antibodies could suppress the immune cellular response even if injected after the antigen (Möller and Wigzell, 1965). However, the degree of suppression decreased with increasing time-intervals between the injection of antigen and antibody. Suppression was maximal with antibody given before or at the same time as the antigen and pronounced suppression could also be achieved with antibody injected 24 and 48 hours after the antigen injection (Fig. 1). The differences between these findings and those obtained with other antigens (Dixon, Jacot-Guillarmod and McConahey, 1967) may be caused by differences in metabolism of the antigens or in the kinetics of the primary immune responses to the two antigens.

With the plaque technique the suppressive effect of antibodies did not show itself during the first 40–48 hours after injection (Fig. 1), irrespective of the time-interval between antibody and antigen treatment (Möller and Wigzell, 1965). After this latency period suppression was revealed as a termination of the exponential increase of the number of PFC. Thus, the actual numbers of PFC observed at 48 and 72 hours after antibody injection were approximately equal.

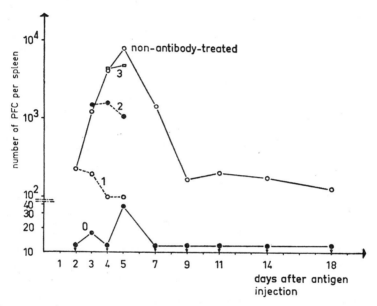

Fig. 1. Development of plaque-forming spleen cells (PFC) in A.BY mice after injection with sheep red cells alone or followed by mouse anti-sheep serum at the indicated days. Each point represents the mean value of two mice.

The possibility exists that antibodies directly suppress the antibody-synthesizing cells. Different opinions have been expressed on this; Rowley and Fitch (1964, 1966) have suggested that antibodies can interact directly with immunologically competent cells and suppress their ability to produce antibodies after antigenic stimulation. This view was based on the demonstration (Rowley and Fitch, 1964) that normal rat lymphoid cells which had been brought into contact with humoral antibodies against sheep red cells *in vitro* or *in vivo* were incompetent to produce antibodies after transfer to irradiated recipients subsequently injected with sheep red cells. However, conflicting results obtained by Möller (1964) indicated that treatment of normal mouse spleen and lymph node cells with humoral antibodies to

Salmonella H antigens did not suppress the ability of the lymphoid cells to carry out a primary immune response against the corresponding antigen after transfer to irradiated recipients. Analogous findings have been obtained with thoracic duct lymphocytes in rats (McCullogh and Gowans, 1966), and with spleen cells in mice (Wigzell, 1967), using sheep red cell antigens. It is possible that the findings of Rowley and Fitch may be explained by adsorption of antibodies to the non-immune lymphoid cell population, which is likely to contain a proportion of macrophages known to fix antibodies to their surfaces; these antibodies may subsequently have been transferred and exerted a suppressive effect on the induction of the primary immune response by interaction with the injected antigen. This possibility is strengthened by the demonstration that only a small amount of antibody is necessary to suppress the immune response against certain antigens.

(b) *Antibody suppression of the 7S response.* The ability of passively transfused antibodies to suppress the primary immune response has also been studied with regard to synthesis of 7S antibody (Möller and Wigzell, 1965; Wigzell, 1966; Möller, 1968). The PFC which are detected by the indirect method using a heterologous anti-gammaglobulin serum will be referred to as 7S or IgG PFC below. It has been repeatedly demonstrated that injection of humoral antibodies simultaneously with the antigen results in marked suppression of 19S antibody and of 7S production as well. Antibodies injected after the antigen often suppress 19S synthesis, although the latent period of suppression allows the 19S titres and the number of PFC to increase for some time. As a rule 7S production is also completely suppresssed by antibodies given during the first three days after the antigen. However, with increasing intervals between the antigen and antibody injection, 7S synthesis becomes more difficult to suppress and serological studies suggest that detectable suppression of 7S production cannot be achieved by antibodies given during the early exponential phase (6–7 days after antigen injection) of synthesis of 7S antibody against sheep red cells (Möller and Wigzell, 1965). However, the serological techniques used for these studies would not detect less than a 50 per cent suppression of the antibody synthesis. It has been generally assumed, however, that 7S synthesis is much less dependent upon the continuous presence of antigen or requires lower concentrations of antigen for its maintenance than 19S synthesis (Uhr, 1964; Svehag and Mandel, 1964). Nevertheless the repeated demonstrations of efficient antibody-mediated suppression of 7S production during the first three days after antigen injection clearly indicate that the events leading to 7S antibody synthesis are antigen-dependent at this early stage.

Recent experiments on suppression of cellular 7S synthesis by antibody have been made with the indirect agar plaque technique (Wigzell, 1966). It was demonstrated that passive transfer of antibodies against sheep red cells and chicken red cells from hyperimmune animals caused a marked decrease in the number of 7S PFC against the specific antigens even when the antibody was transferred several weeks after antigen injection. Suppression was observed after a latency period of 48–72 hours and required large amounts of passively transfused antibodies for a detectable effect. The suppressing efficiency of the antibodies increased with the number of antigen injections of the donor and hyperimmune sera were essential for a demonstrable effect. Suppression appeared to be immunologically specific since antibodies against chicken red cells were ineffective against sheep red cells and *vice versa*. Studies on the serum levels of 2-mercaptoethanol-resistant 7S agglutinins also demonstrated suppression of antibody synthesis.

(c) *Feedback regulation of the immune response.* It has been suggested (Sahiar and Schwartz, 1966; Möller and Wigzell, 1965) that the termination of 19S synthesis against sheep red cells occurring 4–5 days after the injection of antigen may be caused by a feedback suppression of 19S synthesis by 7S antibodies, which start to appear at this time. However, other events may also be responsible for abrogating 19S synthesis. Thus the dependence of 19S synthesis on antigen stimulation suggests that metabolic degradation and/or excretion of the immunogen may stop synthesis of 19S antibody. This possibility is not supported, however, by the experiments showing (Britton, Wepsic and Möller, 1968) that sheep red cells persist for prolonged periods *in vivo* in an immunogenic form.

Several attempts have been made to study the significance of antibody suppression in the regulation of the primary response. It has been experimentally demonstrated that antibodies synthesized within an animal may suppress the induction of a primary response in the same animal, thus excluding the possibility that suppression is mediated only by passively transferred antibodies (Rowley and Fitch, 1964; Morris and Möller, 1968). Thus, Rowley and Fitch showed that a small sensitizing dose suppressed the response to a subsequent large antigen inoculum. Morris and Möller (1968) inoculated hyperimmune spleen cells into untreated syngeneic animals and at different intervals thereafter injected sheep red cells into the recipients and tested them for production of PFC and serum antibody titres. It was found that 19S-producing PFC were markedly suppressed, whereas there was a pronounced stimulation of 7S PFC. It follows that 7S antibodies produced by the transferred cells efficiently suppressed initiation of 19S synthesis in previously uncommitted cells. The transferred 7S cells themselves reacted with a secondary response, revealed as an increased number of 7S PFC and 7S serum antibodies.

To test whether 7S antibodies produced during the primary response may suppress 19S synthesis the same test system was used but the immune spleen cells, which were transferred to syngeneic recipients, were taken from animals immunized against sheep red cells 5–10 days previously. The antibodies produced by such cells also suppressed the initiation of 19S synthesis.

Strong support for a regulatory role of antibody during the primary response has been obtained in another system (Britton and Möller, 1966,

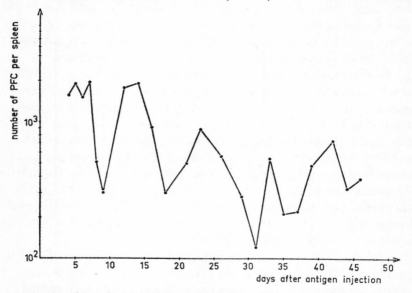

FIG. 2. Regular cyclical fluctuations in the number of plaque-forming spleen cells (PFC) in CBA mice after a single injection of a bacterial vaccine of *E. coli* 055:B5 origin. Each point represents the mean value of five mice.

1968). A lipopolysaccharide antigen from *E. coli* was found to stimulate mice to synthesize only 19S antibody for several months; 7S antibodies were subsequently detected after hyperimmunization. Using the agar plaque technique, sheep red cells coated with the antigen in the form of a bacterial vaccine stimulated the production of 19S PFC, which initially followed the same pattern as observed with sheep red cells. Thereafter, new peaks of PFC appeared at regular 10–15 days intervals, each peak being slightly lower than the preceding one as a rule (Fig. 2).

The regular cyclical fluctuation of cellular 19S synthesis after one injection of antigen was ascribed to a feedback suppression of 19S synthesis by its own product (Britton and Möller, 1968; Möller, Britton and Möller, 1968). According to this hypothesis the antigen would stimulate 19S synthesis in lymphoid cells and the number of antibody-synthesizing cells would increase exponentially during the first four to five days. Eventually the 19S antibodies would reach a sufficiently high concentration to be able to interact with the antigen in such a way as to suppress its ability to stimulate further 19S synthesis. Since no shift to 7S production occurred in this system and since both the 19S-producing cells and the serum antibodies have a short half-life, suppression would be gradually lost. Provided the antigen was still immunogenic it could then initiate a new cycle of 19S-producing PFC.

On this hypothesis it would be possible to suppress any peak of PFC by passive transfusion with specific antibodies shortly before the expected appearance of the peak. Several experiments of this type have been done

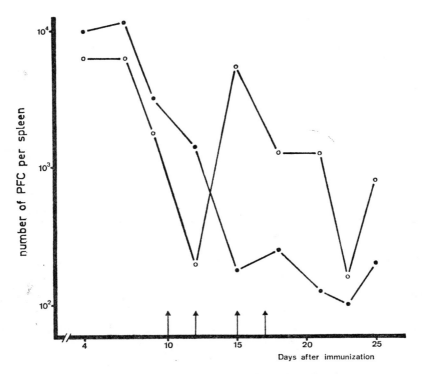

FIG. 3. Plaque-forming cells (PFC) against *E. coli* lipopolysaccharide in CBA mice immunized with *E. coli* bacteria only (o—o) or with bacteria followed by passive transfer of 19S antibodies directed against the antigen at days 9, 13, 15, and 17 (●—●).

(Britton and Möller, 1968) and showed that passively transferred 19S antibodies were competent to suppress the appearance of PFC for about 10 days but thereafter PFC started to appear (Fig. 3). It seemed likely that the progressive metabolism of antibody diminished its concentration below that sufficient to suppress antigenic stimulation of the immune response. Analogous findings have been made previously with toxoid antigens (Uhr and Baumann, 1961).

An essential requirement for this hypothesis is that antigen should persist in an immunogenic form for a long period of time. This has been tested by

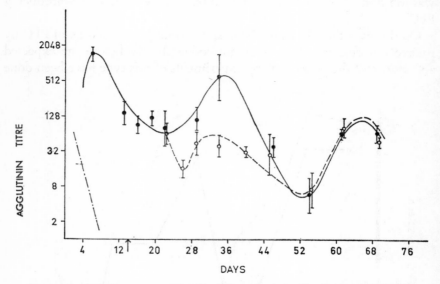

FIG. 4. Agglutinin response against *E. coli* lipopolysaccharide of normal CBA mice (——) and mice irradiated with 900 R at day 14 and then repopulated with non-immune syngeneic bone marrow, lymph node and spleen cells (– – –). The half-life of 19S antibodies is indicated (–.–).

injecting the lipopolysaccharide antigen into mice which thereafter were irradiated and then repopulated with syngeneic bone marrow and lymphoid cells from non-immune animals (Britton, Wepsic and Möller, 1968). It was found that the repopulated animals produced antibodies to the same extent as non-irradiated animals, even if the interval between antigen treatment and repopulation was as long as 70 days (Fig. 4).

An interesting illustration of the efficiency of the feedback suppression of antibody was found in experiments where hyperimmune spleen cells mixed with the specific antigen (sheep red cells) were transferred to non-immune syngeneic and irradiated (600 R) recipients (Möller, 1968). After

Table I

TRANSFER OF SPLEEN CELLS, MIXED WITH SHEEP RED CELLS, FROM CBA MICE HYPERIMMUNIZED AGAINST SHEEP RED CELLS INTO IRRADIATED (600 R) NON-IMMUNE CBA RECIPIENTS

Donor	Number of cells transferred $\times 10^6$	Recipient	Number of cells per recipient spleen* $\times 10^6$	Number of PFC per spleen $\times 10^6$		PFC as percentage of total number of spleen cells	
				19S	7S	19S	7S
CBA, immunized with sheep red cells	50	CBA, 600 R	60.8	0.47	32.0	0.8	52.6
CBA, immunized with sheep red cells	50	CBA, 600 R	60.8	0.37	60.4	0.6	99.3
CBA, immunized with sheep red cells	50	CBA, 600 R	41.6	0.19	24.1	0.5	57.9

* The numbers of nucleated cells and PFC were determined 7 days **after transfer**.

seven days the recipients were tested for the number of 7S PFC in their spleens. Although the spleens were small in size they generally contained more than 10^6 PFC (Table I). The proportion of PFC in the population varied with the number of cells injected and with 10^7 transferred cells it varied between 1 per cent and 7 per cent. By itself the establishment of a population which consists almost entirely of antibody-producing cells is of interest. Transfer of such cells into actively immunized or antibody-treated recipients reduced the number of PFC to that commonly observed in immune and otherwise untreated animals. The proportion of PFC varies between 0·1 per cent and 0·5 per cent as a rule. It seems likely, therefore, that the high proportion of PFC in the irradiated animals is due to the absence of feedback suppression: immune cells stimulated with antigen are transferred to an environment containing no preformed antibodies. Such cells could efficiently interact with the antigen and be stimulated to proliferation. Since they appear to crowd out non-stimulated lymphoid cells, their rate of division must be considerably higher than that of other cells in the inoculum.

Mechanisms of suppression

Although there is no universal agreement on the way in which suppressive antibody works, the evidence cited previously seems to indicate that the first step in suppression is the interaction of antibody with specific antigenic determinants. This conclusion is deduced primarily from the specificity of suppression. The possibility that parts of the antibody molecule other than the combining sites play a role is made less likely by the finding that pepsin-digested antibody is only moderately efficient in suppression (Tao and Uhr, 1966). These observations argue against direct interaction of specific antibody with a nucleic acid informational molecule, or with the cell surface of a potential specific antibody-forming cell, unless antigenic determinants are present on these structures. This latter possibility has been suggested by Rowley and Fitch (1966) who hypothesize that specifically competent lymphoid cells have antigen-like structures on their surfaces as well as specific antibody. Evidence discussed previously (Uhr and Möller, 1968), however, suggests that the experimental basis for this hypothesis is consistent with other, simpler explanations. The concept of interaction between suppressive antibody and antigen is in agreement with an increasing body of evidence suggesting that antigen can persist in the lymphoid tissues for a considerable time and continue to stimulate and influence the immune response (Britton, Wepsic and Möller, 1968). Presumably, passive antibody interacts with such antigen and interrupts the continued stimulation of the immune system.

REGULATION OF QUALITATIVE CHARACTERISTICS OF THE ANTIBODY RESPONSE

The interaction between antigen and the antigen-responding lymphoid cells is generally assumed to result in antibody synthesis either by direct differentiation of the antigen-sensitive cells into antibody-producing cells or by some indirect process, such as the interaction between the stimulated antigen-sensitive cells and other lymphocytes, which subsequently start to produce antibodies. The cellular interaction involved in the latter possibility has recently been established for certain antigens (for review, see *Transplantation Reviews*, vol. 1, 1969). It is obvious that antibody produced in the immune response or passively transferred antibody will interfere with the interaction between antigen and the responding cells, since it will bind to the antigen and thus prevent it from stimulating the antigen-sensitive cells.

It is essential to differentiate between antigen-sensitive cells and antibody-producing cells. The weight of evidence favours the idea that these cells are functionally and physically distinct entities. Thus, antibody-producing cells do not respond to antigen by increased antibody production, nor are they amenable to the induction of tolerance (Britton, 1969; Raidt, Mishell and Dutton, 1968; Möller, 1969a). They appear to be completely insensitive to antigen. Furthermore, indirect evidence cited previously supports the idea that the antibody-producing cells are short-lived and in many experimental situations may survive for only 50–70 hours (Möller and Wigzell, 1965; Britton, 1969). During this period they multiply by mitotic division. In order for antibody synthesis to be maintained new antibody-producing cells must therefore be recruited during the entire immune response. Recruitment is generally considered to occur from antigen-sensitive cells which do not themselves produce antibodies but are equipped with surface receptors for the antigen and when confronted with the antigen, differentiate into antibody-producing cells or, alternatively, trigger other cells to produce antibodies. Presumably the receptors are of antibody nature and it is generally assumed that these have the same characteristics as the antibody produced after stimulation of the corresponding antigen-sensitive cells.

During the exponential phase of the immune response, recruitment of antigen-sensitive cells must occur at an exponential rate in order to account for the exponential increase of antibody-producing cells (for discussion see

Möller, 1967). It seems likely, therefore, that the number of antigen-sensitive cells also increases during immunization. This has also been documented (Raidt, Mishell and Dutton, 1968).

The qualitative changes in the antibodies produced by antibody feedback suppression can be understood in terms of the binding affinity for antigen of serum antibody and of the receptors on the antigen-sensitive cells, respectively. It is commonly assumed that each antigen-sensitive cell bears only one type of receptor for the antigen, characterized by a certain amino acid sequence. Thus, there is a one-cell–one-protein relationship. Furthermore, it is likely that the binding affinity of these receptors varies considerably between different cells. Statistically it is probable that many cells have receptors of low binding affinity and a decreasing number of cells have receptors of high affinity. When antigen is introduced into an animal it will stimulate the antigen-sensitive cells. The antigen will bind most efficiently to the rare high-affinity receptor cells, but is likely to trigger also a large number of cells having a lower binding affinity. When active antibody production is initiated the antibodies with a relatively high binding efficiency will preferentially combine with the antigen. Antigen-sensitive cells having receptors of lower affinity will be selectively suppressed by the higher affinity serum antibodies, whereas cells having a higher binding ability will compete successfully for the antigen with the serum antibodies. The outcome of this competitive interaction will be a gradual increase in the binding affinity of the antibodies during the immune response.

Regulation of serum antibodies

The binding affinity of antibodies may increase 10000-fold during the course of the immune response (Eisen and Siskind, 1964). Although this finding by itself is compatible with the previously outlined scheme of competitive interaction between antigen, antibody and antigen-sensitive cells, other interpretations are possible. Thus, the interaction between antigen and sensitive cells may by itself be sufficient to explain the findings: if the antigen concentration gradually decreases with time after immunization, it is probable that the lower antigen concentration fails to stimulate antigen-sensitive cells with a low binding ability. However, evidence that humoral antibodies participate as a selective force in this respect has been obtained from many groups.

From the concept outlined it would be expected that high-affinity antibody would be more competent to suppress the immune response than low-affinity antibody, since it would interact more efficiently with the antigen. It has been demonstrated that high-affinity antibody against haptens efficiently suppresses antibody synthesis, whereas low-affinity antibody is less competent (Walker and Siskind, 1968). In general agreement with this is the finding that "late" 7S antibody against sheep red cells is required to suppress 7S antibody synthesis long after its induction, whereas "early" 7S antibody is less efficient (Wigzell, 1966). Presumably the "late" 7S antibody was of higher affinity than the early antibody. Analogous results have been obtained by Finkelstein and Uhr (1964) in another experimental system.

It has been well documented that 19S synthesis is easier to suppress by passive antibody than 7S synthesis, if the antibodies are given after the synthesis of the respective immunoglobulin has been initiated. It has been demonstrated that adoptive transfer of immune spleen cells into normal syngeneic recipients, subsequently immunized with the antigen (sheep red cells), leads to an almost complete suppression of 19S synthesis and a vigorous secondary 7S response (Morris and Möller, 1968). Presumably the high-affinity 7S antibodies produced by the transferred cells competed effectively for the antigen with 19S antigen-sensitive cells, whereas they could not efficiently prevent antigen interacting with 7S antigen-sensitive cells carrying high-affinity receptors.

Another consequence of this thermodynamic concept is that passive transfer of antibody into antigen-injected animals should interfere with the normal progressive changes of antibody affinity. Since the passive antibody would suppress stimulation of antigen-sensitive cells bearing low-affinity receptors, the result would be an accelerated rate of increase of affinity for the antigen. This has been demonstrated in haptenic systems: passive transfer of antibody to immunized animals caused a pronounced suppression of the amount of antibody produced, but the binding constant of these antibodies was increased 10-fold over that found in control animals not given antibody (Siskind, Dunn and Walker, 1968).

Regulation of cellular antibody synthesis

A low dose of antigen should favour stimulation of antigen-sensitive cells with high binding affinity receptors whereas a high dose should also be capable of triggering cells less competent to bind the antigen. By determining the degree of DNA synthesis induced *in vitro* by hapten–protein conjugates added to lymphocytes derived from animals immunized with high and low doses of antigen, respectively, this assumption was verified (Paul, Siskind and Benacerraf, 1968). Thus, immunization with a low dose of antigen leads to a cell population capable of being stimulated, on the average, by low concentrations of antigen *in vitro*, whereas stimulation with large antigen doses resulted in a sensitive population requiring a high concentration of antigen for stimulation (Paul, Siskind and Benacerraf, 1968). These findings were correlated with the affinity for the hapten of the serum antibodies produced by the animals and it was found that the low antigen dose caused a higher affinity of the antibody than the high antigen dose.

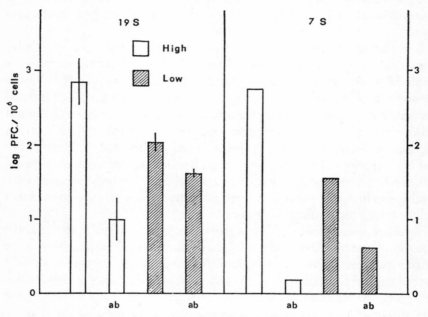

FIG. 5. Number of 19S and 7S plaque-forming cells (PFC) in mice immunized with high and low doses, respectively, of sheep red cells, alone and in mice which in addition received specific antibody to sheep red cells 24–48 hours afterwards. The antibody-treated groups are indicated (ab). Vertical lines indicate standard error of the mean. The figure is based on three experiments using 60 mice.

A consequence of these findings is that the immune response induced by a low antigen dose would be more difficult to suppress by antibody than that stimulated by a high dose, since the high-affinity antigen-sensitive cells appearing after the low dose would compete efficiently with the passive antibody for the antigen. Direct experiments verified this (Möller, 1969b). The 19S and 7S cellular immune response to 10^8 sheep red cells was efficiently suppressed by antibody, whereas that triggered by 10^6 red cells could not be suppressed to the same degree (Fig. 5).

Taken together the findings strongly suggest that actively produced and passively administered humoral antibodies exert a strong selective influence on the binding characteristics of actively produced antibodies. Three components appear to interact in this selective process: the concentration of antigen, the affinity and concentration of serum antibody and the number of antigen-sensitive cells having antigen-capturing receptors of varying degrees of binding affinity for the antigen. Since there is a competition for the antigen between serum antibodies and antigen-sensitive cells, the outcome will be determined by binding affinities and numerical factors in a complex, but in general outline, predictable manner.

IMMUNOLOGICAL ENHANCEMENT

Immunological enhancement is defined as the progressive growth of tumours in histoincompatible recipients which have been treated with antibodies directed against the tumour cell antigens. This is a paradoxical phenomenon since the antibody treatment does not suppress tumour growth, as would be expected, but actually makes growth possible in recipients in which the tumour would otherwise be rejected. The mechanism of this phenomenon has received particular attention and various aspects of it have been reviewed elsewhere (Kaliss, 1958; Batchelor, 1963; Hellström and Möller, 1965; Möller, Britton and Möller, 1968).

Immunological reactions are complex in transplantation systems where the antigen consists of dividing cells. In such situations the effective immune response leading to graft rejection is dependent not only upon the inductive and productive levels of the immune reflex, but also on the efficiency of the effector mechanism, which is dependent on the intensity of cellular and humoral immunity, the sensitivity of the target cells to immune destruction, the growth rate of the target cells and other variables. It is nevertheless clear that the basic principle governing the phenomenon of enhancement is antibody suppression of the immune response, analogous to that studied with non-reproducing antigen.

The basic finding in support of this conclusion is that antibody-treated recipients given histoincompatible tumour cells do not develop a normal humoral or cell-mediated immune response (Snell *et al.*, 1960; Möller, 1963*b*). Of particular importance for enhancement is the failure of cellular immunity to develop efficiently. This can be demonstrated by experiments *in vivo* in which lymphoid cells from the antibody-treated animals are transferred adoptively to other recipients where they fail to reject grafts efficiently (Snell *et al.*, 1960; Möller, 1963*c*). Antibody suppression of both the cellular and humoral antibody responses is highly specific. Thus in situations where tumour cells possess several different antigenic determinants, the immune response is suppressed only with regard to those determinants against which antibodies are present in the host (Möller, 1963a), in analogy with findings in haptenic systems (Brody, Walker and Siskind, 1967).

These and other findings are strong evidence for the importance of antibody suppression of the immune response, usually referred to as afferent inhibition (inhibition of the immunization), in enhancement. However, immunological enhancement can also be demonstrated in animals already sensitized against the tumour cells. In this situation the pre-existing immunity is by itself fully competent to reject the tumour at an accelerated rate. In pre-sensitized animals passive transfer of antibody does not abolish the pre-existing immunity, and therefore additional mechanisms of immunological enhancement must exist. It has been demonstrated that antibody-coated tumour cells are resistant to destruction by immune lymphocytes. If antibody-treated tumour cells are inoculated into one side of an untreated animal they will grow progressively, whereas untreated cells inoculated simultaneously into the other side of the same animal will be rejected (Möller, 1964). Obviously, treatment of the tumour cells with antibody protected them from destruction by sensitized host cells capable of rejecting the untreated tumour cells. An analogous effect can also be demonstrated in tissue culture by adding sensitized lymphoid cells to antibody-coated target cells (E. Möller, 1965). Such cells are protected from destruction, whereas untreated cells are highly vulnerable to destruction by the sensitized lymphoid cells. The protection by antibodies against destruction of tumour cells by sensitized lymphocytes in these experimental situations has been referred to as efferent inhibition. It is presumably caused by humoral antibodies blocking target cell antigenic determinants so that they are no longer available for the attachment of sensitized lymphocytes.

The enhancement phenomenon is not restricted to tumour cell grafts only but can also be demonstrated with various normal tissues, such as skin, ovaries and presumably also kidney grafts. It is obvious that immunological enhancement cannot be demonstrated if the target cells used are vulnerable to the cytotoxic action of humoral antibodies and complement. This is the case with many leukaemias and normal haematopoietic and lymphoid cells. Cells from such tissues are rejected at an accelerated rate as a rule in recipients pretreated with antibody. However, by special procedures aimed at removing or diminishing the cytotoxic effect of humoral antibodies against these cells, they are also amenable to enhancement. Thus, removal of the Fc fragment of the antibodies, leading to a loss of complement fixation, makes leukaemias highly susceptible to enhancement (Chard, French and

Batchelor, 1967). Furthermore, if one uses antigenic systems where humoral antibodies are not cytotoxic to lymphoid or leukaemic cells, immunological enhancement can be demonstrated (Möller, 1963b).

It is particularly important that immunological enhancement is expressed with neoplastic cells in situations where the antigenic discrepancy between the tumour and the host is restricted to that created by the existence of tumour-specific antigens (G. Möller, 1965; Bubenik, Ivanyi and Koldovsky, 1965).

The findings described clearly demonstrate that the basic principle of immunological enhancement is both an afferent antibody-mediated suppression of the immune response, analogous to antibody suppression of non-reproducing antigen, and an efferent protection of antibody-coated target cells from destruction by sensitized lymphocytes. The complexity of the systems used makes it difficult to evaluate the importance and magnitude of the various contributing mechanisms.

SUMMARY

Actively produced or passively administered humoral antibodies may regulate the immune response quantitatively and qualitatively by interacting with the antigen. This interaction inhibits the quantity of antibodies produced, since the continuation of the immune response is antigen dependent. Since antigen-reactive lymphocytes are probably equipped with surface receptors of an antibody nature, there is a competition for the antigen between these cells and humoral antibodies. The antibody with the highest binding affinity to the antigen will preferentially capture antigen. Thus antigen-reactive cells with high-affinity receptors will be preferentially stimulated, whereas cells with low-affinity receptors will not be stimulated because higher affinity humoral antibodies bind to the antigen. This process results in a gradual increase of the binding constant of the antibody during the immune response.

Acknowledgements

This work was supported by grants from the Swedish Medical Research Council, the Swedish Cancer Society and the Damon Runyon Memorial Fund (DRG-954).

REFERENCES

BATCHELOR, J. R. (1963). *Guy's Hosp. Rep.*, **112**, 345.
BRITTON, S. (1969). *Immunology*, **16**, 527.
BRITTON, S., and MÖLLER, G. (1966). In *Genetic Variations in Somatic Cells*, p. 313. Prague: Publishing House of the Czechoslovak Academy of Science.

BRITTON, S., and MÖLLER, G. (1968). *J. Immun.*, **100**, 1326.
BRITTON, S., WEPSIC, T., and MÖLLER, G. (1968). *Immunology*, **14**, 491.
BRODY, N. J., WALKER, G., and SISKIND, G. W. (1967). *J. exp. Med.*, **126**, 81.
BUBENIK, J., IVANYI, J., and KOLDOVSKY, P. (1965). *Folia biol., Praha*, **11**, 426.
CHARD, T., FRENCH, M. E., and BATCHELOR, J. R. (1967). *Transplantation*, **5**, 1266.
DIXON, F. J., JACOT-GUILLARMOD, H., and MCCONAHEY, P. J. (1967). *J. exp. Med.*, **125**, 1119.
DRESSER, D. W., and WORTIS, H. H. (1965). *Nature, Lond.*, **208**, 859.
EISEN, H. N., and SISKIND, G. W. (1964). *Biochemistry, N.Y.*, **3**, 996.
FINKELSTEIN, M. S., and UHR, J. W. (1964). *Science*, **146**, 67.
HELLSTRÖM, K. E., and MÖLLER, G. (1965). *Prog. Allergy*, **9**, 158.
HENRY, C., and JERNE, N. (1968). *J. exp. Med.*, **128**, 133.
JERNE, N. K. (1966). Personal communication.
JERNE, N. K., and NORDIN, A. A. (1963). *Science*, **140**, 405.
KALISS, N. (1958). *Cancer Res.*, **18**, 992.
KENNEDY, J. C., TILL, J. E., SIMINOVITCH, L., and MCCULLOCH, E. A. (1966). *J. Immun.*, **96**, 973.
MCCULLOGH, E. A., and GOWANS, J. L. (1966). Personal communications.
MÖLLER, E. (1965). *J. exp. Med.*, **122**, 11.
MÖLLER, E., BRITTON, S., and MÖLLER, G. (1968). In *Regulation of the Antibody Response*, pp. 141, ed. Cinader, B. Springfield: Thomas.
MÖLLER, G. (1963a). *J. natn. Cancer Inst.*, **30**, 1153.
MÖLLER, G. (1963b). *J. natn. Cancer Inst.*, **30**, 1193.
MÖLLER, G. (1963c). *J. natn. Cancer Inst.*, **30**, 1205.
MÖLLER, G. (1964). *Transplantation*, **2**, 405.
MÖLLER, G. (1965). *Nature, Lond.*, **207**, 1166.
MÖLLER, G. (1967). In *Gamma Globulins, Structure and Control of Biosynthesis*, Nobel Symposium 3, ed. Killander, J. Stockholm: Almqvist and Wiksell.
MÖLLER, G. (1968). *J. exp. Med.*, **127**, 291.
MÖLLER, G. (1969a). In *Immunological Tolerance*, ed. Landy, M., and Braun, W. London: Academic Press.
MÖLLER, G. (1969b). *Immunology*, in press.
MÖLLER, G., and WIGZELL, H. (1965). *J. exp. Med.*, **121**, 969.
MORRIS, A., and MÖLLER, G. (1968). *J. Immun.*, **101**, 439.
PAUL, W. E., SISKIND, G. W., and BENACERRAF, B. (1968). *J. exp. Med.*, **127**, 25.
PLAYFAIR, J. H. L., PAPERMASTER, B. W., and COLE, L. J. (1965). *Science*, **149**, 998.
RAIDT, D. J., MISHELL, R. J., and DUTTON, R. W. (1968). *J. exp. Med.*, **128**, 681.
ROWLEY, D. A., and FITCH, F. W. (1964). *J. exp. Med.*, **120**, 987.
ROWLEY, D. A., and FITCH, F. W. (1965a). *J. exp. Med.*, **121**, 671.
ROWLEY, D. A., and FITCH, F. W. (1965b). *J. exp. Med.*, **121**, 675.
ROWLEY, D. A., and FITCH, F. W. (1966). In *Regulation of the Antibody Response*, pp. 127, ed. Cinader, B. Springfield: Thomas.
SAHIAR, K., and SCHWARTZ, R. S. (1966). *Int. Archs Allergy appl. Immun.*, **29**, 52.
SELA, M. (1967). In *Gamma Globulins, Structure and Control of Biosynthesis*, Nobel Symposium 3, ed. Killander, J. Stockholm: Almqvist and Wiksell.
SISKIND, G. W., DUNN, P., and WALKER, J. G. (1968). *J. exp. Med.*, **127**, 55.
SNELL, G. D., WINN, H. J., STIMPFLING, J. H., and PARKER, S. J. (1960). *J. exp. Med.*, **112**, 293.
STERZL, J., and RIHA, I. (1965). *Nature, Lond.*, **208**, 858.
SVEHAG, S. E., and MANDEL, B. (1964). *J. exp. Med.*, **119**, 21.
TANNENBERG, W. J. K., and MALAVIYA, A. N. (1968). *J. exp. Med.*, **128**, 895.
TAO, T. W., and UHR, J. W. (1966). *Nature, Lond.*, **212**, 208.

Uhr, J. W. (1964). *Science*, **145**, 457.
Uhr, J. W., and Baumann, J. B. (1961). *J. exp. Med.*, **113**, 935.
Uhr, J. W., and Möller, G. (1968). *Adv. Immun.*, **8**, 81.
Walker, J. G., and Siskind, G. W. (1968). *Immunology*, **14**, 21.
Weiler, E., Melletz, E. W., and Brenninger-Peck, E. (1965). *Proc. natn. Acad. Sci. U.S.A.*, **54**, 1310.
Wigzell, H. (1966). *J. exp. Med.*, **124**, 953.
Wigzell, H. (1967). Thesis. Stockholm: Balders tryckeri.
Wigzell, H., Möller, G., and Andersson, B. (1966). *Acta path. microbiol. scand.*, **66**, 530.

DISCUSSION

Ormerod: The mechanism of antibody formation outlined in your paper is rather different from some of the mechanisms suggested in the past; is the selective theory generally accepted now?

Möller: Most scientists accept the clonal selection theory; the alternative is an instructive theory, which implies that the antigen instructs the antibody-forming cell to form a specific product. As a result of more knowledge of protein synthesis very few believe this now. It is implied in clonal selection that cell specificities are present in cells from the beginning. The problem then is to suggest a genetic mechanism whereby this fantastic variability of antibody structure can occur. Suppose that there exist one million antigens, and each individual has only one cell in 10000 which is competent to react with each of the one million antigens. These cells are not expressed before birth. Two possibilities exist: either all genes (10^6) are inherited in the germ line and became derepressed in different cells after birth, or a few genes are inherited and there is a somatic mechanism operating after birth leading to antibody variability. Both possibilities present certain difficulties, which I cannot elucidate here; neither of them is yet proved.

Ormerod: Most of the arguments you put forward in your paper depend on the assumed mechanism by which antibodies are produced.

Möller: If clonal selection in one form or another isn't correct then the interpretations I have given are not valid.

Wolpert: Have you any idea how many molecules of antigen per cell are required to induce tolerance or activation? I am thinking of this in relation to the nerve growth factor. While I feel that the immunological story is not really a model of homeostasis, it may however provide marvellous models of the sort of *cellular* mechanisms that could be involved in cellular interactions and responses.

Möller: Unfortunately this is entirely unknown. If you inject an antigen more than 99 per cent of the molecules are destined to be catabolized in macrophages. It is difficult to trace the relevant antigen in an animal. It is complicated further by the fact that you have little information on the antigen-sensitive cells because they lack markers, since they do not produce antibody. When they are producing antibody, that is, are differentiated, the evidence is rather strong that they do not contain antigen molecules. The basic question—how many molecules are needed on the antigen-sensitive cell to trigger it—cannot be answered yet.

Stoker: Is it not possible to count clones which represent the original cells? Could you count the absolute number of clones arising in the spleen of an animal receiving stimulated cells, in relation to the amount of antigen used?

Möller: You can detect the proportion of antigen-sensitive cells by appropriate *in vivo* experiments.

Stoker: But when you estimate the number of cells which react to the antigen, how does this number correspond to the variation in the dose of antigen?

Möller: That has not been studied.

Allison: You can induce tolerance with most antigens in two zones of dosage, either an extremely low dose or a high dose. This seems to be true even of antigens like bacterial flagellar proteins which are highly immunogenic, and this could imply that very small numbers of antigen molecules can have a biological effect in the induction of tolerance.

Möller: In the flagellar system there are now three doses for tolerance; the lowest dose inducing tolerance corresponds to about 1000 molecules per animal, which is less than the number of lymphocytes (Shellam, 1969). Thus the chance for these molecules to meet the one cell in 20000 that is equipped with a receptor for the particular antigen is extremely low, unless there are either very efficient focusing mechanisms for both antigen and cells, or a magnifying mechanism, which itself must have immunological specificity.

Vernon: You said that an estimate of the total number of antigens would be about a million but of course the total number of possible proteins is astronomically greater than this, so this means that either most of these would not be antigens, or they would be indistinguishable. If you consider a protein with 100 residues, which is a very small protein, and there are 20 amino acids, the number of possibilities is 10^{120}, which probably exceeds the total number of particles in the universe.

Allison: With 20 amino acids, only some of which are known to be antigenic, you are dealing with numbers not of possible proteins, but of antigenic determinants, which may be more limited. One protein will have a number of antigenic sites but many of these will be shared by different proteins having similar configurations. So it is not the number of proteins which is important.

Bergel: Dr Möller suggested that some directing mechanisms will be necessary to bring antigens into contact with the small numbers of cells capable of responding to them, but consider the analogy with drugs, which have to meet specific receptors which may be very few in number, perhaps located in one part of the central nervous system only. We know toxicologically and pharmacologically that they succeed. Consequently, it can't be so difficult for antigens to move to the competent cells, even if they are all over the body.

Möller: If it is a very rapidly moving screening system—for example, antigens sitting on one cell and circulating lymphocytes passing by at quite a rate and becoming fixed only if they recognize the antigen—then even one cell in 20000 is no problem; the spleen would pick it up. But in the thymus–bone marrow cell interaction you have one cell in 20000 in each of *two* populations which have to interact. This creates some intellectual difficulties.

Vernon: What is the distance over which the antigen and cell can actually recognize each other?

Möller: It is likely to be the same as that involved in antigen–antibody interaction.

Vernon: It will then be in the order of 10 or 20 Å. So knowing the size of the particle and the total volume involved, you can work out the collision probabilities, and I am willing to bet that they come out to be very close to zero.

Allison: This is not necessarily true if you have a system like macrophage uptake; the macrophages can move around and present materials to lymphocytes. This is the sort of focusing mechanism that Dr Möller has in mind, I think. And this helps also with the difficulty that although most of the antigen has been degraded, the remaining 10 per cent or so is remarkably efficient in eliciting an immune response.

Möller: In a way it is an artificial system, when the antigen is given after ingestion by macrophages; you don't know that it is efficient *in vivo* when it is taken up in the lysosomes.

Allison: You know the efficiency in the sense that the antibody response can be measured in animals into which macrophages containing known amounts of antigen have been injected.

Möller: I think it is necessary to postulate either a very efficient focusing mechanism of both antigen and cells, or a magnifying mechanism which is specific. The only specific magnifying mechanism which can be postulated must involve antibody itself, because we know of no other specific mechanism in immunity.

Elsdale: Actually cells very seldom get as close as 10 Å to one another; the usual separations are larger than this. However, a focusing mechanism would be provided if cells, because of their specific properties, were relatively unhindered in their approach to within 10 Å of one another. Biophysical theories of cell adhesion suggest that cells separated by this short distance would be firmly held together.

Lamerton: On a different point, Dr Möller, you said that the antigen-sensitive cells, which apparently precede the antibody-producing cells, undergo only a limited number of divisions. How many times do they divide, and what is the evidence for this?

Möller: There is only indirect evidence. The number of divisions of the antibody-producing cells is likely to be 5–10 (Möller, 1967).

Elsdale: Todaro and Green (1964) have grown human cells for 100 generations using a medium supplemented with serum albumin to raise the total protein concentration.

Mason: Dr Möller, is there a difference between cells producing IgM and cells producing IgG? I notice that the increase of IgG occurs rather later, in a primary response. Does this mean that the cells multiply before they produce IgG, whereas the cells producing IgM enter almost immediately into synthesis without necessarily proliferating?

Möller: The problem is whether the same cell is producing IgG and IgM. Earlier it was claimed (Nossal *et al.*, 1964) that there was a shift in production from IgM to IgG in the same cells, but most scientists now think that they are produced by different cells. The question then is: why is there a latency period before the appearance of IgG? It has been claimed (Wei and Stavitsky, 1967) that there is no latency period. However, it seems difficult to get round the fact that IgM and IgG peak at different times. It seems plausible that the IgM represents a very primitive response involving less cell interaction. This is also the response appearing first in ontogenetic development, and it is more difficult to suppress by irradiation and drugs. IgG occurs later phylogenetically and may depend more on cellular interaction and might therefore be more easily disturbed. This is speculation, however.

REFERENCES

MÖLLER, G. (1967). In *Gamma Globulins, Structure and Control of Biosynthesis*, Nobel Symposium 3, p. 473, ed. Killander, J. Stockholm: Almqvist and Wiksell.

NOSSAL, G. J. V., SZENBERG, A., ADA, G. L., and AUSTIN, C. M. (1964). *J. exp. Med.*, **119**, 485–502.
SHELLAM, G. R. (1969). *Immunology*, **16**, 45.
TODARO, G. J., and GREEN, H. (1964). *Proc. Soc. exp. Biol. Med.*, **116**, 688–692.
WEI, M-M., and STAVITSKY, A. B. (1967). *Immunology*, **12**, 431–444.

GENERAL DISCUSSION

IMMUNE REACTIONS AND HOMEOSTASIS

Bergel: May I pose the general question of how far immune reactions, or the presence of immune competence, contribute to the homeostasis of the whole organism. The immune reaction is a protection against foreign proteins and therefore will tend to avert disasters, whether through invasion by microorganisms or by foreign tissue in transplantation. Immunological reactions clearly play a part in the overall preservation of stability which homeostatic systems are supposed to create.

Möller: I agree entirely from the point of view of the multicellular individual. However, one can look at this from the point of view of the unicellular intruder also. Consider microorganisms which are pathogenic in man. These infect the host, and this can result in such extensive multiplication of the microorganisms that the host is killed. This is also a disaster for the infecting organism because it will not have a chance to spread to new hosts. In order to be successful the organism should become antigenic to such a degree that the host can defend itself against it, and therefore survive for such a length of time that the organism has a chance to infect another individual. On the other hand if the organism is too antigenic the immune response will destroy it immediately. Therefore, pathogenic microorganisms should try to reach a state of weak antigenicity, creating a state of chronic infection in order to survive.

Allison: Dr Möller is taking the question at the second level of complexity, but at the first level there are beautiful and instructive examples in the genetic defects of development of the immunological system, including the agammaglobulinaemias and a range of defects involving cell-mediated immunity. These have interesting consequences of which the most obvious are the chronic bacterial and fungal infections, and some virus infections, to which the patients are prone. What is also coming up now is that children with these immunological defects have a much higher incidence of malignancy than other children of the same age, in particular malignancy of the lymphoreticular system. That this is due to the deficiency of the immune system seems to be confirmed by the fact that immunosuppressive treatment, either by drugs or by antilymphocytic serum, in kidney-transplant patients is also associated with a greatly increased risk of lymphoreticular malignancy. This is being well documented by R. A. Good, T. Starzl and others. The findings suggest that cell-mediated immunity really is an important homeostatic control mechanism, not only against infection, but also against malignancy in man.

This is certainly true in experimental animals. For instance, we have found that antilymphocytic serum and also thymectomy have increased the incidence of certain virus-induced tumours in mice (Allison and Law, 1968), and can allow tumour development even when adult animals are exposed to viruses.

Möller: One theory of chemical carcinogenesis is based on the idea that the carcinogens are extremely potent immunosuppressive agents; thus methylcholanthrene would cause a pronounced decrease in the effector mechanism of cellular immunity at the site of injection, so that any antigenic variant which is neoplastic could not be rejected as it normally would be. This could also explain why antilymphocytic serum doesn't work with these chemical carcinogens, since they themselves would be so strongly immunosuppressive.

Roe: A dose measured in micrograms of 3-methylcholanthrene or 7,12-dimethylbenz(a)anthracene is enough to induce tumours in a mouse, if given during the first few days of life. Although the dose is given subcutaneously, tumours may arise remotely in a wide variety of tissues and organs, often after a very long latent interval. Existing methods of measurement of immune competence do not reveal general impairment following such low doses. In this case it is difficult to argue that local impairment might have been sufficient since tumours may arise at so many different locations.

Möller: The argument is based on a local effect, at the site of application. It is very difficult to test this possibility. A strong argument against this hypothesis would be the demonstration of chemical carcinogenesis in tissue culture, where there are no immunological surveillance mechanisms. This has actually been done by Sachs and Heidelberger who succeeded in obtaining neoplastic transformation in tissue culture by chemical carcinogens (Huberman and Sachs, 1966; Heidelberger and Iype, 1967).

Stoker: Is there not some relationship between the degree of antigenicity of a tumour and its growth rate?

Möller: I don't think there is any proof for such a simplified view. What factors affect tumour growth *in vivo*? For simplicity we can only consider growth rate, antigenicity (immunosensitivity) and immunogenicity. These three components can interact in a highly complex manner. Thus there is an antagonistic influence between humoral antibodies and cellular immunity. This is illustrated by the process of self-enhancement, in which a tumour stimulates the immune response of the host and provokes production of circulating antibodies, which coat the antigenic determinants of the tumour cells, but are not capable of killing them if the tumour is immunoresistant. The tumour is no longer recognized as antigenic and will therefore grow. If the spleen, where humoral antibody is synthesized, is removed, the tumour will be rejected. Thus an immune response may facilitate tumour growth without any additional pretreatment. Thus a strongly immunogenic but immunoresistant tumour may grow faster *because* it stimulates an immune response. There is no simple correlation between tumour growth rate and antigenicity.

O'Meara: We discussed cell receptors for antigens earlier; is anything to be gained by the study of species differences in receptors? Are there instances in which certain proteins or other materials are antigenic in one species and not in another? There certainly are instances where toxic antigens have been used and there has been no response to the toxicity of the antigen by one species and a response in others; the guinea pig is highly susceptible to diphtheria toxin, for example and the rat is almost completely insusceptible. The insusceptibility might be a failure on the part of the animal to possess receptors to this particular antigen. A similar instance occurs with tetanus toxin in cold-blooded animals, such as the tortoise, by comparison with warm-blooded animals. The toxin can circulate for days in cold-blooded animals without giving rise to any symptoms. This would suggest that there may be a lack of receptors for antigens in certain instances.

Möller: Usually immunologists have wanted to get an antibody response and if an animal did not respond they disregarded this. Now they are equally interested in non-responding animals because of the potential theoretical significance. Inbred strains of mice have been found which can or cannot respond to certain antigens (McDevitt and Sela, 1965). If this were due to lack of cellular receptors it would be strong evidence for the germ-line hypothesis of antibody variability; it would be difficult to explain on a somatic hypothesis. By appropriate transfer tests it was shown that cells from responding animals were fully competent to respond in a non-responding environment (McDevitt and Tyan, 1968). The reverse transfer has not yet been done. So far it is only in agammaglobulinaemia that a lack of cellular receptors is likely to exist.

Allison: In work going on at Harvard it has been shown that diphtheria toxins are not taken up by cells from an insusceptible animal (the mouse), but if a small amount of basic protein is added to diphtheria toxin it can be taken up and is extremely toxic and damaging to mouse heart (Moehring *et al.*, 1967; H. J. P. Ryser, unpublished observations). In this case it does seem to be a question of whether there is specific uptake of the toxin.

Möller: Non-responsiveness can be caused by a variety of factors. The two-receptor hypothesis of Jerne (1967) and Mitchison (1967) illustrates this elegantly. According to this hypothesis it is a prerequisite that the antigen can bind to two antibody receptors. An antigen with only one determinant cannot interact with two cells, each carrying one receptor of a particular specificity, and cannot stimulate an immune response. If a hapten is attached to protein A a good immune response is obtained both to the hapten and to the protein. If the same hapten is attached to another protein a secondary response to the hapten cannot be obtained. Although the hapten can bind to the cellular receptor there are no cells available for the new carrier. If tolerance is induced to the carrier protein an immune response to the hapten cannot be obtained.

Bergel: Dr Möller mentioned in his paper the selectivity of antigens in terms of the *kind* of antibody produced. Is it correct that transplantation immunity seems to be dependent upon a different type of immune response from immunity against infections? It has been stated that it is the 19S fraction of immunoglobulin, instead of the 7S fraction, which is effective in graft rejection.

Möller: 19S antibody is most efficient in all serological tests, including cytotoxicity, and therefore it is the type likely to be more responsible for graft rejection. On the other hand there is no clear evidence that these antibodies are involved at all in graft rejection. The reason why certain antigens stimulate certain types of immune responses is entirely unknown.

Bergel: There appears also to be variation in the immunosuppressing effect by different suppressors, such as 6-mercaptopurine or cyclophosphamide, in organ transplantations when compared with the problem of resistance to infection or lack of it in the patient. I gather that these immunosuppressors prevent different types of antibody formation; they are not simply suppressing the whole range of immunoglobulin from 7S up to 19S.

Möller: Immunosuppressive drugs are rather selective in this action. It also depends when they are given (before or after the antigen). Certain immunosuppressive agents such as irradiation may actually increase the immune response quite dramatically if given shortly after the antigen. The most radiosensitive step, except for division itself, appears to be initiation of antibody synthesis. IgG synthesis seems to be particularly sensitive to immunosuppressive treatment.

Wolpert: May I make a more general point here? I hope I have not been misunderstanding the tenor of the discussion, but I have a feeling that the word homeostasis is being used in such general terms as to lose all meaning. I get the impression that immunological behaviour is being considered as a homeostatic mechanism for cellular proliferation. One should always be very careful to specify homeostasis with respect to what. One can have homeostasis with respect to blood pressure, or temperature or size and so on. If one does not specify carefully one ends up saying nothing other than that the body is somehow being kept constant, which I don't think is very helpful.

Stoker: I concur in this, because it strikes me that for example antibody inhibition is the exact opposite of homeostasis, because it is a way of picking out a *new* cell system, in this case a very high affinity antibody-producing cell, and of changing from one steady state to a new state altogether. This is to define the thing in a particular way: you can call the whole immunological mechanism homeostatic if you look at it as a surveillance mechanism. But as Dr Möller described it, it is the opposite of homeostasis.

Möller: Of course antibody synthesis is an adaptive response to a changing environment. At the cell level this is the opposite to homeostasis but at the level of the individual it is a homeostatic mechanism, since its purpose is to keep the individual undisturbed by the environment. Furthermore, the cellular events in antibody synthesis are carefully controlled in a feedback system, aiming at homeostasis of cell proliferation after the initial increase of antibody-producing cells.

Bergel: We shall have to consider what we mean by, or how we use, this word "homeostasis". It is used, as Professor Wolpert indicated, in a number of situations. So far in this meeting we have used it very loosely, which has been appropriate to a symposium which we are treating like a spoken piece of research rather than one with a specific problem to answer. In fact we are talking about whether there *is* a problem. But we shall need to consider formulating a definition of homeostasis which is applicable to at least a majority of observations.

Möller: Burnet's idea (Burnet, 1967) is that the immunological mechanism developed in evolution as a homeostatic mechanism for controlling the cellular *milieu intérieur*, to remove aberrant cells. Later the same mechanism has been used as a defence against intruders; but its basic function is to control the internal environment. What Dr Allison has said about antilymphocytic serum in immunosuppression shows that the immune system has a very fundamental role; if you lack that mechanism you may succumb to a neoplastic development.

Allison: I want to extend the interesting remarks made earlier by Professor Wolpert on the question of time-scale (p. 105). It seems to me that we have considered at least three different types of homeostasis: biochemical homeostasis using pre-existing enzymes, which has a very short time-scale (seconds or minutes); an intermediate type involving adaptive responses of cells (synthesis of new enzymes and so on, which has a time-scale of hours); and thirdly long-term homeostatic mechanisms which are concerned with the well-being of the whole organism, for example replacement of lost blood or liver, or immune responses preventing tumour formation (which takes days or weeks). So perhaps the key distinction is not so much the event itself as the time-scale on which it happens.

Subak-Sharpe: There is also a fourth type of homeostatic mechanism, that of the population, and at this level the mutation rate exerts an important control.

Vernon: They all have something in common; they all describe the organism's tendency to return to a steady state.

Stoker: I would modify Professor Vernon's definition to say a *fairly* steady state.

Lamerton: Because neoplasia represents a qualitative change in biological organization in tissues and not merely a quantitative disturbance of normal function, it should perhaps be considered as a unique type of perturbation, requiring to be dealt with by a very special type of homeostatic mechanism.

Allison: This distinction is not confined to neoplasia. In a number of diseases, cell-mediated immune responses are critical. This is true, for instance, of leprosy and leishmaniasis. The nature of the disease produced by these organisms depends on the immune reactions elicited. With a good immune reaction the patient will survive, but without it the patient develops lepromatous leprosy or kala azar. In other words the importance of cell-mediated immunity isn't confined to neoplasia: it is a long-term response to perturbation from the normal.

Wolpert: One unifying criterion for homeostatic mechanisms may be that they all involve some kind of continual communication; they need some sort of flow of information or continual monitoring. A cell in a tissue may not be dividing and yet it may be capable of division. But the fact that it is not dividing need not mean that it is under homeostatic control. It may just not be dividing for some other reason: something is turned off. You can get the cell to divide by doing something, but you wouldn't say necessarily that that was a breakdown of homeostatic control. In another situation there may be continual monitoring and the cell doesn't divide for that reason; it starts dividing when you interfere with its homeostatic mechanism. We should consider the distinction between these two things.

Bergel: Homeostasis is a two-way traffic, slowing things down and speeding them up. We should look at it dynamically. The words hormone, or stimulator, and chalone, or inhibitor, fit together. We may eventually have to use the word chalone for a lot more substances than at present.

Subak-Sharpe: Dr Allison distinguished homeostatic mechanisms according to the time-scale on which the events happened; in addition one should do so according to where they happen, so that there is a division in time and in space. On the spatial scale we would distinguish homeostatically regulated events within an organelle, within a cell, within an organ, within an individual; and possibly at the last level, within a population. The two scales do not necessarily measure the same thing, but may complement one another.

Roe: Perhaps neoplasia should be regarded as outside the subject of homeostasis, because homeostasis surely implies that what is being controlled is itself otherwise normal. There is plenty of evidence that the cells of which most cancers are composed are abnormal. Therefore a factor that controls their proliferation, or fails to control it, cannot be regarded as homeostatic.

Möller: I am afraid that if we leave out neoplasia we leave out the most essential cellular homeostatic mechanism.

O'Meara: A disease which is illustrative is pernicious anaemia; it has relevance also to what Dr Johns spoke about earlier (p. 128), insofar as in this condition there is a failure of normal development of the precursors of the red cells so that at a stage corresponding more or less to the erythroblastic stage the cells contain haemoglobin. At this stage the nuclei still have the potential to divide, but they don't divide; they are extruded from the cell so that a macrocyte is formed and is set free into the blood stream. There is a massive proliferation of the bone marrow which gives an appearance almost of being a tumour; incidentally the spleen becomes enlarged and transformed and goes over to haemopoiesis. With these changes therefore you have something which is very close to a tumour in general appearance, but if vitamin B12 is given it all clears up. It may be interesting from the point of view of extrusion of the nucleus from the haemoglobiniferous erythroblasts.

Johns: It would certainly be of interest to examine those nucleated, haemoglobin-containing cells, if sufficient quantities could be obtained.

REFERENCES

ALLISON, A. C., and LAW, L. W. (1968). *Proc. Soc. exp. Biol. Med.*, **127**, 207.
BURNET, F. M. (1967). *Lancet*, **1**, 1171.
HEIDELBERGER, C., and IYPE, P. T. (1967). *Science*, **155**, 214–217.
HUBERMAN, E., and SACHS, L. (1966). *Proc. natn. Acad. Sci., U.S.A.*, **56**, 1123.
JERNE, N. K. (1967). *Cold Spring Harb. Symp. quant. Biol.*, **32**, 591.
MCDEVITT, H. O., and SELA, M. J. (1965). *J. exp. Med.*, **122**, 517.
MCDEVITT, H. O., and TYAN, M. L. (1968). *J. exp. Med.*, **128**, 1.
MITCHISON, N. A. (1967). *Cold Spring Harb. Symp. quant. Biol.*, **32**, 431.
MOEHRING, T. J., MOEHRING, J. M., KUCHLER, R. J., and SOLOTOROVSKY, M. (1967). *J. exp. Med.*, **126**, 407.

THE STRUCTURE OF MAMMALIAN CELL SURFACES

J. A. Forrester

Department of Cell Biology, Chester Beatty Research Institute, London

The cell surface is the external aspect of the complex of structures which form the cell-limiting membrane. In *in vivo* situations when it may be juxtaposed to other cell surfaces, or to connective tissues, or where macromolecular components of the body fluids may be in reversible association with it, the definition of the cell surface becomes difficult. For washed cells in free suspension in a simple electrolyte solution matters are easier, and an electrophoretic surface may be defined as the envelope through those charge centres which contribute to the cell ζ-potential. However, before I go on to discuss the information which cell electrophoretic investigations have yielded concerning this restricted part of the subcellular anatomy, I would like to consider briefly the possible molecular architecture of the membrane as a whole.

The lipid bilayer model which originated in the paper of Gorter and Grendel (1925) has in recent years received a succession of criticisms in which alternative orientations of the polar lipid in membranous structures have been suggested (e.g. Korn, 1966; Green and Perdue, 1966). In 1961 there was produced for the first time an artificial lipid membrane in which the orientation of the lipid could be reasonably assumed to be that suggested by Gorter and Grendel for biological membranes. Since that time many groups of workers have measured the properties of such "black" lipid membranes prepared from a wide variety of polar lipids and mixtures of them. In Table I, I have gathered together some of the published values for some physical and electrical properties of these membranes and compared them with corresponding values for cellular membranes also taken from the literature. The ranges include values obtained for artificial membranes of widely varying composition and for biological membranes from many species and fulfilling a wide range of functions. I think that the general agreement is impressive and is good circumstantial evidence that the bimolecular lipid leaflet could provide the essential mechanical and electrical properties of biological membranes. I would, however, like to discuss in a little more detail the electrical resistances recorded in this table.

The artificial bilayers are characterized on first formation by a very high ohmic resistance; 10^8 or 10^9 ohms/cm² are typical figures. These are many orders of magnitude above the measured values for biological membranes and may be changed by subsequent modification of the membrane by the incorporation of, for example, certain cyclic antibiotics, which probably function as cation carriers, or specific protein preparations (Mueller and Rudin, 1968). These bring about a reduction of the membrane resistance by a factor of from a thousand- to a million-fold into good agreement with the measured values for biological membranes. The ohmic resistance of a membrane is one of its most characteristic properties and these results strongly indicate that the bimolecular lipid layer cannot exist in an unmodified form over the whole of the cell surface. However, only a small proportion of the surface membrane need be altered to account for this

Table I

COMPARISON OF SOME MECHANICAL AND ELECTRICAL PROPERTIES OF NATURAL MEMBRANES AND BLACK LIPID MEMBRANES

	Natural membranes	Black lipid membranes
Thickness, Å		
Electron microscope	40–130	60–90
X-ray diffraction	40–85	–
Optical methods	–	40–80
Capacitance	30–150	40–130
Refractive index	~1·6	1·55–1·66
Interfacial tension (dynes/cm)	0·03–3·0	0·2–6·0
Water permeability (μm/sec)	0·25–60	2·3–25
Resistance (ohms/cm²)	10^2–10^5	10^3–10^9
Capacitance (μm/F/cm²)	0·5–1·3	0·3–1·3
Resting potential (mv)	10–90	0–140
Breakdown voltage (mv)	~100	100–550

discrepancy, the remainder having the classical bilayer structure. There is other evidence (e.g. Korn, 1966; Hoffman, 1968) suggesting on chemical grounds that there is insufficient lipid in the erythrocyte to form a complete bilayer over the whole surface. However, whatever is the true situation as to the orientation of the membrane lipid core, it matters little from the point of view of cell electrophoresis unless the method used for preparing cells for measurement drastically alters the membrane resistance to render it a comparatively conducting barrier. The results of Loewenstein and Kanno (1967) suggest that this is not the case.

The widely used technique of cell electrophoresis (Abramson, Moyer and Gorin, 1942; Ambrose, 1965) gives information about the nature and amount of fixed charge groups in the plane of shear of suspended cells; in practice, at physiological ionic strength, they will constitute a shell 10–15 Å thick at the exterior of the cells. The first generalization that I think may be made from the available results is that most mammalian cells seem to have a highly hydrophilic surface coat in whose structure carbohydrate plays an important role. This accords well with recent views on the nature of lipid–protein interactions in lipoproteins and membranous structures, in which increasing weight is being given to hydrophobic bonding between the lipid and the protein components. Glycoproteins taking part in this type

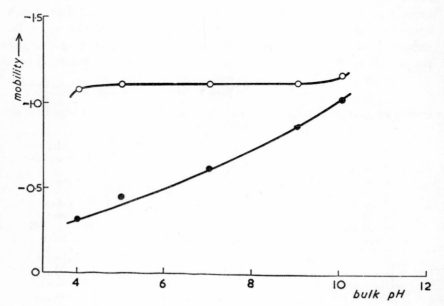

FIG. 1. Electrophoretic mobility (μm/sec/v/cm) of EL4 ascites tumour cells as a function of pH of the suspending medium (0·145M-NaCl) before (o——o) and after (●——●) coating with heterologous immune globulin. (Forrester, Dumonde and Ambrose, 1965.)

of interaction would be expected to orient themselves with their more hydrophobic portions directed towards the lipid core and their hydrophilic, carbohydrate portions directed towards the aqueous external solution. Fig. 1 (Forrester, Dumonde and Ambrose, 1965) supports this view. The washed EL4 ascites tumour cells display a pH–mobility relationship typical of many mammalian cells (e.g. Cook, Heard and Seaman, 1961) including erythrocytes (Heard and Seaman, 1961), in that over a wide range of pH they show little or no change in electrophoretic mobility. Proteins typically show a continuous variation of mobility with pH (Abramson, Moyer and Gorin, 1942) and, as the figure shows, coating the surface of these cells with immune globulin from a heterologous antiserum confers upon them a comparable mobility–pH variability. The importance of carbohydrates at the surface of the cell is supported by the chemical studies of Davies (1962) on mouse histocompatibility antigens and of Morgan and Watkins (1959) on human blood group substances.

When it comes to making generalizations about absolute values of cell ζ-potentials, matters become a little more difficult, however. For example,

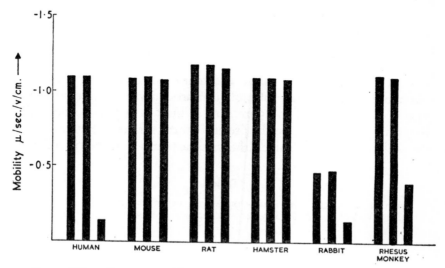

FIG. 2. Effect of treatment with neuraminidase on the electrophoretic mobilities of various species of erythrocyte. First pillar, normal untreated value; second, control with heat-inactivated enzyme; third, treatment with active enzyme. Suspending medium: 0·145 M-NaCl, pH 7·0.

even if we confine our attention to a cell type as specialized as the erythrocyte, very great differences exist between species (Brinton and Lauffer, 1959). The red cell of the dog, for example, possesses a ζ-potential nearly four times as great as that of the rabbit at physiological ionic strength and pH. But the red cells of the majority of species so far examined fall in a fairly narrow band between these two extremes, having electrophoretic mobilities which cluster between -0.90 and -1.25×10^{-4} cm^2 sec^{-1}v^{-1} and corresponding ζ-potentials of 14–19 mv. In many cases it has been possible to demonstrate that a large proportion of this negative charge is derived from the ionization of sialic acid residues present in the surface coat. Fig. 2 shows the electrophoretic mobilities of the red blood cells from six species before and after treatment with neuraminidase from *Vibrio cholera*.

As may be seen, the rat, mouse and hamster erythrocytes retain their electrophoretic mobilities intact after incubation under conditions where the mobility of human erythrocytes is reduced to about 20 per cent of its initial value. From a physical chemical point of view, however, human and mouse red blood cells are indistinguishable, having identical electrophoretic mobilities and pH–mobility curves. Since acid hydrolysis will release comparable amounts of sialic acid from erythrocyte stroma from both these species, I am inclined to the opinion that the negative charge at the surface of the mouse red cell also stems largely from sialic acid residues but that these are in a linkage with their parent macromolecules which is not cleaved by the *V. cholera* enzyme.

Two other favoured objects of study by the cell electrophoretic technique have been cells from tissue culture and murine ascites tumour cells. These cells also show a clustering of mobilities around values corresponding to negative ζ-potentials of 12–19 mv, and in most cases a proportion of this charge has been ascribed to the presence in the surface of neuraminidase-labile sialic acid groups. Table II shows some results I obtained in collabora-

Table II

EFFECT OF TREATMENT WITH NEURAMINIDASE ON THE ELECTROPHORETIC MOBILITIES (μm/sec/v/cm) OF NORMAL AND TRANSFORMED CLONES OF HAMSTER KIDNEY FIBROBLASTS

Clone	Untreated	After enzyme treatment
C13 (normal)	-1.02 ± 0.05	-0.68 ± 0.06
	-1.01 ± 0.05	-0.64 ± 0.05
N (transformed)	-1.30 ± 0.05	-0.65 ± 0.06
	-1.27 ± 0.06	-0.65 ± 0.05

tion with Professor Stoker using hamster fibroblasts and polyoma-transformed variants of them from tissue culture. The transformed clone N had an initially higher mobility than the normal C 13 but after exposure to the enzyme both cell types had indistinguishable reduced mobilities, suggesting that the enhanced charge of the transformed clone could be accounted for by the appearance in the surface of shear of extra sialic acid residues.

Experiments with proteolytic enzymes suggest the manner in which sialic acid residues may be bound in the cell surface. Cook, Heard and Seaman (1960) have described a sialopeptide released from the surface of human erythrocytes by treatment with trypsin, and Langley and Ambrose (1964) have described a similar material derived from the surface coat of the Ehrlich ascites tumour cell. We have extended these observations to other tumours and to hamster fibroblasts from culture, and we find that between

30 and 70 per cent of the total sialic acid in a cell may be brought into solution in the form of glycopeptides by the use of proteolytic enzymes, leaving the cells still intact as far as permeability to lissamine green is concerned (Holmberg, 1960). These results suggest that sialic acids are present as components of structural glycoproteins at the surfaces of a variety of mammalian cells.

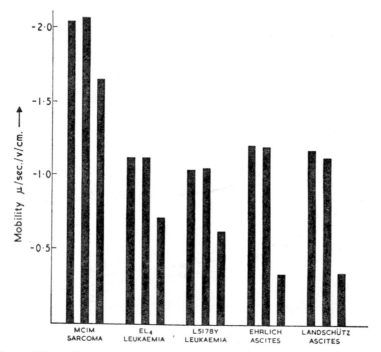

FIG. 3. Effect of treatment with neuraminidase on the electrophoretic mobilities of various mouse ascites tumours. Other details as Fig. 2.

In Fig. 3 are displayed the electrophoretic mobilities of five ascites tumours before and after treatment with bacterial neuraminidase. All show a loss of charge varying up to about 75 per cent, and the two lymphomas and the two forms of the Ehrlich tumour display original values in the range mentioned earlier. The MC1M sarcoma, however, displays some peculiar characteristics and is worth considering in more detail since it shows how cell electrophoresis may be used to analyse the ionogenic groupings at the cell surface.

Purdom, Ambrose and Klein (1958) first showed that the various sublines of this tumour possessed different and characteristic mobilities. These are reflected in the points at zero calcium concentration in Fig. 4. The M_{AA} ascites subline has an electrophoretic mobility more than twice as great as the parent M_{SS} solid sarcoma when measured in a simple saline medium containing no multivalent cations. Inclusion of calcium in the suspending medium brings about a pronounced decrease in the mobility of the

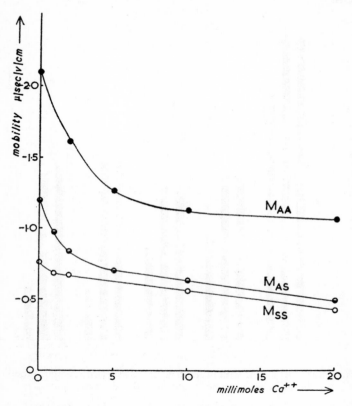

FIG. 4. Effect of increasing calcium ion concentration at constant ionic strength (0·15) and pH (7·0) on the electrophoretic mobilities of sublines of the MC1M sarcoma.

ascites subline which is effectively complete at a calcium concentration of 0·01 M; the effect on the solid subline is very small. This ability to bind calcium reversibly from concentrations of the order of a few millimolar may be due to the presence of phosphate groups in the electrophoretic shear plane, as I have elsewhere suggested (Forrester, Dumonde and Ambrose, 1965). In any case, Fig. 5 shows that the calcium-binding groups may be distinguished from the neuraminidase-labile charge groups.

After treatment with the enzyme, which removes about 20 per cent of the surface charge, the calcium-binding capacity of the cells is unimpaired. Thus at least three anionic types have been demonstrated semi-quantitatively in the surface of the M_{AA} subline: those removed by neuraminidase treatment, those which bind calcium, and a residual charge, not further charac-

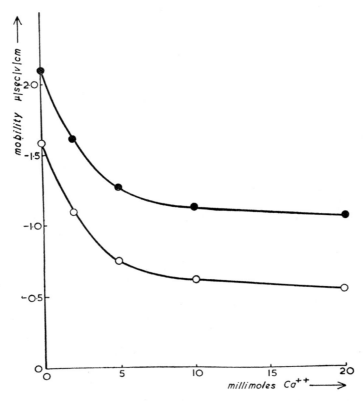

FIG. 5. Effect of increasing calcium concentration on the electrophoretic mobility of the M_{AA} ascites subline of the MC1M sarcoma before (●——●) and after (o——o) treatment with neuraminidase.

terized, necessary in order to account for the original ζ-potential. Incidentally, the small reduction in mobility after treatment of this cell line with neuraminidase is in agreement with Gasic and Berwick's histochemical finding (1963) that this tumour line has a very sparse acid mucopolysaccharide coat.

The various roles that the whole cell membrane complex may play in controlling cell interactions and thus the formation of organized tissues will be discussed in greater detail by Professor Wolpert in the following paper (p. 241), but I would like at this point to discuss whether information about cell ζ-potentials is of use when considering cell-to-cell contacts and adhesion.

In any physical chemical treatment of the approach of two surfaces, a curve of potential energy against separating distance may be constructed in which the repulsive force between the surfaces stems wholly from coulombic repulsive forces dependent upon the surface ζ-potentials, except at very close approaches when essentially mechanical interaction forces dominate. The attractive forces may come from a variety of sources (Pethica, 1961). Thus, if it is valid to apply the theoretical approach of Verwey and Overbeek (1948) to cell surfaces, the ζ-potential of the cells must play a central role in determining the parameters of the potential-against-distance curve. The assumption of an ascites habit of growth together with a much enhanced ζ-potential by the MC1M sarcoma M_{AA} subline could be interpreted in this way. However, the experiments with erythrocytes indicate that other factors are also important. Neuraminidase-treated human erythrocytes show as little inclination to stick to each other or to a solid substrate as do untreated ones despite having a ζ-potential of only about 4 mv compared with the normal value of about 17 mv. In this instance the comparative stiffness of the red cell membrane is all-important and although point contacts are possibly more easily formed, the non-deformability of the membrane may not allow these to spread and form permanent adhesion over a large area. Furthermore, experiments that I performed with Professor Stoker (Forrester, Ambrose and Stoker, 1964) show that although the loss of contact inhibition and reduction in adhesiveness associated with viral transformation of hamster cells was sometimes accompanied by an increase in ζ-potential this was not always the case.

It has been suggested by Bangham and Pethica (1960) that in the formation of intercellular adhesions the tips of fine pseudopodia of small radius of curvature would be at a considerable energetic advantage in the initiation of contacts compared with two approaching surfaces of large radius of curvature. Their treatment applies to cells approaching a position of equilibrium in the primary minimum of the potential energy curve, not to secondary minimum effects (Verwey and Overbeek, 1948; Curtis, 1962). Observations by Taylor (1966) on isolated chick and mouse embryonic cells and by Gustafson and Wolpert (1967) on echinoderm embryos have demonstrated that in these systems contacts are indeed initiated by the tips of very fine pseudopodia with diameters down to about 0·1 μm. If Bangham and Pethica's arguments are correct, this suggests that the cell surfaces are coming into molecular contact at the point of adhesion. Pethica (1961) has indicated the variety of attractive forces which may be acting in these conditions. Unfortunately cell electrophoresis is essentially an integrating measurement giving a value for the ζ-potential which is an average over the whole of the cell surface and can supply no information about the existence of atypical areas or domains which may have specific adhesive properties. The results I have shown earlier clearly provide plenty of opportunity for specialized regions of the cell surface to be differentiated in the magnitude of the local ζ-potential and the type of grouping generating it. The presence of positive charge centres, which have been demonstrated in several cell types (Ward and Ambrose, 1969), is a further complicating component in evaluating attractive forces.

It must, I think, be said that despite the fact that cell electrophoresis is a powerful tool of great specificity and precision as far as the region of the cell structure which may be investigated by its use is concerned, there have been far too few analyses in detail, such as I have described above, of cells which have a well-documented and interesting social life. This in part reflects the availability of materials. Many of the cells in which, for example, the experimental embryologist is interested, and whose surface composition, particularly as a function of time, would be of the greatest interest, are available in numbers too few for the present techniques of cell electrophoresis. Furthermore, it is often extremely difficult to devise means whereby cells from organized tissues can be brought into suspension and presented to the instrument in such a way that the information derived from them is meaningful in their context as tissue components.

REFERENCES

ABRAMSON, H. A., MOYER, L. S., and GORIN, M. H. (1942). *Electrophoresis of Proteins.* New York: Rheinhold.
AMBROSE, E. J. (ed.) (1965). *Cell Electrophoresis.* London: Churchill.
BANGHAM, A. D., and PETHICA, B. A. (1960). *Proc. R. phys. Soc. Edinb.,* **28,** 43.
BRINTON, C. C., and LAUFFER, M. A. (1959). In *Electrophoresis,* pp. 427–492, ed. Bier, M. New York: Academic Press.
COOK, G. M. W., HEARD, D. H., and SEAMAN, G. V. F. (1960). *Nature, Lond.,* **188,** 1011–1012.
COOK, G. M. W., HEARD, D. H., and SEAMAN, G. V. F. (1961). *Expl Cell Res.,* **28,** 27–39.
CURTIS, A. S. G. (1962). *Biol. Rev.,* **37,** 82–129.
DAVIES, D. A. L. (1962). *Biochem. J.,* **84,** 307–317.
FORRESTER, J. A., AMBROSE, E. J., and STOKER, M. G. P. (1964). *Nature, Lond.,* **201,** 945–946.
FORRESTER, J. A., DUMONDE, D. C., and AMBROSE, E. J. (1965). *Immunology,* **8,** 37–48.
GASIC, G., and BERWICK, L. (1963). *J. Cell Biol.,* **19,** 223–228.
GORTER, E., and GRENDEL, F. (1925). *J. exp. Med.,* **41,** 439.

GREEN, D. E., and PERDUE, J. F. (1966). *Proc. natn. Acad. Sci. U.S.A.*, **55**, 1295–1302.
GUSTAFSON, T., and WOLPERT, L. (1967). *Biol. Rev.*, **42**, 442–498.
HEARD, D. H., and SEAMAN, G. V. F. (1961). *Biochim. biophys. Acta*, **53**, 366–374.
HOFFMAN, J. F. (1968). *J. gen. Physiol.*, **52**, no. 1, part 2, 185s.
HOLMBERG, B. (1960). *Expl Cell Res.*, **22**, 406–414.
KORN, E. D. (1966). *Science*, **153**, 1491–1498.
LANGLEY, O. K., and AMBROSE, E. J. (1964). *Nature, Lond.*, **204**, 53–54.
LOEWENSTEIN, W. R., and KANNO, Y. (1967). *J. Cell Biol.*, **33**, 225–234.
MORGAN, W. T. J., and WATKINS, W. M. (1959). *Br. med. Bull.*, **15**, 109–112.
MUELLER, P., and RUDIN, D. O. (1968). *J. theoret. Biol.*, **18**, 222–258.
PETHICA, B. A. (1961). *Expl Cell Res.*, Suppl. 8, 123–140.
PURDOM, L., AMBROSE, E. J., and KLEIN, G. (1958). *Nature, Lond.*, **181**, 1586–1587.
TAYLOR, A. C. (1966). *J. Cell Biol.*, **28**, 155–168.
VERWEY, E. J. W., and OVERBEEK, J. T. G. (1948). *Theory of the Stability of Lyophobic Colloids.* New York: Elsevier.
WARD, P. D., and AMBROSE, E. J. (1969). *J. Cell Sci.*, **4**, 289–298.

For discussion of this paper, see pp. 259–263].

THE CELL MEMBRANE AND CONTACT CONTROL

L. WOLPERT AND D. GINGELL

Department of Biology as Applied to Medicine, Middlesex Hospital Medical School, London

WE are largely ignorant of the mechanisms of cell contact and their effect on cellular interactions. Nevertheless, cell contact has been invoked in the explanation of a wide variety of biological phenomena related to homeostasis, and these can be loosely grouped together under the headings of cell movement, cell growth, developmental interactions, and immunological responses. In relation to cell movement, contact is clearly involved in contact inhibition (Abercrombie, 1961, 1967), morphogenetic movements (Gustafson and Wolpert, 1963, 1967; Trinkhaus, 1966) the aggregation and sorting out of cells (Holtfreter, 1939; Moscona, 1968; Steinberg, 1964) and the invasiveness of tumour cells (Abercrombie, 1967; Carter, 1967). The term contact inhibition of growth (Stoker, 1967) has been used in relation to certain density-dependent growth phenomena of cells in culture, and contact seems to be involved in the anchorage dependence of cells (Stoker, 1968). Under the more general heading of developmental interactions, contact has been invoked in relation to inductive phenomena (Weiss, 1947; Grobstein, 1961), and more recently has received considerable attention in view of the finding of functional coupling between the cells of developing embryos (Furshpan and Potter, 1968; Loewenstein, 1966) and its absence in certain tumour cells (Loewenstein and Kanno, 1967). Cell contact appears to play an important role in several intercellular recognition processes involved in the immune response (Dresser and Mitchison, 1968).

In this paper we shall first consider some model mechanisms by which cell contact could provide a basis for control of cellular behaviour and then consider some of the above phenomena in the light of these models. The definition of cell contact raises some problems (Grobstein, 1961) but in this paper we shall consider cell contact to occur when there is some physical or chemical interaction between the apposed surfaces.

Any consideration of cell contact ought to take into account the factors determining whether cell contact can be established, how it is maintained, and how it may be broken. It is beyond our scope to consider such problems here (see Curtis, 1967; L. Weiss, 1967; Moscona, 1968, for reviews). It must nevertheless be very strongly emphasized that quite different factors may be operating in these three aspects of contact (L. Weiss, 1967). Moreover, it should be remembered that contact phenomena will be dependent not only on the interfacial properties of the apposed surfaces but also on their geometrical form, the forces generated by the cell, and the mechanical properties of the membrane (Gustafson and Wolpert, 1967). As regards adhesion, one school of thought, most notably represented by Moscona (see his 1968 paper for further references) has argued for "the existence of specific factors associated with the cell surface and intercellular spaces which function as cell ligands and mediate histogenetic attachment". This conception of an intercellular cement must be contrasted with the view that the colloidal properties of the cell surface are responsible for cell adhesion (Curtis, 1967). The colloidal properties of cell surfaces must clearly play a role in cell contact (Pethica, 1961; Curtis, 1967). The approach of two cell cell surfaces will be opposed by the electrical double layer, since all tissue cell surfaces carry a net negative charge of the same order of magnitude, and this clearly can be a limiting factor (Born and Garrod, 1968). However, it does not follow that the colloidal forces of the London–van der Waals type are responsible for holding cells together. This is an area of considerable controversy. Several recent investigations have, for example, failed to find a satisfactory correlation between surface potential and cell adhesion (L. Weiss, 1968; Gingell and Garrod, 1969). It is becoming increasingly clear that factors other than colloidal forces are acting in the making, maintaining and breaking of cell contacts. Whether this involves cell-specific ligands is still open to question. The problem of specificity is returned to below.

In so far as cell contact is involved in control mechanisms we are primarily concerned with how such interactions may occur and how they could modify cellular behaviour. In considering possible model mechanisms of contact control we shall distinguish between the functions of the cell membrane acting as a sensor, a transducer and a channel. These relate to the degree and nature of the communication between the cell and its environment.

THE CELL MEMBRANE AS A SENSOR

When the cell membrane is acting as a sensor the properties of both the membrane and the cytoplasm remain unchanged in different environmental situations, but the cell surface provides a means for sensing environmental variations, such as in adhesiveness. For example, a cell moving by means of random extension and retraction of pseudopods over a surface of differential adhesiveness might be expected to move in the direction where the pseudopods make the stronger contacts: there would in a sense be a tug-of-war between pseudopods and those with the better adhesion could determine the direction of movement (Gustafson and Wolpert, 1963, 1967). The cell is responding to variations in adhesiveness but there need be no change in membrane properties nor need there be any cytoplasmic change. This type of mechanism is best illustrated by Steinberg's (1964) theory of cell sorting out, based on adhesive strength. It is also implied in Carter's (1967) model of cell movement. A sensor mechanism could also operate with a high degree of specificity in the cell-to-cell or cell-to-substratum contact. If, for example, molecular contact does occur and the adhesive sites are chemically specific, stable intercellular contact should reflect this specificity (see p. 248 for discussion of molecular contact and specificity).

Another situation in which the cell membrane could act as a contact sensor arises when major topographical variations exist in its environment. Random pseudopodal activity could lead, for example, to a cell tending to remain at a corner, where the number of possible contact points is greater than on a plane (Gustafson and Wolpert, 1963).

THE CELL MEMBRANE AS A TRANSDUCER

In contrast to its role as a sensor, when it is acting as a transducer the cell membrane provides a dynamic means whereby changes in the external environment are communicated to the interior of the cell through some change in a property of the membrane (Gingell, 1967; Gingell and Wolpert, 1969; Wolpert and Gingell, 1968). We have suggested that modification of the outer surface potential of large free-living cells is responsible for permeability changes in the initiation of pinocytosis and other cellular responses. In relation to cell contact this implies that contact brings about a change in a property of the membrane, and we must now consider how this could happen. P. Weiss (1947, 1962) has given considerable attention to this aspect of membrane dynamics. The general question is how one cell membrane could, as it were, know that another membrane were close by; just how close by it would need be is a major question as well. Molecular contact between membranes could clearly provide a mechanism for changing membrane properties. We have given attention to the less obvious question of how interaction could occur in the absence of molecular contact. This requires a consideration of the physical effects on each membrane resulting from the approach of two membranes.

If the cell surface membrane bears an excess of fixed negative charges (Heard and Seaman, 1960; Seaman and Heard, 1960; Cook, 1968) it can be calculated that approach in physiological saline to within 30 Å can

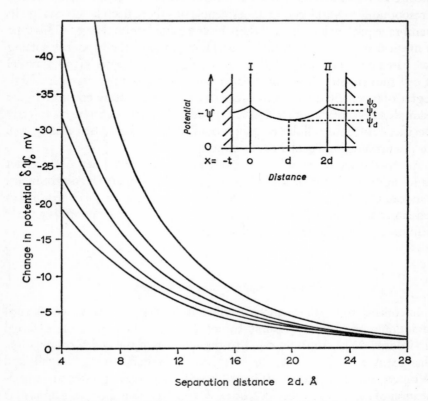

FIG. 1. Change in surface potential ψ_0 at the planes bearing fixed charges, resulting from the approach of membrane double layers in physiological saline. Thickness of the surface layer on each membrane is t and its degree of impenetrability is α. (1) $\alpha = 1.0$ (no layer); (2) $\alpha = 0.8$, $t = 10$ Å; (3) $\alpha = 0.8$, $t = 20$ Å; (4) $\alpha = 0.4$, $t = 10$ Å; (5) $\alpha = 0.4$, $t = 20$ Å. Surface potential, ψ_0 at infinite separation (10^{10} Å) is 25 mV in every case. The insert shows a schematic representation of the electrical potential profile between identical interacting plane parallel double layers under conditions of constant surface charge density during interaction. Fixed charges reside in planes I and II and are separated from a surface, impenetrable to solute ions, by a partially penetrable layer of thickness. Midpoint between planes I and II is d.

result in a significant increase in negative electrostatic surface potential (Gingell, 1967, 1968). Fig. 1 shows curves of potential change against the distance separating membranes. The degree to which counterions can penetrate behind the fixed charges of each membrane is represented by the constant α, while the thickness of this penetrable region at the surface is given by t. One can calculate that at distances corresponding to a unit membrane separation of 50 Å as seen in the electron microscope it is reasonable to expect a potential change of about 5 mv ($t=20$ Å, $\alpha=0\cdot8$). It is important to realize that a surface coat, here represented by the partially penetrable layer, could play a role in the transducer mechanism by transmitting the signal deeper into the membrane. It is clear that surface potential rises might provide a basis for a change in membrane properties leading to a cellular response. Changeux and his co-workers (1967) have argued that a small environmental change might bring about radical conformational changes in membrane macromolecules and it is significant that Hill (1967) has envisaged the stimulus for conformational change in terms of a change in electric field strength. Such a concept appears attractive in the light of work on the dependence of cation permeability on surface potential in artificial lipid micelles (Bangham, Standish and Watkins, 1965). These workers found that an increase in surface negativity from 20 mv to 25 mv increased the potassium permeability by 80 per cent. Mueller and Rudin (1968) have also elegantly demonstrated that the cation permeability of artificial excitable lipid membranes is delicately controlled by the interfacial potentials. It is common knowledge that the permeability state of excitable biological membranes is a function of the difference in potential across the membrane (Hodgkin and Huxley, 1952) so that it might be anticipated that a change in either inner or outer surface potentials could be responsible for a change in membrane properties (Frankenhaeuser and Hodgkin, 1957; Hodgkin and Chandler, 1965).

It should not be thought that environmental information can reach the cytoplasm only via a change in membrane permeability. A modification of the membrane–cytoplasmic interface might be produced, such as enzyme activation, which could effectively transfer information (P. Weiss, 1962). Indeed, this appears to be the mode of action of many hormones which activate adenyl cyclase in membranes, resulting in synthesis of cyclic AMP which may act as the cytoplasmic messenger (Sutherland, Robison and Butcher, 1967).

A somewhat different transducer mechanism might involve the mechanical deformation of the cell membrane. Stretching of the cell surface could lead to change in permeability properties or to exposure of new groups at the cytoplasmic interface. Some evidence to support this view comes from studies of muscle cells. Tension in the membrane appears to cause an increase in the rate of Na^+ extrusion (Harris, 1954) and a rise in metabolic rate (Clinch, 1968). It also seems that the response of the Pacinian corpuscle is associated with a change in permeability in response to membrane deformation (Gray and Sato, 1953).

THE CELL MEMBRANE AS A CHANNEL: FUNCTIONAL COUPLING

We have pointed out above that cell contact could lead to a permeability change. The concept of channel to be considered here is intended to emphasize the possibility that this permeability change might facilitate the flow of information between cells. It is thus a special case of the transducer mechanism described above. Such a concept is made necessary by the work of Furshpan and Potter (1968) and Loewenstein (1966) who have shown that cells in contact may have localized regions of increased ionic permeability: the so-called functional coupling phenomenon. This has now been described in a variety of excitable tissues (Furshpan and Potter, 1959; Pappas and Bennett, 1966; Bennett, 1966). More important from our point of view, such functional coupling has been described in a variety of non-excitable tissues including insect salivary gland; toad bladder; mammalian liver; amphibian epidermis; tissue cells in culture and squid embryo (reviewed by Furshpan and Potter, 1968); chick embryo (Sheridan, 1966); and amphibian embryos (Ito and Hori, 1966; Slack and Palmer, 1969).

In some cases where the ultrastructural basis of functional coupling in vertebrates has been investigated, "tight" junctions (those places where no intercellular gap is visible in the electron microscope) have been found between the cells (Furshpan and Potter, 1968). In invertebrate material fewer cases have been investigated, but coupling may be associated with septate desmosomes (Bullivant and Loewenstein, 1968).

Loewenstein (1967) describes the onset of functional coupling between like cells of disaggregated sponges of *Haliclona* and *Microciona* species. Cells of the same species spontaneously aggregated in artificial seawater, electrical coupling being established within 1–40 minutes of apparent contact. When cells of the same species were pressed together in artificial seawater with lowered Ca^{++} and Mg^{++} concentrations, under which conditions the cells are non-adhesive, no communication was detected even after some hours, but on addition of divalent cations to restore their normal concentrations coupling frequently became detectable. Cells of different species are not mutually adherent nor do they develop electrical communication in the conditions used. This suggests that the factors necessary for establishing adhesive contact are also necessary, and may even be a prerequisite, for the genesis of functional coupling. That such contacts can, under physiological conditions, be relatively easily broken is shown by Furshpan and Potter's

(1968) observations on the relationship between functional coupling and apparent contact between moving tissue cells.

Loewenstein (1967) has put forward a hypothesis to account for the genesis of functional coupling, based on the fact that loss of Ca^{++} from cell surface membranes increases their permeability. He suggests that in close apposition the future junctional region becomes isolated from the external pool of Ca^{++} by a sealing element and the Ca^{++} in the compartment formed is sequestered into the cells, with the result that the junctional membrane becomes permeable.

There are several difficulties with this suggestion. In the first place it does not explain how the calcium is removed from the junctional compartment. Secondly, if all surface membrane is potentially capable of forming a junctional complex (Loewenstein, 1967) and all pumps out calcium, the calcium pump must be stopped in the region where coupling occurs, unless the calcium in the compartment is removed more rapidly than it is pumped in. We would suggest that the feature lacking in this analysis is the initial spontaneous localized change in membrane properties brought about at contact. The membrane appears to be acting as a transducer.

As pointed out above, the close approach of membranes can lead to a rise in surface potential. Large potential rises are predicted at small separations. This effect might conceivably be the first significant event in the genesis of functional coupling. One result might be simply a permeability change due to increased negative surface potential having altered the field in the membrane. It is possible that the permeability increase caused by exposure of cell surface membranes to low concentrations of Ca^{++} is at least partly due to the increased negative surface potential which results from detachment of Ca^{++} from anionic groups in the membrane (see Frankenhaeuser and Hodgkin, 1957).

It is of importance to consider not only what sort of information could pass through a channel provided by functional coupling, but how such information could be used by the cell. At present it is not clear what substances other than small ions can pass across functionally coupled junctions. Loewenstein (1966) has claimed that proteins of molecular weight up to about 60000 can pass across the junctions of salivary gland cells. Pappas and Bennett (1966) found the crayfish electronic junctions permeable to fluorescein (molecular weight 332) while Furshpan and Potter have reached a similar conclusion for fibroblasts in culture. However, in the blastula of the amphibian embryo, Slack and Palmer (1969) have found such junctions to be impermeable to fluorescein. It will clearly be of great importance to determine the maximum size of molecule that can pass across the junction.

The potential information-carrying capacity of macromolecules, if they pass through a channel between dissimilar cells, is clear. What is less obvious is how the channel could provide information when the cells are similar. A particularly interesting possibility whereby small ions could provide the basis for positional information in a field of similar cells has been put forward by Goodwin and Cohen and will be discussed later, under developmental interactions (p. 256).

SPECIFICITY

Many aspects of cell contact bring up the question of specificity and central to this problem is whether or not specific molecular interactions occur at the site of cell contact. Do cell contact phenomena depend on a lock and key mechanism at the site of contact, of the type known to provide a high degree of specificity in enzyme–substrate and antigen–antibody interactions? This is a controversial area but it is extremely important to realize that the apparent specificity of cellular responses associated with cell contact does not in itself imply molecular specificity at the site of contact (see for example the sorting out of cells discussed below). One may ask whether direct molecular contact occurs at the site of contact. Unfortunately, we are unable to answer this question except for the case of tight junctions. Here the membranes are in very intimate contact; no gap is observable: we may thus assume that molecular contact is possible. It is more difficult to know whether there is molecular contact at close junctions in which the extracellular space seen in the electron microscope is about 100 Å or less (Hay, 1968), and is particularly difficult for gaps of 200 Å. Because of our uncertainty of the nature of cell adhesion, of where the cell surface ends, and whether or not a surface coat or cell-specific ligands are present (Moscona, 1968), it is not yet possible to decide whether or not molecular contact is occurring.

Much of the relevant evidence comes from electron microscopy which has failed to resolve such problems and might not be expected to reveal occasional sites of macromolecular contact. If there really is a structureless gap between the surfaces the possibilities for specific interaction must be very seriously reduced. It should be very strongly emphasized however that molecular contact could easily occur between occasional macromolecules bridging an apparent intercellular gap of say 50 Å. Such bridging would not be easy to detect.

It is important to point out that the specificity of cellular response associated with cell contact when the membrane is acting as a sensor will largely depend upon the specificity of the adhesive mechanism of the cell. This need not be the case when the membrane is functioning as a transducer, where the specificity may arise at any point in the transducing mechanism. For example different cells could respond with different changes in membrane properties to the same contact stimulus. Moreover, even the cellular responses to similar changes in membrane properties could be different. For example, the formation of cyclic AMP by membranes as a result of the action of several different hormones leads to a characteristic cellular response, depending on the cell type (Sutherland, Robison and Butcher, 1967). If molecular contact occurred the possibilities for specificity would be very greatly increased when coupled with a transducing mechanism.

EXAMPLES OF CELL CONTACT CONTROL AND POSSIBLE INTERPRETATIONS

CELL MOVEMENT

We have suggested that the movement of tissue cells includes the extension, attachment and retraction of pseudopods (Gustafson and Wolpert, 1963, 1967; Wolpert and Gingell, 1968), a mechanism particularly evident in the movement of the mesenchyme of sea urchin embryos. Such a mechanism is also supported by electron microscope observations of chick mesenchyme (Hay, 1968) and also by Ingram's (1969) recent observations on tissue cells in side view. The movement of cells by this type of mechanism poses some important problems of cell contact (Wolpert and Gingell, 1968) since it requires the repeated making and breaking of contact with the substratum over which the cell is moving. We have almost no idea of the mechanism whereby the cell adheres to the substratum or how this contact is broken, and it should be recognized that the processes of making and breaking adhesive contacts might be very different. For example, L. Weiss (1962) has suggested that the breaking of contact is not limited by the adhesion between cell and substratum but by the cohesive strength of the cell surface, loss of adhesive contact thus meaning rupture of the cell surface.

The mechanical properties of the membrane may play a role in controlling the movement of cells which spread on the substratum to which they are attached. Abercrombie (1961) has pointed out that there may be competition between pseudopods and we have suggested that if the membrane of flattened cells is stretched as a result of pseudopodal activity the cell may not be able to form a new pseudopod unless the tension is released by withdrawal of other pseudopods. In this situation the membrane is providing a transducer mechanism for measuring the degree of cell spreading.

Contact inhibition and contact paralysis

One of the most important aspects of cell movement and cell contact arises from Abercrombie's work on contact inhibition (Abercrombie, 1961, 1967) which shows that when a cell comes into contact with the surfaces of surrounding cells, its movement may be inhibited. Contact inhibition appears to be absent from certain tumour cells. Contact inhibition as defined by Abercrombie refers to the overall behaviour of the cell, and occurs when one "cell will not use another cell as a substrate for its locomotion" (Abercrombie, 1965). We have introduced the term contact paralysis to refer to the local inhibition in pseudopodal activity (Gustafson and Wolpert, 1967; Wolpert and Gingell, 1968) which usually occurs when contact is made (Abercombie, 1961). Contact paralysis may be but one aspect of contact inhibition and may be operative even when contact inhibition is not manifested. For example, cells with relatively long independent pseudopods, such as the primary mesenchyme of sea urchin embryos, exhibit contact paralysis at their points of contact, but do not show contact inhibition. Contact inhibition is a multifactorial phenomenon depending on a variety of factors including adhesive forces, mechanical properties of the cell, the shape of the cell, the disposition of pseudopods, the forces generated and contact paralysis (Gustafson and Wolpert, 1967). It is possible that contact paralysis is a normal component of cell movement when this occurs by pseudopodal retraction, since the contact between pseudopod tip and substratum may result in cessation of extension followed by contraction of the pseudopod. It is a matter of some importance to know whether any mechanism of this type operates in either normal locomotion or contact inhibition. The mechanism may involve either, or both, the sensor and transducer properties of the cell membrane. The cessation of activity could result from a local immobilization of the surface by the adhesive forces of the adjacent cell (Abercrombie, 1961, 1967), in which case the membrane is acting as a sensor, detecting differential adhesiveness between the substratum over which the cell is moving and the adjacent cell. On the other hand, Weiss (1961) has suggested that a more subtle mechanism than mere mechanical immobilization may be involved. We envisage that this could involve a transducer mechanism: as pointed out above, the approach of two membranes can lead to an increase in surface potential which in turn may lead to conformational changes in the membrane, and this in turn could affect pseudopodal activity. Such a conformational change might involve, for example, the activation of adenyl cyclase leading to local production of cyclic AMP. This in turn could alter the local calcium concentration (Rasmussen and Tenenhouse, 1968) and so affect a cytoplasmic system responsible for motility (Gingell and Palmer, 1968). There is good evidence for fibrillar and/or microtubule-like structures in the small pseudopods of tissue cells and there is also evidence that Ca^{++} may be involved in their formation, stability, and the mechanism whereby they generate forces.

That lack of contact inhibition does not imply lack of functional coupling has been shown by Potter, Furshpan and Lennox (1966) using transformed fibroblasts. Recent electron microscopic evidence of the contacts made by migrating cells indicate that their pseudopods make tight or close junctions at their site of attachment (Hay, 1968). Loewenstein and Penn (1967) have made the particularly interesting observation that during wound healing of the urodele epidermis, functional coupling between the migrating cells at the edge of the wound is lost and then re-established when cells from opposite sides of the wound make contact and cell movement ceases. From this experiment it cannot be ascertained whether the loss of electrical coupling initiates pseudopodal activity or whether both result from the release from contact paralysis by the removal of the cells in the wounded area.

Although it is obvious that transmission of signals between dissimilar cells could effect contact paralysis and modulate other aspects of cell behaviour, it is less clear in the case of cells of similar type.

Cell movement and cell contact in pattern formation

Cell movement and cell contact are clearly concerned in several phenomena in which cells, initially randomly dispersed, move to take up well-defined patterns. This is clearly a feature of considerable importance in morphogenesis. For example, during the early development of the sea urchin embryo the primary mesenchyme cells after entering the blastocoele are randomly distributed, but after a few hours move by pseudopod extension and retraction to take up a well-defined pattern (Gustafson and Wolpert, 1963, 1967). Holtfreter (1939) and more recently Steinberg (1964) have demonstrated differences in contact affinities among embryonic tissues which may be related to their normal morphogenetic movements. In this connexion, particular attention has in recent years been given to the sorting out of cells in mixed aggregates (Moscona, 1968). In this sorting out one type of cell usually becomes covered by cells of another type. In contrast to this rather simple inside-outside pattern formation it has recently been demonstrated that cells from the imaginal discs of insects when reaggregated tend to take up positions similar to those they occupied before disaggregation. As Garcia-Bellido (1966) has put it, "the determined cell possesses some kind of spatial information which characterizes it as a cell in a certain place in a given pattern". Rather specific cell movements and cell contact have also been implicated in the development of the central nervous system, particularly regeneration of the optic nerve (reviewed by Jacobson, 1966).

These are clearly cell contact phenomena of considerable relevance to developmental homeostasis. The questions we would like to pose are (1) to what extent could the membrane acting as a sensor provide a basis for such phenomena? (2) how much specificity is required? (3) are the transducer and channel properties of the membrane involved?

A common misconception is that the sorting out of cell types in mixed aggregates necessarily implies a highly specific mechanism for cell recognition: that the cell contact requires stereochemical molecular interactions. The mechanism suggested by Steinberg (1964) for the sorting out of cells may require only sensor properties, since he postulates that the final arrangement of randomly motile cells will be that of minimal potential energy, which will be a function of the different energies of adhesion between the cells involved. He has provided evidence that this mechanism, relying only on differences in adhesive energy, could operate for the cases in which the final pattern is more or less radially symmetrical. However, some doubt has been cast on this interpretation by Roth and Weston (1967) who found that the cell types used by Steinberg invariably preferentially aggregated with like cells rather than unlike cells. Nevertheless, it is possible to explain pattern formation by the mesenchyme of the sea urchin in terms of the cells moving over a substratum to take up positions where their pseudopods make the most stable contacts (Gustafson and Wolpert, 1967). Although this type of mechanism does not necessarily require specificity at the sites of contact, it does not preclude it.

It seems possible, but not easy, to extend the concept of a sensor mechnism to more complex cases in which different cells on the same substratum move in different directions. This is shown by the behaviour of grafts of retina on to the optic tectum by De Long and Coulombre (1965). Regardless of the site of the graft of retina on the tectum the direction taken by outgrowing retinal fibres was correlated with the site of origin of the retinal graft. One might imagine independent gradients in particular chemical groups on the surface of the tectum, and that cells from different portions of the retina are sensitive to one type of group only. This type of specificity for a sensor mechanism has yet to be demonstrated, but could provide a means for cells to move to very specific sites on a substratum. Even with a mechanism of this type it is not easy to explain those situations in which cells, when dissociated, move so as to take up very specific positions in a two-dimensional array. The experiments of Garcia–Bellido (1966) suggest that each cell is uniquely specified, including at least two cellular polarities. This means that the cell surface must include a specification of the left and right sides and also the up and down faces: in effect, the cell would be like a piece in a jig-saw puzzle. One way out of this rather unlikely set of requirements would be for the cells to make use of positional information, and this will be discussed below (p. 256). When looking at these more complex patterns it is possible to invoke the transducer and channel properties of cell contacts. If functional coupling were rapidly established at each point of pseudopodal contact it would be possible, in principle at least, for signals to be passed between the cells. These signals could determine whether the cells continued to move or not. The cells could, in a sense, be measuring the positional information.

CELL GROWTH

Two striking features of the growth of cultured freshly isolated diploid cells are that they are unable to grow in suspension but require contact with a substratum—they are anchorage-dependent (Stoker, 1968)—and they tend to stop growing on a substratum when the monolayer becomes fully confluent (Stoker, 1967). This latter statement requires modification since perfusion can lead to multilayering (Kruse and Miedema, 1965). These two phenomena present something of a paradox since contact seems to be required for growth on the one hand, yet can inhibit it on the other. We suggest that both may be related to a transducing mechanism in which there is mechanical deformation of the cell membrane. As pointed out above there is some evidence to support the view that this could lead to a permeability change. Cells in suspension are macroscopically rounded but the cell surface may be highly convoluted, as seen in BHK cells (Wolpert and Gregory, 1969). When the BHK cells spread such convolutions in the surface are pulled out, resulting in the spread cell having a more or less smooth surface (see Fig. 2). The apparent increase in surface area could possibly be provided by the pulling out of such folds. This implies that cell spreading involves forces tending to stretch the membrane, since there is little reason to believe that in cell spreading or cell movement new surface membrane is formed (Wolpert and Gingell, 1968). It is tempting to think that an associated permeability change could result in cell growth and cell division which could possibly account for anchorage dependence.

The evidence that contact inhibition of growth is a cell-to-cell contact-mediated phenomenon is not persuasive, since if inhibition is solely dependent on cell-to-cell contact, multilayering should not occur. Nevertheless there seems evidence that some inhibition occurs at about the time the monolayer approaches confluence. 3T3 cells seem particularly sensitive to contact inhibition of growth (Todaro, Lazar and Green, 1965) and it is of particular interest that localized removal of cells from a stationary monolayer culture led to cell divisions in that region even when the culture medium was agitated. "It is clear that the removal of cells adjacent to an arrested cell is sufficient *per se* to induce that cell to divide." It thus seems possible that a similar mechanism to that proposed for anchorage dependence may be operative, since contact inhibition of movement may, when the monolayer becomes confluent, prevent cell spreading and thus inhibit growth. It would thus be of great interest to know the relationship between cell morphology and cell growth in multilayers. For example, growth and division might be confined to those cells capable of spreading on the glass. In this respect, it is interesting that Miedema (1968) has found that alkaline phosphatase activity is, in multilayered cultures, found only in the layer adherent to the glass.

The possibility that some alteration of the cell surface occurs in malignancy is receiving increasing attention (Wallach, 1968). Tumour cells are less subject to anchorage dependence and contact inhibition of growth (Stoker, 1967). It is of particular interest that Loewenstein and Kanno (1967) found no coupling in liver tumour cells whereas the normal or regenerating cells were coupled. Similar results have now been reported for thyroid epithelial tumours (Jamakosmanovic and Loewenstein, 1968) and human stomach cancer epithelial cells (Kanno and Matsui, 1968). The site of the change in all these cases is probably in the membrane but the details are entirely unknown.

IMMUNOLOGICAL RESPONSES

Contact between cells seems to be an essential feature of certain immunological phenomena. For example the interaction between macrophages with non-specifically adsorbed cytophilic antibody and allogeneic cells leads to mutual adhesion and death of both cell types. It is possible that macrophages, which are unable to synthesize immunoglobulins, may thus be enabled to interact with allogeneic cells in delayed hypersensitivity with a high degree of molecular specificity, cytophilic antibody acting as a recognition factor (Dumonde, 1967; Nelson and Boyden, 1967). Contact with previously sensitized lymphocytes is also thought to be a major factor in the destruction of allogeneic cells.

It is possible that contact-mediated transfer of immunological information between cells might play an even wider role in intercellular recognition

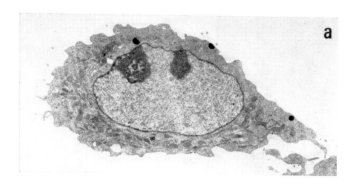

Fig. 2. Electron micrographs of BHK cells to show the disappearance of convolutions in the surface membrane when the cell spreads on a Millipore filter. In (a) the cell has just begun to spread and has been on the filter for 3 hours. In (b), after 48 hours, the cells are well spread and the surface is now smoother. Both pictures are × 4000.

processes. The hypothesis has been put forward (see Dresser and Mitchison, 1968) that whereas the interaction of humoral antigens with appropriate "committed" small lymphocytes may result in immunological paralysis, or tolerance, antigen conveyed by macrophages results in division of committed lymphocytes to form primary cells with antibody-like receptors at the surface, responsible for the immune response. Dresser and Mitchison consider it likely that the role of the macrophage lies in presenting antigen to lymphocytes in a suitable steric form, rather than recognizing antigen and then transferring to the lymphocyte a message lacking antigenic structure.

Again, the mixed lymphocyte reaction in tissue culture can be inhibited by an antiserum directed against the light chain of the immunoglobulin. This suggests that the recognition of histocompatibility antigens on other cells is mediated through molecules with light chain determinants, presumably some form of antibody (Greaves, Torregiani and Roitt, 1969). This implies direct molecular contact between cells of a lock and key type.

Interactions of the type described may well involve transducer mechanisms, but this is purely speculative. Permeability changes have been reported in cells exposed to humoral antibody and complement (Preto, Kornblith and Pollen, 1967) and this may be of more general significance. However, the contact interactions occurring in immunological systems may provide us with models of how highly specific responses could occur between isogenic cells.

DEVELOPMENTAL INTERACTIONS

The importance of cell contact in relation to morphogenetic movements has been considered above. Here we are more concerned with the role of cell contact in controlling developmental interactions, particularly communication between parts of the system. Twenty years ago, Paul Weiss proposed what is essentially a transducer mechanism for embryonic induction (Weiss, 1947). This postulated that close contact was essential for induction and brought about a reorientation of the surface molecules of the responding cell. It is still not clear whether or not induction requires cell contact (Saxen and Toivonen, 1962). The discovery of functional coupling in developing systems referred to above (p. 246) has given considerable impetus to research into the role of cell contact in intercellular communication. It is of interest to consider whether pattern formation, which clearly requires intercellular communication, involves cell contact (Wolpert, 1969). Very little evidence is available but Saunders and Gasseling (1963) have investigated the problem in the developing chick wing. Reversal of the apical zone of the wing results in mirror-twin duplication of the hand. This effect still occurred if Millipore filters were placed between the reversed graft and the rest of the wing, preventing cell-to-cell contact, but the effect was attenuated as the pore size of the filter was decreased and was completely blocked by a non-porous barrier. Direct cell-to-cell contact does not in this case seem necessary.

It has recently been suggested that pattern in a morphogenetic field requires the specification of positional information: that is, cells have their position specified with respect to one or more points in the system. This positional information is thought largely to determine the cell's molecular differentiation (Wolpert, 1969). An ingenious model whereby positional information could be specified has been proposed by Goodwin and Cohen (1969), making use of periodic phenomena in cells. If there were an autonomously periodic event in each cell of a population it can be shown that certain cells would become entrained to the period of the fastest cell—the dominant cell—if a brief synchronizing signal could be transmitted, perhaps by small ions, from cell to cell. The propagation of a second periodic event, with appropriate intercellular delays, would result in a phase angle difference between the two periodic events, increasing with distance from the dominant cell, and thus providing each cell with positional information. This mechanism shows how small ions could provide a means of communication between similar, but not identical, cells. A mechanism of this type could be involved in those situations referred to above where cells move to take up rather complex patterns. For example, in the movement of the primary mesenchyme cells of the sea urchin embryo we have suggested that the final pattern reflects the sites where their pseudopods form the most stable contacts. Rather than depending on an adhesive mechanism the stability of the contact may depend on local positional information being communicated to the interior of the cell.

An amusing possibility whereby cell-to-cell communication occurs without passage of any molecules across the membrane could also be envisaged. Changeux and his colleagues (1967) have, for example, suggested that if the cell surface membrane were made of protomeric units, localized interaction with small ligands could bring about a conformational change in the whole membrane. On this basis it seems possible that such a change might be propagated between cells in contact, resulting in signal transmission analogous in some ways to the nervous impulse, except that a memory of the event in terms of changed surface membranes would remain. A different ligand–protomer interaction might then trigger off a propagated return to the initial state. It is possible even to envisage that a membrane in state A would induce state B in the membrane of the adjacent cell which could induce state C in its neighbour and so on down the line. Such a system coupled with the reversible exchange of molecules between the membrane and the interior could provide a remarkable communication system and could even provide a mechanism for the specification of positional information.

SUMMARY

We remain largely ignorant of the mechanisms involved in cell contact and their effect on cellular interactions. An attempt has been made to analyse possible contact control phenomena in terms of the behaviour of the membrane as a sensor, transducer and channel, in relation to cell movement, cell growth, developmental interactions and immunological responses. The transducer concept emphasizes the possibility of changes in membrane properties which may result from contact and lead to responses in the interior of the cell. Particular attention has been given to the possible changes in surface potential that may occur when cells come into contact, and the formation of high permeability junctions. It has been suggested that both contact inhibition of growth and anchorage dependence of cells may be accounted for by a change in permeability of the membrane when it is stretched. With the exception of certain immunological reactions, there does not seem to be any persuasive evidence for the existence of specificity at the molecular level at the site of contact. However, such specificity cannot be precluded. Our present state of knowledge does not even permit a decision to be made about whether or not cells can make direct molecular contact, except in the case of tight junctions. In general it would be surprising if future work did not reveal a greater role for the membrane in contact control, which makes use of its dynamic properties.

REFERENCES

ABERCROMBIE, M. (1961). *Expl Cell Res.*, Suppl. 8, 188–198.
ABERCROMBIE, M. (1965). In *Cells and Tissues in Culture*, vol. 1, p. 77, ed. Willmer, E. N. New York: Academic Press.
ABERCROMBIE, M. (1967). *Natn. Cancer Inst. Monogr.*, **26**, 249–277.
BANGHAM, A. D., STANDISH, M. M., and WATKINS, J. C. (1965). *J. molec. Biol.*, **13**, 138.
BENNETT, M. V. L. (1966). *Ann. N.Y. Acad. Sci.*, **137**, 509–539.
BORN, G. V. R., and GARROD, D. (1968). *Nature, Lond.*, **220**, 616–618.
BULLIVANT, S., and LOEWENSTEIN, W. R. (1968). *J. Cell Biol.*, **37**, 621–632.
CARTER, S. B. (1967). *Nature, Lond.*, **213**, 256–260.
CHANGEUX, J. P., THIÉRY, J., TUNG, Y., and KITTEL, C. (1967). *Proc. natn. Acad. Sci. U.S.A.*, **57**, 335–341.

CLINCH, N. F. (1968). *J. Physiol., Lond.*, **196**, 397–414.
COOK, G. M. W. (1968). *Biol. Rev.*, **43**, 363–391.
CURTIS, A. S. G. (1967). *The Cell Surface: Its Molecular Role in Morphogenesis*. New York: Academic Press.
DE LONG, G. R., and COULOMBRE, A. J. (1965). *Expl Neurol.*, **13**, 351–363.
DRESSER, D.W., and MITCHISON, N. A. (1968). *Adv. Immun.*, **8**, 129.
DUMONDE, D. C. (1967). *Br. med. Bull.*, **23**, 9–14.
FRANKENHAEUSER, B., and HODGKIN, A. L. (1957). *J. Physiol., Lond.*, **137**, 218–244.
FURSHPAN, E. J., and POTTER, D. D. (1959). *J. Physiol., Lond.*, **145**, 95–127.
FURSHPAN, E. J., and POTTER, D. D. (1968). In *Current Topics in Developmental Biology*, vol. 3, ed. Moscona, A. A., and Monroy, A. New York: Academic Press.
GARCIA-BELLIDO, A. (1966). *Devl Biol.*, **14**, 278–306.
GINGELL, D. (1967). *J. theoret. Biol.*, **17**, 451–482.
GINGELL, D. (1968). *J. theoret. Biol.*, **19**, 340–344.
GINGELL, D., and GARROD, D. R. (1969). *Nature, Lond.*, **221**, 192–193.
GINGELL, D., and PALMER, J. F. (1968). *Nature, Lond.*, **217**, 98–102.
GINGELL, D., and WOLPERT, L. (1969). In preparation.
GOODWIN, B., and COHEN, M. (1969). *J. theoret. Biol.*, in press.
GRAY, J. A. B., and SATO, M. (1953). *J. Physiol., Lond.*, **122**, 610–636.
GREAVES, M., TORREGIANI, G., and ROITT, I. (1969). *Nature, Lond.*, **222**, 641–645.
GROBSTEIN, C. (1961). *Expl Cell Res.*, Suppl. 8, 234–245.
GUSTAFSON, T., and WOLPERT, L. (1963). *Int. Rev. Cytol.*, **15**, 139–214.
GUSTAFSON, T., and WOLPERT, L. (1967). *Biol. Rev.*, **42**, 442–498.
HARRIS, E. J. (1954). *J. Physiol., Lond.*, **124**, 242–247.
HAY, E. D. (1968). In *Epithelial Mesenchymal Interactions*, pp. 31–35, ed. Fleischmajer, R., and Billingham, R. Baltimore: Williams and Wilkins.
HEARD, D. H., and SEAMAN, G. V. F. (1960). *J. gen. Physiol.*, **43**, 635–654.
HILL, T. L. (1967). *Proc. natn. Acad. Sci., U.S.A.*, **58**, 111–114.
HODGKIN, A. L., and CHANDLER, W. K. (1965). *J. gen. Physiol.*, **48**, 27–30.
HODGKIN, A. L., and HUXLEY, A. F. (1952). *J. Physiol., Lond.*, **117**, 500–544.
HOLTFRETER, J. (1939). *Arch. exp. Zellforsch.*, **23**, 169.
INGRAM, V. T. (1969). *Nature, Lond.*, **222**, 641–645.
ITO, S., and HORI, N. (1966). *J. gen. Physiol.*, **49**, 1019–1027.
JACOBSON, M. (1966). In *Major Problems in Developmental Biology*, pp. 315–383, ed. Locke, M. New York: Academic Press.
JAMAKOSMANOVIC, A., and LOEWENSTEIN, W. R. (1968). *Nature, Lond.*, **218**, 775.
KANNO, Y., and MATSUI, Y. (1968). *Nature, Lond.*, **218**, 775–776.
KRUSE, P. F., and MIEDEMA, E. (1965). *J. Cell Biol.*, **27**, 273–279.
LOEWENSTEIN, W. R. (1966). *Ann. N.Y. Acad. Sci.*, **137**, 441–472.
LOEWENSTEIN, W. R. (1967). *Devl Biol.*, **15**, 503–520.
LOEWENSTEIN, W. R., and KANNO, Y. (1967). *J. Cell Biol.*, **33**, 225–234.
LOEWENSTEIN, W. R., and PENN, R. D. (1967). *J. Cell Biol.*, **33**, 235–242.
MIEDEMA, E. (1968). *Expl Cell Res.*, **53**, 488–496.
MOSCONA, A. A. (1968). *Devl Biol.*, **18**, 250–277.
MUELLER, P., and RUDIN, D. O. (1968). *Nature, Lond.*, **217**, 713–719.
NELSON, D. S., and BOYDEN, S. V. (1967). *Br. med. Bull.*, **23**, 15–20.
PAPPAS, G. D., and BENNETT, M. V. L. (1966). *Ann. N.Y. Acad. Sci.*, **137**, 495–508.
PETHICA, B. A. (1961). *Expl Cell Res.*, Suppl. 8, 123–140.
POTTER D. D., FURSHPAN, E. J., and LENNOX, E. S. (1966). *Proc. natn. Acad. Sci. U.S.A.*, **55**, 328–336.
PRETO, A., KORNBLITH, P. L., and POLLEN, D. A. (1967). *Science*, **157**, 1185–1187.
RASMUSSEN, H., and TENENHOUSE, A. (1968). *Proc. natn. Acad. Sci. U.S.A.*, **59**, 1364–1370.

DISCUSSION 259

Roth, S. A., and Weston, J. A. (1967). *Proc. natn. Acad. Sci. U.S.A.*, **58**, 974–980.
Saunders, J. W., and Gasseling, M. T. (1963). *Devl Biol.*, **7**, 64–78.
Saxen, L., and Toivonen, S. (1962). *Primary Embryonic Induction*. New York: Academic Press.
Seaman, G. V. F., and Heard, D. H. (1960). *J. gen. Physiol.*, **44**, 251–268.
Sheridan, J. D. (1966). *J. Cell Biol.*, **37**, 650–659.
Slack, C., and Palmer, J. F. (1969). *Expl Cell Res.*, **55**, 416–419.
Steinberg, M. S. (1964). In *Cellular Membranes in Development*, pp. 321–366, ed. Locke, M. New York: Academic Press.
Stoker, M. (1967). In *Current Topics in Developmental Biology*, vol. 2, pp. 107–129, ed. Monroy, A., and Moscona, A. A. New York: Academic Press.
Stoker, M. (1968). *Nature, Lond.*, **218**, 234–238.
Sutherland, E. W., Robison, G. A., and Butcher, R. W. (1967). *Circulation*, **37**, 279–306.
Todaro, G. J., Lazar, G., and Green, H. (1965). *J. cell. comp. Physiol.*, **66**, 325–334.
Trinkhaus, J. P. (1966). In *Major Problems in Developmental Biology*, pp. 125–176, ed. Locke, M. New York: Academic Press.
Wallach, D. F. H. (1968). *Proc. natn. Acad. Sci. U.S.A.*, **61**, 868–874.
Weiss, L. (1962). *J. theoret. Biol.*, **2**, 236–250.
Weiss, L. (1967). *The Cell Periphery, Metastasis, and other Contact Phenomena*. Amsterdam: North Holland.
Weiss, L. (1968). *Expl Cell Res.*, **51**, 609–625.
Weiss, P. (1947). *Yale J. Biol. Med.*, **19**, 235–278.
Weiss, P. (1961). *Expl Cell Res.*, Suppl. 8, 260–281.
Weiss, P. (1962). In *The Molecular Control of Cellular Activity*, pp. 1–72, ed. Allen, J. M. New York: McGraw-Hill.
Wolpert, L. (1969). In *Towards a Theoretical Biology*, vol. 3, ed. Waddington, C. H. Chicago: Aldine. In press.
Wolpert, L., and Gingell, D. (1968). *Symp. Soc. exp. Biol.*, **22**, 169–198.
Wolpert, L., and Gregory, M. (1969). Unpublished data.

DISCUSSION

Stoker: I would like to have clarified what is meant by molecular contact. When one considers the molecular conformation of a cell surface, whatever the model of membrane structure, there is likely to be a series of molecules as one goes out from the apparent surface, starting where molecules are very stably attached and very rarely exchanged with the environment and proceeding outwards to less firmly attached molecules in a kind of cloud. The next cell, too, will have a cloud of molecules constituting its "surface". One could suppose that these molecules vary in their probability of interchange with the extracellular environment. What does one mean by "contact" in those circumstances?

Forrester: The problem is a difficult one and related to that of defining the cell surface which I mentioned (p. 230). Moving from the inside of a washed cell outwards one would pass through regions of increasingly hydrophilic character, finishing up with an array of sugar residues covalently bonded in macromolecules which are themselves integral parts of the membrane.

In a medium containing adsorbing and desorbing macromolecules the situation becomes more difficult still. What I understand as "molecular contact" between such surfaces would exist when they were linked by forces of the order of magnitude of salt linkages. I would include in this definition cases where the linkage is effected by ion triplet formation or where the bonding from each surface is to an interposed polymer. This corresponds to the primary minimum in the colloid theory treatment and to close approach to distances of the order of 5–10 Å. I would not describe the situation that Professor Wolpert discussed, when he considered interactions over a distance of 50 Å, as molecular contact, albeit Gingell's calculations indicate that there may be potential changes sufficient to produce conformational changes in the juxtaposed membranes at this distance.

Wolpert: I would agree. One is also concerned about molecular contact because of the degree of specificity possible. If we are going to invoke specificity in stereochemical and chemical terms, there must be molecular contact.

Möller: If immunized lymphocytes are added to target fibroblasts growing in a monolayer the target cells will be killed. However, if the fibroblasts are coated with antibodies and then exposed to killer lymphocytes, no cytotoxicity occurs, presumably because the lymphocytes cannot make contact with the fibroblast receptors. So the distance here is that of one molecule on the target cells.

If lymphocytes are non-specifically made competent to kill by, for example, coating them with antilymphocytic serum, they can exert their killing effect only if the antilymphocytic serum can also react with the target cells and thus form a bridge between the targets and the lymphocytes. If such a bridge cannot be formed the single layer of antilymphocytic serum molecules present on the lymphocytes prevents the close contact necessary for killing.

The phenomenon of allogeneic inhibition may also be mentioned here. If tumour cells of genotype A are injected into $F_1(A \times B)$ hybrids of the same species, the hosts cannot react immunologically against the tumour A, but in spite of this tumour growth is retarded by comparison with cells injected into the syngeneic A strain. The transplantation antigens are on the cell surface and in the $A \rightarrow (A \times B)$ situation there is probably an interaction between two cells having different configurations of transplantation antigens which suppresses growth.

Wolpert: This may be our best model for real specificity of molecular contact at the surface between two cell types.

Stoker: The other example is that shown by Borek and Sachs (1966), which also seems to invoke contact.

Iversen: Professor Bergel enquired earlier, apropos histones and DNA, whether they were married or simply lived together. We have been talking very much about cells that meet and form brief contacts, but at least in the epidermis the cells are really married! In such systems you find desmosomes with tonofibrils going into the cells. Now why is this system there? Is it also used for information exchange of some sort? We must not forget that in an organism, cells live closely together in tissues and organs; they do not meet and go away again as in some cell cultures.

Wolpert: The functions of desmosomes and of tight junctions are rather confused at the moment. There is increasing evidence in vertebrates that the so-called tight or close junction is the site of the low ionic permeability.

O'Meara: When the epidermis becomes cancerous certain interesting changes take place in the behaviour of the cells. As Professor Iversen points out, the cells of the normal epidermis are in contact with one another in a special way through the intercellular bridges. These bridges are maintained when an epidermis has become malignant and infiltration has taken place. In a proportion of tumours, 10 per cent or so, as time goes on a change takes place in the behaviour of these cells; they become dissociated, almost as if a three-dimensional type of growth has changed to a two-dimensional form. Not only do the cells round off as dedifferentiation occurs but they become separated, no longer united by intercellular bridges. Furthermore quite a number of these tumours, instead of growing in plaques, begin to grow in chains in which the cells are quite well separated from one another (what I call streptocytic growth), almost as if cell repulsion had developed in place of a cell adhesion. I wonder whether there may not be repulsive forces at work as well as attractive forces and whether the end result may not be a combination of the two.

Forrester: It is precisely by considering the balance between attractive and repulsive forces that the potential energy curve between two approaching surfaces is derived. Dr B. Larson, from Aarhus, working in my laboratory, obtained some results which indicate how the magnitude of the coulombic repulsive interaction may determine whether cells will come into contact. DEAE-dextran of molecular weight about two million adsorbs strongly to erythrocytes and to ascites tumour cells, causing agglutination and profoundly altering their electrophoretic mobilities to positive values. To our surprise we found that dextran sulphate of similar molecular size, although it does not adsorb to erythrocytes (confirming other evidence on the unavailability of positive charge centres in the electrophoretic plane of shear of those cells), does adsorb strongly to ascites cells, increasing their electrophoretic mobilities up to two-fold. However, no agglutination is observed. Presumably the polymer is interacting with the small proportion of positive charge centres in the surfaces of these cells, but the increased repulsive forces between the polymer-coated cells does not permit them to approach sufficiently close for bridging by the polyelectrolyte to take place.

Stoker: A relevant piece of information is in the discrepancy between different reports concerning the difference in coupling between normal and tumour cells. Loewenstein and Kanno (1967) reported that hepatomas show a loss of coupling compared to normal tissues whereas D. D. Potter and E. J. Furshpan find no difference in coupling between normal cells and tumour virus transformed cells (Furshpan and Potter, 1968). This discrepancy may be due to the fact that the results of Loewenstein and his colleagues were obtained with tumours that had developed *in vivo*, whereas Potter and Furshpan's data were obtained with cells that had been transformed *in vitro*. Uncoupling may have a selective advantage *in vivo* but not *in vitro*. This could be tested by putting transformed cells, which show coupling, back into an animal and seeing whether the tumours that develop are uncoupled.

Allison: I am concerned that people might conclude that the close cell contacts which inhibit mitosis are the only mechanism by which growth inhibition and stimulation can occur. There must be situations, however, as in a liver after partial hepatectomy, where the social relations of the cells are not changed in any obvious way and yet the cells are stimulated to divide. Again, removal of circulating red cells or injection of erythropoietin cannot have any effect on the social relations of the cells in the bone marrow, yet division is stimulated there. We must envisage other mechanisms which are separate from density-dependent inhibition and can overcome it.

Roe: On a rather different point, we have not considered tumours which involve two types of tissue *ab initio*, such as a benign fibroadenoma or a scirrhous carcinoma of the breast. None of the homeostatic mechanisms that have been discussed at this meeting relates to the occurrence of complex neoplasms such as these.

Stoker: One cell can however give rise to tumours of enormously varying transplantability.

Roe: I agree, but this can hardly be the explanation of mammary fibroadenoma induction, as by Huggins' technique (Huggins, Briziarelli and Sutton, 1959). The fibroadenoma of the breast is an organized tumour involving two embryologically different types of tissue, mesoderm and ectoderm. The smallest tumours, like normal mammary tissue, contain both elements, and both elements proliferate in concert as tumours grow.

Allison: Where a genetic marker in tumour cells has been studied, for example glucose 6-phosphate dehydrogenase (Linder and Gartler, 1965), it has so far suggested a clonal origin of each tumour from a single cell or small number of cells.

Bergel: May I consider an even lower level of organization? Dr Forrester mentioned coulombic forces, and I take it that van der Waals and London forces also might act at certain distances. A point that is almost lost sight of when one comes to larger biological integrated units is that the energy level of attracting or repelling forces is of a different order. When one reaches ionic interaction one is relatively speaking in a different kind of world from that in which polar or London effects are the dominant forces. It may be useful to consider the Oppenheimers' (Oppenheimer, Oppenheimer and Stout, 1948) incorporation of completely inert surfaces of cellophan or nylon films and the subsequent production of sarcomas. It is not impossible that this is due to irregularities or disturbances of a homeostatic system regulating that part of the tissue. I believe that the study of relatively simple chemical events, particularly in the field of surface chemistry, might throw light on the biological problems of cell surface interaction and contact inhibition.

Jacques: The interest of surface physics and chemistry is further illustrated by the fact that cellular locomotion ceases not only when the mobile cell meets another cell (contact inhibition), but also when it simply adsorbs inducers of pinocytosis (Chapman-Andresen, 1957, 1958) or when it sticks to and spreads on a suitable substratum in surface phagocytosis (Robineaux and Pinet, 1960).

REFERENCES

BOREK, C., and SACHS, L. (1966). *Proc. natn. Acad. Sci. U.S.A.*, **56**, 1705.
CHAPMAN-ANDRESEN, C. (1957). *Expl Cell Res.*, **12**, 397.
CHAPMAN-ANDRESEN, C. (1958). *C.r. Trav. Lab. Carlsberg, ser. Chim.*, **31**, 77.
FURSHPAN, E. J., and POTTER, D. D. (1968). In *Current Topics in Developmental Biology*, vol. 3, ed. Moscona, A. A., and Monroy, A. New York: Academic Press.
HUGGINS, C., BRIZIARELLI, G., and SUTTON, H. (1959). *J. exp. Med.*, **109**, 25.
LINDER, D., and GARTLER, S. M. (1965). *Science*, **150**, 67.
LOEWENSTEIN, W. R., and KANNO, Y. (1967). *J. Cell Biol.*, **33**, 225–234.
OPPENHEIMER, B. S., OPPENHEIMER, E. T., and STOUT, A. P. (1948). *Proc. Soc. exp. Biol. Med.*, **67**, 33.
ROBINEAUX, R., and PINET, P. (1960). *Ciba Fdn Symp. Cellular Aspects of Immunity*, pp. 5–43. London: Churchill.

REGULATING SYSTEMS IN CELL CULTURE

M. G. P. STOKER

Imperial Cancer Research Fund, London

THIS paper will be concerned with growth regulation in cultured fibroblasts. Such cells and their transformed derivatives have been extensively used, particularly by virologists and oncologists, whose greed has led to excellent techniques for rapid and unregulated growth of the cells. In recent years however there has been an increasing interest in the control of cell division, and there are now some useful experimental systems for inhibiting and initiating growth at will, although their relevance to the homeostatic systems operating *in vivo* is as yet unknown.

The systems I shall discuss are (1) density dependent, or contact inhibition, (2) serum factor dependence, and (3) anchorage dependence. It is possible, and indeed likely, that these are all aspects of the same control system and their relationship will be considered.

Though it is cell division which is of final relevance in the study for example, of developmental biology or cancer, it now seems clear that the principal controls operate very early in the cell cycle, soon after the previous division. With few exceptions, non-growing but competent cells are in the G_1 phase (or perhaps the G_0 phase, discussed elsewhere in this symposium, p. 9). Release from inhibition first initiates the series of synthetic processes, including DNA synthesis, which finally leads to the production of two daughter cells. RNA and protein synthesis are reduced in density-dependent inhibition at least (Eagle and Levine, 1967), and release leads first to RNA synthesis (Todaro, Lazar and Green, 1965): control may therefore be at the transcription level.

DENSITY DEPENDENT INHIBITION (CONTACT INHIBITION) OF GROWTH

Fibroblasts grow on rigid surfaces of glass or suitable plastic, but the rate of cell division is markedly dependent on the initial density of cells or the density achieved as multiplication proceeds. With typical cell types, in conventional culture conditions, growth is slow at less than 10^4 cells per cm^2, maximal between 10^4 and 10^5, and inhibited at more than about 3×10^5 cells per cm^2. This suggests that the effects of density are complex, but we shall restrict consideration to the changes at the higher densities which are obviously more likely to be related to *in vivo* conditions. The density at which growth is inhibited is referred to as the "saturation density" and it often corresponds to a confluent monolayer with all cells in mutual contact. Since contact is known to inhibit movement, it is suggested that growth is also subject to contact inhibition. When the density is reduced growth recommences, to regain the saturation density; so the inhibition is fully reversible. With certain cell lines, such as the specially selected 3T3 line, mutual contact and inhibition of growth is achieved with thin sheets of cells at much lower densities, of about 5×10^4 cells per cm^2 (Todaro and Green, 1963).

Under standard conditions of medium supply most tumour cells and cells transformed *in vitro* will grow to higher saturation densities than normal cells, and so are apparently less sensitive to the control mechanism. It was shown in our laboratory however that certain transformed cells remain sensitive to inhibition when in contact with non-growing normal cells (Stoker, 1964; Stoker, Shearer and O'Neill, 1966): hence the suggestion that transformed cells could receive and respond to some sort of inhibitory signal which they could not themselves initiate or transmit. It has subsequently been shown however that this sensitivity to normal cell inhibition is not universal but varies between different types of transformed cell (Eagle, 1965; MacIntyre and Pontén, 1967; Pontén and MacIntyre, 1968). Recently Pollack, Green and Todaro (1968) have isolated mutants from both normal and virus-transformed 3T3 cells to form a whole hierarchy with a wide variation in saturation density (irrespective of monolayer formation) and also variation in sensitivity to inhibition by co-cultivation with non-growing cells. This suggests that the inhibition is not a simple "touch and stop" phenomenon but must be expressed in quantitative terms.

The view that contact was the critical requirement for inhibition of growth was originally supported by reports that replenishment of the medium in dense inhibited cultures did not restimulate cell division and that old medium from such cultures was adequate for growth of cells at lower density. More recent work however has shown that contact is not the sole determining factor because the saturation density is, after all, dependent upon the supply of available medium (Stoker and Rubin, 1967). By increasing the relative volume of the medium per unit area of culture, or the rate of change of the medium, multilayers of cells can be obtained. Using perfusion cultures, for example, Kruse and Miedema (1965) were able to produce normal fibroblast cultures 5–6 cells thick. The continued growth is now thought to be due to factors in the serum, which will next be discussed.

SERUM FACTOR DEPENDENCE

In their early work with 3T3 cells Todaro, Lazar and Green (1965) found that addition of fresh serum to non-growing confluent cultures restimulated DNA synthesis in a proportion of cells, and they attributed this to a serum factor, which released cells from contact inhibition. The effect of serum on 3T3 cells has more recently been investigated by Holley and Kiernan (1968) who have reported that the number of cells produced is directly proportional to the amount of serum present, and is not affected by contact. They also reported that medium could be depleted of the factor responsible by exposure to growing cells. This last result is at variance with earlier reports that medium was not depleted but the experimental conditions are not easily comparable, and some of the discrepancy may be due to the complication that cultured cells may themselves release serum-like proteins into the medium (Rubin and Williams, 1968). Holley and Kiernan have isolated a pronase-sensitive and rather heat-resistant fraction corresponding to a substance of about 100000 molecular weight, which is present in serum, and which they believe to be responsible for limited growth in cultured normal cells.

Holley and Kiernan therefore suggest that the inhibition of growth in confluent cultures of 3T3 cells is due not to contact but to the accident that conventional medium containing 10 per cent serum contains enough growth factor for a cell yield corresponding to a monolayer of cells but not more.

SHORT-RANGE EFFECTS

These experiments at least show that the serum factor (or factors) is of utmost importance and must be considered in all growth control phenomena. Nevertheless, there remains strong evidence that some sort of short-range interaction between the cells limits growth, and that alterations in the medium through depletion of growth factors or addition of inhibitory factors cannot explain all the data.

In the earlier work on transfer of inhibition to transformed cells from confluent mouse embryo cultures covering a limited area of surface, it was found that the inhibitory effect was lost if the added transformed cell lay on the bare surface, even if it was only a few cell diameters distant from the edge of the inhibitory layer of normal cells and in the same medium. The inhibitory region was not extended beyond the edge of the confluent cell sheet by flow of medium in either direction (Stoker, Shearer and O'Neill, 1966).

Much of the more recent work with 3T3 cells has also emphasized this strong topographical effect. For example, a scratch in a non-growing confluent layer of 3T3 cells is followed by emigration of cells into the bare area, and Todaro, Lazar and Green (1965) have shown that thymidine incorporation is mostly limited to these free cells, in contrast to those in the undisturbed confluent sheet exposed to the same medium.

Recently Dulbecco and I examined the events in scratched 3T3 cell sheets by combining time-lapse cinematography with autoradiography, in order to determine the prior history and particularly the topographical features of particular cells in which thymidine incorporation was later observed to begin (Dulbecco and Stoker, 1969). Briefly we found that the cells started to move from the edge of the original sheet as a mass, remaining in contact and preserving their approximate relationship with one another, until they had moved about 100 μm from the edge of the original edge, when they became free and moved individually and randomly. Cells showing thymidine incorporation and mitosis eventually accumulated in the centre of the scratch among the free cells, as reported by Todaro, Lazar and Green (1965).

Thymidine pulse experiments however showed that DNA synthesis began 12 to 16 hours after the cells migrated over the original edge, whether or not they were finally free moving, or still in the zone of mass movement and in mutual contact. We concluded that the primary event which leads to induction of growth occurs at or soon after the first movement of cells from their original position into the bare area. Since the cells initially move *en masse*, this does not require complete loss of contact or communication (coupling) with neighbouring cells, or even gross alteration of topographical relationship.

Convection of the medium was rapid and there was no evidence of a static layer at, or close to the cell surface, so it is unlikely that movement through a free extracellular diffusion gradient was responsible for the induction. A concentration gradient through which the cell moves may still be important but it must be attached to the surface or be intracellular through cytoplasmic connexions.

Sections of confluent monolayers of 3T3 cells show considerable overlapping of cells (Todaro, Green and Goldberg, 1964). The inhibition and its release may depend on a change in the relative area of cell–cell contact compared to cell–medium contact, or to a change in the nature of the contact or type of connexion.

These experiments were carried out with unchanged medium under conditions in which the serum factor had been reduced to the extent that DNA synthesis was low in the confluent cell sheet. The interrelationship of serum factor and cell interaction will be considered later.

ANCHORAGE DEPENDENCE

We turn now to a type of growth regulation which at first appears to be unrelated to those just considered. As already described, normal fibroblasts will grow when spread on the rigid surface of a glass or plastic culture vessel under suitable conditions of density and medium supply. They will not grow however when suspended in fluid or semi-solid medium under conditions which nevertheless permit the growth of many virus-transformed or tumour cells. We have therefore studied the inhibition of suspended normal cells trapped in agar, or in methyl cellulose gels, from which they can be quantitatively recovered (Stoker et al., 1968).

When exponentially growing normal cells from surface cultures are first placed in suspension they remain as spheres. The mean cell volume increases for 24 hours, at about the same rate as in transformed cells, and up to 40 per cent of cells may divide once. During this time thymidine incorporation also continues in nearly half the cells. Thereafter growth and DNA synthesis slow down or stop. Nevertheless the viability of the cells, judged by their ability to grow into clones after reattachment to a surface, remains unaltered for at least a week in suspension, so they are competent. When the cells are allowed to anchor to a surface in the presence of thymidine and colchicine, incorporation begins again after about 8 hours and is followed by mitosis commencing 10–12 hours later. All the mitoses are labelled and so the inhibition is assumed to occur between mitosis and the S phase, probably in the G_1 phase of the cycle. We assume that the continued growth of some cells for the first 24 hours in suspension is due to the fact that the asynchronous population contains individuals which have passed the block in the cycle.

Inhibition might operate on each cell individually, or might be a property of the culture as a whole; for example, the suspending medium might be inhibitory, or the intercellular distance a critical factor. To investigate this, small lengths of glass fibre were introduced into suspension cultures. Cells which were close enough to attach spread on the fibres and grew into spindle-shaped colonies. Neighbouring cells in the same medium which did not attach remained inhibited. This shows that there is no general inhibition in the culture and that each cell behaves independently of the others. It also shows that elongation on a rigid surface (in this case about 100 μm long and 24 μm diameter) is associated with induction of growth, and might be causal. For example, spreading is likely to alter the membrane by changing the surface area or configuration and this might alter membrane function.

There are several ways in which suspended normal cells may be induced to grow without anchorage. For example, the BHK21 line of hamster cells which we have studied are induced to grow by infection with a high multiplicity of polyoma virus, irrespective of stable transformation (Stoker, 1968). We assume that there is a product of the viral genome which directly affects the regulation mechanism.

Macpherson (1968) has shown that growth in suspension is induced by large doses of insulin, and this is of interest because of the enhancement of glucose uptake. Transformed cells which can grow in suspension without added insulin have also a high glucose uptake.

Finally, growth is also induced by increasing the concentration of serum (Clarke and Stoker, 1969). Preliminary studies indicate that foetal serum is more active than post-natal serum and the limited diffusion suggests that the factor responsible is associated with a large molecule. Insulin itself in serum is insufficient in quantity to account for the effect unless its activity is enhanced by association with other molecules. The relationship to Holley's factor, or to fetuin, studied by Puck and his colleagues (Fisher, Puck and Sato, 1958), is not yet known.

INTERRELATIONSHIP OF GROWTH CONTROL PROCESSES

It is abundantly clear that the three phenomena discussed must be interrelated in some way. Apart from anything else the sensitivities of cells to density, to serum factor, and to suspension are all affected by the very limited genetic information specified by a small DNA tumour virus, such as polyoma or SV40. It is likely that one gene product is involved and produces all these effects through a common pathway.

The attached cells at high density in surface cultures and spherical suspended cells do not at first sight have much in common. However, lowering the cell density to induce growth provides more available substrate for each cell. This may lead to increased spreading with a change in relative surface area or in the proportion of surface exposed to medium or substrate. This extension of spreading may be similar to that found when suspended cells spread freely on a surface. A similar change in cell shape is probably common to both types of induction.

To link the serum factor with density inhibition one could suppose that close packing of cells at high density interferes with uptake, by obstruction of sites at cell-cell interfaces. However, it is impossible to distinguish between a passive effect of this type, and an alternative active change, initiated by cell interaction, which itself changes the uptake of or response to a serum factor. Similar considerations apply to suspended cells, which may be unable to grow through passive obstruction of uptakes sites, perhaps by membrane folding, or alternatively through an active process which raises the requirement for serum factor. In either case growth could be initiated either by increasing serum factor concentration with no change in cell shape or topography, or by increasing the rate of uptake through a change in the cell with no addition of fresh serum.

Further progress will be limited until the factors in serum which affect growth can be isolated and identified. Ultimately a solution may require an understanding of membrane structure and function. At present the data are not available to allow the general simplification which we would all desire for this complex business.

SUMMARY

The growth of cultured fibroblasts may be controlled experimentally in several ways:

(1) Growth is inhibited at high density. The operative interaction between the cells occurs over a very short distance and probably involves physical contact. It is not known whether such contact involves the initiation of an active mechanism or merely inhibits by a passive process such as blockage of sites or metabolite uptake.

(2) Growth may also be controlled by a protein factor or factors present in sera. It has been shown that in certain cell systems such factors become limiting under conventional culture conditions, but they may play an important, though hitherto unrecognized role in density dependence and other forms of regulation.

(3) Most normal cells are unable to grow in suspension as spheres but depend on anchorage to a rigid surface, with consequent spreading of the cell and probable change in conformation of the plasma membrane.

In all three forms of growth regulation, inhibited cells are probably held in the G_1 phase of growth. Moreover, they are all affected by infection of the cell with small DNA-containing tumour viruses with very limited genetic information. It is therefore probable that a common mechanism is involved, which may also operate for the regulation of cell growth *in vivo*.

REFERENCES

CLARKE, G. D., and STOKER, M. G. P. (1969). Unpublished.
DULBECCO, R., and STOKER, M. G. P. (1969). Unpublished.
EAGLE, H. (1965). *Israel J. med. Sci.*, **1**, 1220–1228.
EAGLE, H., and LEVINE, E. M. (1967). *Nature, Lond.*, **213**, 1102–1106.
FISHER, H. W., PUCK, T. T., and SATO, G. (1958). *Proc. natn. Acad. Sci. U.S.A.*, **44**, 4–10.

Holley, R. W., and Kiernan, J. A. (1968). *Proc. natn. Acad. Sci. U.S.A.*, **60**, 300–304.
Kruse, P. F., and Miedema, E. (1965). *J. Cell Biol.*, **27**, 273–279.
MacIntyre, E., and Pontén, J. (1967). *J. Cell Sci.*, **2**, 309–322.
Macpherson, I. A. (1968). Unpublished.
Pollack, R. E., Green, H., and Todaro, G. J. (1968). *Proc. natn. Acad. Sci. U.S.A.*, **60**, 126–133.
Pontén, J., and MacIntyre, E. H. (1968). *J. Cell Sci.*, **3**, 603–613.
Rubin, H., and Williams, J. (1968). Unpublished.
Stoker, M. (1968). *Nature, Lond.*, **218**, 234–238.
Stoker, M. G. P. (1964). *Virology*, **24**, 165–174.
Stoker, M., O'Neill, C., Berryman, S., and Waxman, V. (1968). *Int. J. Cancer*, **3**, 683–693.
Stoker, M. G. P., and Rubin, H. (1967). *Nature, Lond.*, **215**, 171–172.
Stoker, M. G. P., Shearer, M., and O'Neill, C. (1966). *J. Cell Sci.*, **1**, 297–310.
Todaro, G. J., and Green, H. J. (1963). *J. Cell Biol.*, **17**, 299–313.
Todaro, G. J., Green, H., and Goldberg, B. D. (1964). *Proc. natn. Acad. Sci. U.S.A.*, **51**, 67–73.
Todaro, G. J., Lazar, G., and Green, H. (1965). *J. cell. comp. Physiol.*, **66**, 325–334.

DISCUSSION

Abercrombie: You mentioned the problem of applying this kind of work to *in vivo* situations; it might be worth mentioning some work which seems to bear on this. J. S. Santler and I (Santler and Abercrombie, 1957) studied a system in which we could prepare tissues *in vivo* (we used peripheral nerve) at different cell densities and apply to them pulses of growth stimulator in different amounts. We found that the final density of the population depended on the amount of stimulator put into the system, which seems to parallel the effect of the serum factor *in vitro*. However, there was also a density effect; in essence, the denser the initial population, the lower the specific growth rate, which again seems to parallel what one finds in tissue culture.

Mason: Healy and Parker (1966) have found two serum factors necessary for good growth of newly explanted mouse embryo cells. One was a relatively low molecular weight (40–50000) α_1-acid-glycoprotein or orosomucoid which was taken up by the cells and exhausted in the medium. The other was a high molecular weight substance, an α_2-macroglobulin (molecular weight 850000), which also decreased in the medium but was not specific; it could be replaced by substances like dextran or Ficoll. Could these observations explain two effects of serum: (1) an effect which one might call metabolic which can be exhausted in the medium; (2) an effect through cell contact which is less specific and physical in nature?

Stoker: There must be many factors in the serum. The Holley factor seems to be a large molecule.

Elsdale: We have investigated the effects of serum in a slightly different way to Holley, using human diploid fibroblasts (unpublished observations). Plated out at about two-thirds confluent in Petri dishes in serum-free medium, the cells remain non-growing and healthy for about a week. After two or three days in serum-free medium the cells were stimulated with different amounts of serum at low concentrations ranging from 0·2 to 5 per cent. The growth obtainable without further medium change was measured. The amount of growth was related to the amount of serum made available to the cells. The growth-promoting material was excluded by G-50 Sephadex.

Regarding Holley and Kiernan's (1968) suggestion that inhibition of growth is due not to contact but to the traditional use of 10 per cent serum, we find that using 10 per cent serum and cell lines established from human foetuses, lung fibroblasts routinely grow to about 10×10^6 cells/50 mm dish, muscle fibroblasts to around 5×10^6, and gut fibroblasts to low densities, sometimes to no more than 2×10^6. What is controlling growth? Mixed cultures initiated with equal numbers of densely growing lung fibroblasts and sparsely growing gut fibroblasts grow to densities only a little higher than the gut cells alone. From general experiments of this type it is clear that sparsely growing cells inhibit dense growers.

Stoker: This agrees with the findings of Pollack, Green and Todaro (1968) that variants of 3T3 with higher saturation densities are inhibited by those with lower saturation densities. Those with the highest densities, however, are not subject to this inhibition.

Möller: This is in agreement with the existence of humoral factors. There is probably also cellular recognition, however. A few findings do suggest that; for instance the induction of tumours by plastic films introduced subcutaneously. The mechanism is unlikely to be chemical induction or an effect on soluble circulating factors. If one simply separates fibroblasts from each other, neoplasia is caused to develop. Another example is allogeneic inhibition, which has been demonstrated only when cells in suspension have been inoculated; it cannot be demonstrated with skin, which is grafted as a continuous sheet and where the only contact between donor and host is the outer layer. This suggests that cells have recognition mechanisms for identity, which operate when there is close contact with something identical. Non-identities lead to growth inhibition. Another example of an analogous situation is amoebae which don't eat each other but do eat other foreign substances and micro-organisms. Thus they can distinguish themselves from others.

Iversen: On this question of cells recognizing each other and not eating other cells of the same type, Bessis (1963) irradiated small areas of white blood cells *in vitro* with micro-beams of ultraviolet light; the cells then became sick and died. Then macrophages immediately started eating the dead cells. Some change must have occurred to enable them to recognize which was a dead cell and which was a living one. But even before the cells were dead, when they were only damaged, the macrophages collected around them, as if they could "smell" the sick cells and detect which were going to die and which they could expect to eat in the near future. There must thus be a mechanism by which the cells not only recognize each other but also recognize whether they are healthy or not.

Jacques: Although macrophages show a definite predilection for old, sick or dead cells it should not be concluded that healthy cells are never taken up by larger ones. In cultured thymus for instance, active thymocytes are frequently engulfed by epithelioid cells, move around in their cytoplasm and usually escape alive and well after some time (Törö, 1968).

Abercrombie: It seems to me that the crucial observation on plastic film carcinogenesis is that roughening of the surface by scratching it greatly diminishes the incidence of cancer (Bates and Klein, 1966), and that it isn't a matter of the cell disliking a non-self material but disliking a non-self surface in a particular conformation, which in this case is a flat plane.

Bergel: But is it not true that even the powders of some of these plastics when introduced into the animal are still carcinogenic, though weakly so, according to Druckrey and Schmahl, and to Oppenheimer and co-workers (see Clayson, 1962)?

Roe: Many polymers, although they consist mainly of chemically inert materials, are contaminated with a wide variety of chemical agents, some of which are carcinogens or closely related to carcinogens. Not all polymers are inactive when powdered or broken up into smaller pieces. It is important, therefore, that those who base arguments on data derived from the induction of cancer by polymers should be sure that the implanted materials are active because of their physical rather than chemical characteristics.

Lamerton: What happens *in vivo* to the shape of the cells in contact with a smooth plastic surface?

Allison: In one report the cells in immediate contact with the plastic formed a sheet along its surface, which is exactly what you would expect.

Stoker: You could say that it is very similar to the anchorage-dependent situation, if you assume that the unattached cells, although not necessarily rounded, are still unanchored.

Allison: In another experiment holes were made in the plastic sheet, and this greatly diminished its effectiveness as a carcinogen.

Wolpert: Dr Elsdale and Dr Jones reported a long time ago that embryonic myoblasts would not form muscle unless they could spread (Elsdale and Jones, 1963).

O'Meara: To go back to Professor Stoker's observations, the system of repair that occurs when you wound your multilayered culture of cells is very similar to what takes place in repair of the epidermis. When a portion of the epidermis is denuded the cells which multiply are a little back from the edge, and are pushed out over the intervening space to cover up the area that has been denuded. This seems to be a very similar process to what you find in culture.

On the effect of serum: serum of course has been used traditionally in cell cultures. In bacterial cultures, serum may either be nutritional in its effect or may have the effect of neutralizing agents which are antagonists to cell growth.

Another point which arises in relation to the addition of high molecular weight proteins to media, especially when silicates and glass are also introduced, is the fact that the materials adsorb proteins on to their surfaces and thereby become altered. Not only does the surface become altered, but the agent which is adsorbed on to it is changed. This is known as "contact activation". So we must recognize the fact that we can coat surfaces and the coat may actually be altered by the contact.

Stoker: Unfortunately this doesn't explain the effect of difference in the size of the particles.

O'Meara: Not unless it is a question of total amount. In that event it might. You say that insulin will allow cells to grow in free suspension. Could this possibly be an effect of zinc, because I don't think you can get insulin that is free of zinc?

Stoker: I don't know.

Bergel: An effect on blood plasma by surfaces such as glass has been demonstrated by Keele (1957); he brings plasma in contact with glass surfaces and then injects the fluid intradermally or applies it to the base of a blister, and pain is produced. This indicates a possible alteration of the plasma itself, through an activation of a proteolytic enzyme, according to Keele.

REFERENCES

BATES, R. R., and KLEIN, M. (1966). *J. natn. Cancer Inst.*, **37**, 145.
BESSIS, M. (1963). *Ultraviolet irradiation of white blood cells.* A film produced by Science Film, 22 Bd. Victor Hugo, Paris.

Clayson, D. B. (1962). *Chemical Carcinogenesis*, pp. 101–109. London: Churchill.
Elsdale, T., and Jones, K. (1963). *Symp. Soc. exp. Biol.*, **17**, 257–273.
Healy, G. M., and Parker, R. C. (1966). *J. Cell Biol.*, **30**, 359–553.
Holley, R. W., and Kiernan, J. A. (1968). *Proc. natn. Acad. Sci. U.S.A.*, **60**, 300–304.
Keele, C. A. (1957). *Proc. R. Soc. Med.*, **50**, 477.
Pollack, R. E., Green, H., and Todaro, G. J. (1968). *Proc. natn Acad. Sci. U.S.A.*, **60**, 126–133.
Santler, J. S., and Abercrombie, M. (1957). *J. cell. comp. Physiol.*, **50**, 429.
Törö, I. (1968). In *Cell Biology*, p. 85, ed. Dustin, P., Stebbing, N., and Bostock, C. J. Amsterdam: Excerpta Medica Foundation, International Congress series, **166**.

METABOLIC COOPERATION BETWEEN CELLS

J. H. SUBAK-SHARPE

Institute of Virology, University of Glasgow

WHEN cultured mammalian cells come into contact their individual metabolism is modified by the exchange of elaborated materials. A description of this phenomenon, which has been termed "metabolic cooperation" (Subak-Sharpe, Bürk and Pitts, 1966, 1969), will now be given and its nature examined. First the evidence is presented that the genetically variant cells used cannot carry out certain biosynthetic steps. Next follows a detailed account of metabolic cooperation. Various aspects of the phenomenon and possible explanations at the molecular level are then discussed and finally some wider implications are mentioned.

THE CELLS IN PURE CULTURE

The origin of the hamster fibroblast line BHK21 clone C13 has been described (Macpherson and Stoker, 1962), as has the derivation of the polyoma virus transformed cell line PyY (Stoker and Macpherson, 1964) and its biochemical variant PyY/ *TG1* (Subak-Sharpe, 1965), which lacks detectable inosinic pyrophosphorylase (E.C. 2.4.2.8. IMP, GMP: pyrophosphate phosphoribosyl transferase) activity.

From PyY and PyY/ *TG1* a number of other variant cell lines have been isolated, as shown in Fig. 1.

Cell lines used for the present study have been tested for their ability to incorporate into nucleic acid various preformed purines, nucleosides and also nucleotides supplied in the medium. The experimental procedure adopted was as follows: 10^5 monodisperse cells were seeded on to 60 mm glass Petri dishes containing three or four 13 mm glass coverslips under 5 ml of EFC medium (9 parts modified Eagle's medium, 1 part foetal calf serum). Then either 10 μc of [^{3}H]-labelled or 1 μc of [^{14}C]-labelled purine etc. was added and the dish incubated for 24 hours in a gassed incubator at 37°C. The coverslips were then fixed, extracted with trichloroacetic acid and processed through radioautography as described previously (Subak-Sharpe, Bürk and Pitts, 1969). The results of one typical experiment, which are summarized in Table I, are best considered in the light of the relevant purine metabolic pathways in mammalian cells, shown in simplified form in Fig. 2.

Table I demonstrates that PyY cells, which in this context can be regarded as the parental wild type, have the enzymic equipment to incorporate any

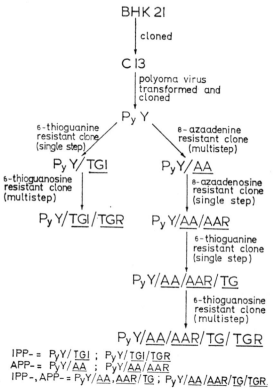

Fig. 1. The derivation of biochemically marked lines from BHK 21 clone C13.

of the purine compounds into their nucleic acid. PyY/*TG1*/*TGR* and PyY/*AA*/*AAR*/*TG*/*TGR* cells were unable to incorporate any of the hypoxanthine or guanine-containing precursors except trace amounts where shown.

Thus both cell lines lack detectable inosinic-guanylic pyrophosphorylase activity and inosine-guanine kinase activity. Similar observations have been made with human Lesch-Nyhan cells (Friedmann, Seegmiller and Subak-

Sharpe, 1969) which had previously been shown to lack virtually all inosinic-guanylic pyrophosphorylase activity (Seegmiller, Rosenbloom and Kelley, 1967). PyY/*TG1*/*TGR* cells can therefore be referred to as IPP⁻.

PyY/*AA*/*AAR* and PyY/*AA*/*AAR*/*TG*/*TGR* cells could not incorporate adenine or adenosine. When adenylic acid was supplied a few silver grains

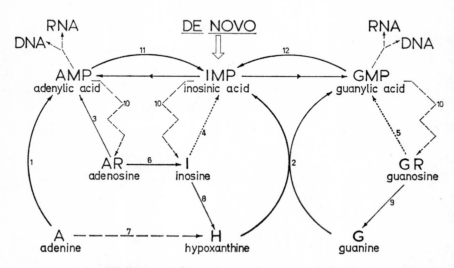

FIG. 2. A simplified diagram of the pathways of purine metabolism in mammalian cells.

The enzymes mediating the various numbered steps are:
1. adenylic pyrophosphorylase (2.4.2.7).
2. inosinic-guanylic pyrophosphorylase (2.4.2.8).
3. adenosine kinase (2.7.1.20).
4, 5. inosine and guanosine kinase (2.7.1.f). It is still uncertain whether a separate kinase or both kinases exist in mammalian cells.
6. adenosine deaminase (3.5.4.4).
7. adenine deaminase (3.5.4.2).
8, 9. hypoxanthine-guanine phosphorylase (2.4.2.1).
10. 5′-nucleotidase (3.1.3.5).
11. 5′-adenylic deaminase (3.5.4.4).
12. guanylic reductase.

above background levels were observed above the tested PyY/*AA*/*AAR*/*TG*/*TGR* cells, indicating that only trace amounts had been incorporated into nucleic acid. Both cell lines thus lack adenylic pyrophosphorylase activity and adenosine kinase activity; and the PyY/*AA*/*AAR*/*TG*/*TGR* cells are devoid of all the enzyme activities looked for. For convenience PyY/*AA*/*AAR* cells will be referred to as APP⁻ and PyY/*AA*/*AAR*/*TG*/*TGR* cells as IPP⁻, APP⁻.

Table I

CAPABILITY OF VARIOUS CELL LINES TO INCORPORATE PREFORMED PURINES, SUPPLIED IN THE MEDIUM, INTO RNA OR DNA

Label	^3H	^3H	^3H	^3H	^3H	^3H	^3H	^{14}C
Compound	G	GR	GMP	H	I	A*	AR	AMP
Specific activity mc/mM	70	500	410	634	1050	3920	2700	17
Cell line								
PyY	+++	+++	+++	+++	+++	+++	+++	+++
PyY/TG1/TGR	—	—	— trace	—	—	+++	+++	+++
PyY/AA/AAR	+++	+++	+++	+++	+++	—	—	not done
PyY/AA/AAR/TG/TGR	—	—	— trace	—	—	—	—	— trace

* 100 μM cold hypoxanthine + 10 μM cold adenine were also added.
Abbreviations used: Guanine, G; guanosine, GR; guanylic acid, GMP; hypoxanthine, H; inosine, I; adenine, A; adenosine, AR; adenylic acid, AMP.
+++ heavy incorporation; — no incorporation; — trace, very few silver grains above background.

The virtual failure to incorporate guanylic acid by IPP$^-$ and adenylic acid by APP$^-$ cells confirms that nucleotides cannot normally enter mammalian cells from the medium (Leibman and Heidelberger, 1955). This is further supported by the following experiment. Cells were fed freshly sonicated IPP$^+$ cell extract and labelled hypoxanthine, but the cells failed to incorporate [^{3}H]hypoxanthine (Subak-Sharpe, unpublished observations). Apparently nucleotides are split at the cell membrane, probably by 5'-nucleotidase or possibly by 5'-nucleotide hydrolase, to purine or nucleoside which can then enter the cell and be incorporated into nucleic acid, provided that the cell is genetically capable (that is, possesses pyrophosphorylase activity). The relevance of this point will become apparent after the phenomenon of metabolic cooperation has been described.

THE CELLS IN MIXED CULTURE

Experiments using mixtures in the ratio of 300 IPP$^-$ cells to one IPP$^+$ cell have revealed that hypoxanthine was not only incorporated into nucleic acid by the rare IPP$^+$ cells as expected, but also, though to a lesser extent, by IPP$^-$ cells which were *in direct or indirect contact* with these IPP$^+$ cells. IPP$^-$ cells *not in contact* with IPP$^+$ cells, nor with IPP$^-$ cells themselves in direct or indirect contact with IPP$^+$ cells, did not incorporate hypoxanthine (Subak-Sharpe, Bürk and Pitts, 1966, 1969). This phenomenon was called "metabolic cooperation" and is defined as the process whereby the metabolism of cells in contact is modified (perhaps controlled) by exchange of elaborated materials. Metabolic cooperation is not to be confused with the quite different phenomenon of cell fusion described by Harris and his co-workers (1966).

By exposing IPP$^+$ and IPP$^-$ cell mixtures to labelled hypoxanthine at various times after plating (0–12, 12–24, 24–36, and 36–48 hours) and scoring the observed numbers of heavily labelled cells and lightly labelled cells in contact with them, it could be shown that the former were IPP$^+$ cells and the latter IPP$^-$ cells, and that metabolic cooperation could take place without being accompanied by cell division. The detailed experimental results have already been given (Subak-Sharpe, Bürk and Pitts, 1969) and will not be reproduced here.

Evidence has now been obtained that metabolic cooperation is not confined to cells lacking inosinic pyrophosphorylase, but is similarly observed in cells lacking adenylic pyrophosphorylase and adenosine kinase.

APP$^-$ cells and APP$^+$ cells were grown separately in pure culture and also mixed in the proportion of 300:1, using the experimental procedure

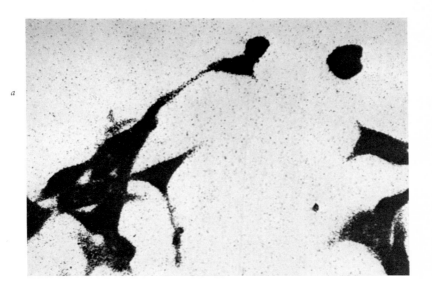

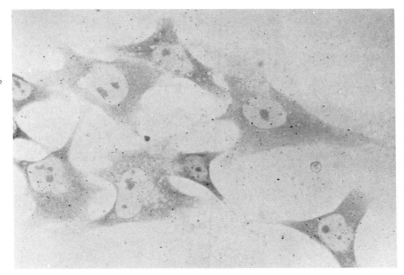

FIG. 3. Radioautography of (a) APP+ cells. Note the heavy silver grain deposition over the cells. (b) APP− cells.

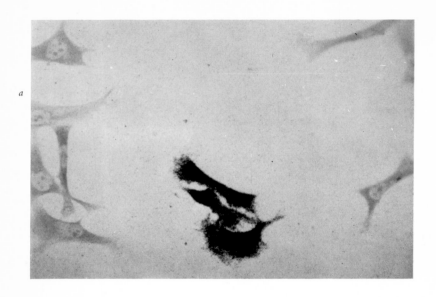

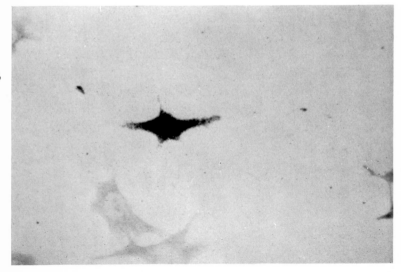

FIG. 4 (*a* and *b*). Radioautography of mixed APP⁺ and APP⁻ cells not in contact.

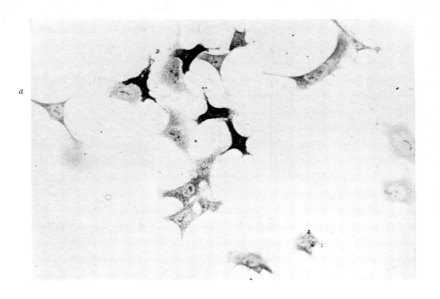

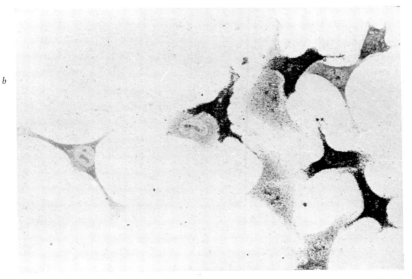

Fig. 5 (a). Radioautography of metabolic cooperation: a clone of APP⁺ cells in direct and indirect contact with nine APP⁻ cells. Seven further APP⁻ cells are not in contact and have not incorporated label. Low power.

(b) A higher magnification of the same group of cells.

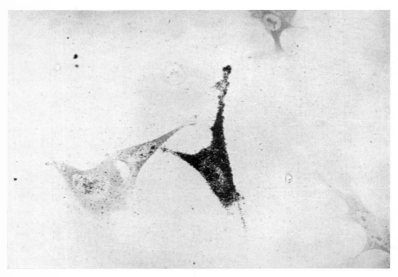

FIG. 5 (c). A clone of two APP+ cells in very localized contact with two APP− cells. Compare with the nearby APP− cells which have not made contact.

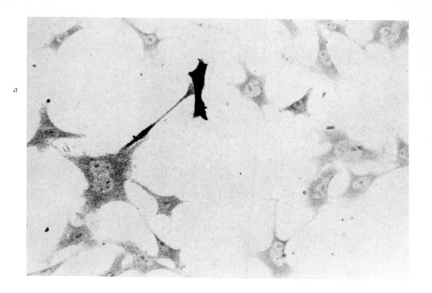

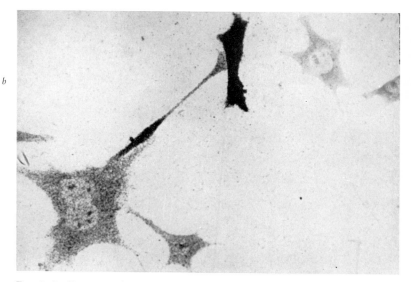

FIG. 6. Radioautographs illustrating the metabolic cooperation gradient.
(a) A clone of three APP+ cells in direct and indirect contact with several APP− cells. Compare with the APP− cells which are quite close but not in contact.
(b) A higher magnification of the same group of cells.

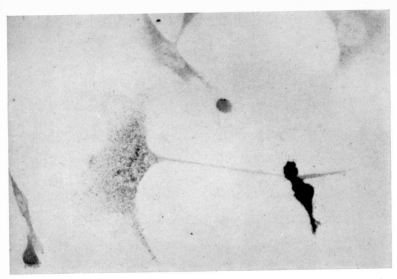

Fig. 7. Radioautography of metabolically cooperating cells which have moved a considerable distance apart, but a long thin cytoplasmic process still remains, proving that they have been in contact.

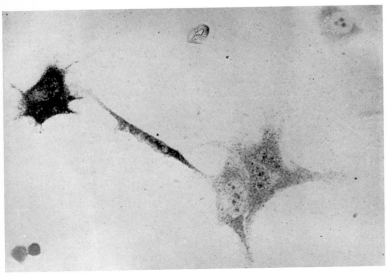

Fig. 8. Radioautography of metabolic cooperation. As in Fig. 6 note the distances between cooperators. Sometimes cells like the connecting APP⁻ cell fall off the glass. Note three nearby APP⁻ cells.

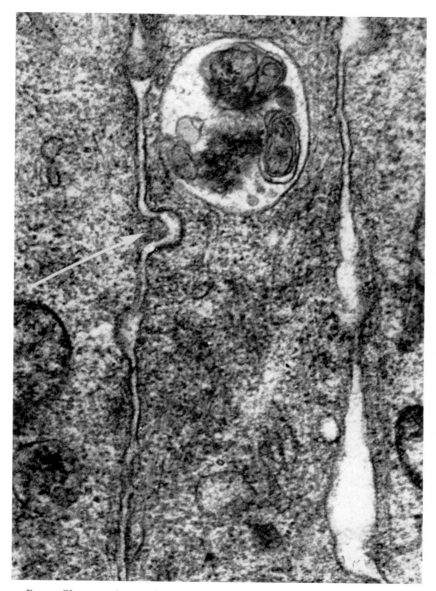

Fig. 9. Electron micrograph of a section through portions of two adjacent BHK C13 cells. Note the protrusion of one cell surface into a cup-like invagination of the adjacent cell (arrow). Fixed in 1·5 per cent glutaraldehyde and 1 per cent osmium tetroxide. ×80000.

FIG. 10. Electron micrograph of two adjacent BHK C13 cells demonstrating a possible mechanism of cytoplasmic exchange by a process akin to phagocytosis (see arrows). ×48000.

described earlier except that 10 μc of [³H]adenine (Radiochemical Centre, Amersham; specific activity, 3920 mc/mM) was the labelled purine added. Film was developed after 14 days' exposure at 4°C.

Fig. 3a shows APP⁺ cells; all have incorporated a great deal of label into their nucleic acid, as is demonstrated by the heavy deposition of silver grains above each cell. Just as observed previously with hypoxanthine, there is a slight variation in the actual amount of isotope incorporated per cell, which is probably at least partly due to their lack of synchronization. Fig. 3b shows APP⁻ cells which have not incorporated the labelled adenine into their nucleic acid. Silver grain deposition over the cells does not exceed the background level. These cells were clearly devoid of any detectable adenylic pyrophosphorylase activity.

Fig. 4a and b shows two typical examples of APP⁺ and APP⁻ cells in mixed culture in areas where contact has not been made between the APP⁺ cells which are heavily labelled and the several nearby APP⁻ cells which are totally devoid of label. Provided that cell contact has *not* taken place between APP⁺ and APP⁻ cells the phenotypes always correspond to the genotypes. Cross-feeding does not take place despite the relatively close proximity of the APP⁺ cells (Fig. 4a and b).

A totally different situation is found when APP⁻ cells are in microscopically observable contact with APP⁺ cells (Fig. 5a, b and c). The APP⁻ cells in contact with APP⁺ cells are now seen to have incorporated label into their nucleic acid, though always to a lesser extent. Moreover APP⁻ cells in *indirect* contact with APP⁺ cells (Fig. 5b)—that is, in contact with other APP⁻ cells which are themselves in contact with APP⁺ cells—also incorporate label. Under favourable circumstances a gradient of incorporation of label can be seen to go over many cells (Fig. 6a and b), but such a gradient is never steep—a finding which remains quite unexplained. Where the cells are allowed to grow for 24 hours before label is added, small clones of APP⁺ cells are frequently observed in contact with clones of APP⁻ cells, as is shown in Fig. 5a. Both cell types have divided once or twice and in addition the cells have moved so that sister cells have often become separated. Furthermore, new cell contacts are made and old ones unmade as time passes. Fig. 7 shows a case where two metabolically cooperating cells have moved apart but are still joined by a long and thin cytoplasmic process. This cell movement can make interpretation more difficult. In some instances a linking cell may come off the glass and be lost, leaving previously connected cooperators without any remaining, observable connexion. Fig. 8 illustrates what might well have been such a case, only the linking cell has fortunately not fallen away.

These findings of metabolic cooperation are exactly as previously observed with IPP+ and IPP− cells (Subak-Sharpe, Bürk and Pitts, 1966, 1969).

GENERALITY OF OCCURRENCE, RECIPROCITY AND SOME QUANTITATIVE ASPECTS

These observations establish that metabolic cooperation is not confined to the activity of a single enzyme, for adenylic pyrophosphorylase and inosinic pyrophosphorylase, though not biochemically purified, are genetically distinct. Of course it is possible that both enzymes belong to a special class—they may for example be membrane bound. (I am indebted to Dr F. Gros for the information that a collaborator of Dr J. Senez has found that the enzyme orotidylic pyrophosphorylase is membrane bound in bacteria.) In the case of inosinic pyrophosphorylase activity, metabolic cooperation has been demonstrated between lines of transformed hamster cells (Subak-Sharpe, Bürk and Pitts, 1966, 1969), between transformed and non-transformed hamster cells (Subak-Sharpe, Bürk and Pitts, 1966, 1969), between transformed hamster and normal mouse embryo cells (Stoker, 1967), and between human fibroblasts from normal individuals and from patients suffering from Lesch-Nyhan disease (Friedmann, Seegmiller and Subak-Sharpe, 1968). Thus metabolic cooperation is not a phenomenon peculiar to tumour cells or cells of unlimited growth potential, but is a normal characteristic of fibroblasts.

It has been shown by two types of experiment that metabolic cooperation is a true phenomenon of reciprocal exchange and not a one-way donor-to-recipient relationship.

First, a 1:1 mixture of $10^{4.7}$ IPP− and $10^{4.7}$ APP− cells was seeded and given either [^{3}H]adenine or [^{3}H]hypoxanthine for 24 hours and then processed for radioautography. Irrespective of which isotopically labelled purine had been used, metabolic cooperation was observed when cells were in contact, as described earlier (R. A. Roosa and J. H. Subak-Sharpe, unpublished results).

Secondly, a 1:1 mixture of $10^{5.7}$ IPP− and $10^{5.7}$ APP− cells was grown for 24 hours under conditions of close cell contacts, and then given either [^{3}H]adenine or [^{3}H]hypoxanthine for 2 hours. After radioautography, silver grains above the individual cells in randomly chosen fields were counted. Total average incorporation per 3000 cells was also calculated from cell counts and measurements of disintegrations/minute were made on three coverslips for each purine. Table II, extracted from the data of Bürk,

Table II
RADIOACTIVITY INCORPORATED BY CELLS ON COVERSLIPS MEASURED IN TWO WAYS*

	Disintegrations/minute per 3000 cells		Grains per cell	
	[³H]Adenine	[³H]Hypoxanthine	[³H]Adenine†	[³H]Hypoxanthine
APP+, IPP−	1122	5	31·9 (8995/282)	0·6 (166/262)
APP−, IPP+	7	762	0·6 (146/229)	51·0 (13981/274)
1:1 mixture [APP−, IPP+ / APP+, IPP−]	807	308	27·0 (7232/268)	24·9 (6687/269)
The half with low count			16·2 (2167/134)	12·3 (1666/135)
The half with high count			37·8 (5065/134)	37·5 (5021/134)
Excess of mix over 1:1 expectation	242	−76	10·7	−1·0

* Calculated from the data in Table I and Fig. 1 of Bürk, Pitts and Subak-Sharpe (1968).

† The fraction in brackets shows total grains over total cells counted.

Pitts and Subak-Sharpe (1968), summarizes the findings, which show clearly that virtually *all* cells had incorporated label into their nucleic acid irrespective of which isotopically labelled purine had been provided.

The results summarized in Table II also provide tentative answers to the following three questions. (1) Was the total isotope incorporated into nucleic acid by mixed cells only equal to that expected from the proportion of competent and variant cells, or was there an increase in the total amount incorporated by the mixed cells? (2) What was the relative amount of labelled purine incorporated in cooperator cells compared to that incorporated by the genetically competent cells in the mixed culture? (3) Was there an effect on the amount of label incorporated by competent cooperating cells?

The answer to the first question differs with the enzyme activity studied. In the case of inosinic pyrophosphorylase, the amount of [^{3}H]hypoxanthine incorporated by all the mixed cells is just about equal to that which the competent IPP$^+$ cells alone would have been expected to incorporate. In contrast the adenylic pyrophosphorylase activity of the mixed cells was very much higher than expected (66 per cent). At present it is not known why the two pyrophosphorylase systems should differ in this respect though one might speculate that the former may already be fully induced while the latter is not.

The answer to the second question is that cooperator cells incorporated 33 per cent (IPP) and 43 per cent (APP) of the labelled purine incorporated by the genetically competent cells in the mixed culture. Of course in 1:1 mixture conditions like these almost every non-competent cell has established contacts with several competent cells. When metabolic cooperation is observed between cells plated at high dilution the relative amount incorporated by cooperator cells appears to be much lower.

The answer to the third question again depends on the actual system. In the case of inosinic pyrophosphorylase the genetically competent IPP$^+$ cells in mixed culture incorporated about 75 per cent of what they did in pure culture, the cooperating IPP$^-$ cells accounting for the remaining 25 per cent. But with adenylic pyrophosphorylase the situation was different. The competent APP$^+$ cells in mixed culture incorporated 118 per cent—that is, nearly 20 per cent more—of what they did in pure culture, and the cooperating APP$^-$ cells accounted for a further 51 per cent. Thus metabolic cooperation in the APP system was accompanied by a considerable increase in total adenylic pyrophosphorylase activity.

These are preliminary answers as far as precise quantitation is concerned, for they depend on a number of necessary assumptions: that the cells plated with equal efficiency (if they did not, the actual relative numbers must be known); that the distributions of cooperator and competent cells touched but did not overlap (if they did, then the answer to question 2 is too low); and that low and high grain counts over cells were made equally accurately (if not, then the high grain counts were probably underestimates).

POSSIBLE MOLECULAR EXPLANATIONS

It is pertinent to consider what the molecular basis of metabolic cooperation could be. For the IPP cells, the observations are compatible with explanations based on the entry into IPP$^-$ cells of five different classes of molecule.

The first is radioactively labelled nucleotide *synthesized in the IPP$^+$ cells* (in other words, labelled IMP or GMP). This explanation is less attractive than appears at first sight, for neither AMP nor GMP appear to be able to enter these cells from the medium without loss of the phosphate group. Thus some unknown special mechanism of nucleotide transfer from cell to cell, either involving actual contact or necessitating very close proximity, would have to be postulated.

The second is radioactively labelled nucleic acid—any RNA or DNA—again synthesized in the IPP$^+$ cells.

The third is informational nucleic acid—messenger RNA or episomal DNA—actually coding for inosinic-guanylic pyrophosphorylase. This nucleic acid may or may not be labelled when it first enters the IPP$^-$ cells. It is important only that it is biologically active.

The fourth is polypeptide—this may be the enzyme inosinic-guanylic pyrophosphorylase itself, or only that polypeptide chain (if the enzyme is made up of subunits) which is inactive or absent in the IPP$^-$ cells.

The fifth class would comprise "regulating" molecules specific for inosinic-guanylic pyrophosphorylase, for it is not known at present whether the structural enzyme in IPP$^-$ cells is intrinsically non-functional or destabilized or totally repressed.

The first and second explanations require the synthesis of radioactive products from the supplied precursor within (or at the membrane of) those cells which are always genetically competent to do this. The first in addition ascribes a protective function to contact, which enables the passage of nucleotides from one cell to another. The third, fourth and fifth alternative explanations, however, introduce a new concept. It is implicit in all three that direct or indirect contact with a genetically capable cell can endow a different cell with the ability to perform a biosynthetic step *for which its genome lacks the genetic information.* Thus in a monolayer (and presumably *in vivo* in a tissue) a cell's metabolic capability would not necessarily be absolutely defined by its genotype, but rather by the overall gene pool of all the cells with which it is directly or indirectly in contact.

At present the evidence that cell-to-cell contact is a necessary prerequisite for metabolic cooperation rests on observations made with the light microscope. Whether actual contact is made at the level resolved by the electron microscope has not yet been investigated. However I have held the notion for some time that cells *in contact* may exchange small amounts of cell membrane and cytoplasm, and my colleague Dr R. Goldman has recently made observations on C13 cells which appear to favour this possibility. Dr Goldman has kindly permitted me to reproduce two of his electron micrographs which suggest that cells in contact might exchange small quantities of membrane and cytoplasm by a process akin to phagocytosis (Figs. 9 and 10). Rose (1963) has also observed, by time-lapse analysis, that microscopically observable intercellular exchange of pieces of cytoplasm and even of whole mitochondria can take place.

It is also conceivable that small short-lived cytoplasmic bridges may mediate transfer either of membrane, or of cytoplasm, or both following cell-to-cell contact. There are other possibilities, but at this stage of our knowledge it seems unjustifiable to speculate further.

SOME IMPLICATIONS OF METABOLIC COOPERATION

Whatever molecular explanation may finally prove correct, one must revise one's view of the cell (at least in tissue culture) as a self-contained and self-delimited genetical unit of biosynthetic as well as reproductive function. The phenomenon of metabolic cooperation clearly suggests that in some instances a cell's biosynthetic ability may be communicated to, or be enhanced by communication from, genetically different cells with which the cell is (or was) in contact. It is at present unknown whether the biosynthetic ability gained by metabolic cooperation ceases when contact is broken, or whether it decays with a measurable half-life. If explanation 3, 4 or 5 is correct then decay with a measurable half-life would be expected in many cases.

The consequences of metabolic cooperation, if the phenomenon is widespread, must be considered in designing experiments for the selective recovery of genetic variants of mammalian cells in culture. If cells are plated in concentrations which make contact between neighbouring cells almost inevitable or at least very frequent, then the usual selection procedures will mainly favour changed cells whose genotype is not masked by metabolic cooperation with neighbouring unchanged cells. Dominant changes would be picked up whereas many recessive changes would be lost. The simple remedy is to plate cells at concentrations sufficiently low to minimize the probability of early cell-to-cell contact. Should it be found that the consequences of metabolic cooperation decay with a half-life of several hours, then selective conditions had better not be applied until sufficient time has elapsed to allow the molecules obtained by previous cooperative contact to become exhausted.

If metabolic cooperation occurs normally also *in vivo* and is not just confined to a few biosynthetic situations, it is highly pertinent to the problem of cancer and in particular to attempts at the selective destruction of malignant cells. Where it operates the phenomenon implies that tumour cells, though themselves genetically sensitive to a therapeutic agent, may be protected by the genetically resistant surrounding normal cells. In other instances tumour cell characteristics may be expressed by surrounding normal cells, causing them to respond to the chemotherapy as though they also were malignant cells. In either case it may prove advantageous and important if conditions can be found which reduce the amount of metabolic cooperation at least temporarily, as this should bring the target cells into sharper focus for chemotherapy.

SUMMARY

(1) Cell lines are described which lack inosinic pyrophosphorylase activity (IPP^-) or adenylic pyrophosphorylase activity (APP^-). These genetic variants have been used to confirm that preformed nucleotides do not normally enter mammalian cells.

(2) The phenomenon of metabolic cooperation is described. APP^- and APP^+ cells were grown in mixed culture in the presence of labelled adenine and the results studied by radioautography. If not in direct or indirect contact with APP^+ cells, APP^- cells do not incorporate the supplied label into their nucleic acid. APP^+ cells always do incorporate the label, and APP^- cells in direct contact with APP^+ cells or indirectly in contact with them through other APP^- cells also incorporate the supplied adenine. Cell-to-cell contact appears to be essential for APP^- cells to gain the ability to incorporate. This was also previously found in the $IPP^+:IPP^-$ system.

(3) The relative amounts of label incorporated by genetically competent cells and by metabolically cooperating cells are discussed. The IPP and APP systems do not behave identically in this respect. It is also shown that

metabolic cooperation is reciprocal cooperation and not a unidirectional donor–recipient relationship.

(4) The possible molecular basis of metabolic cooperation is believed to be one of five different classes of molecule.

(5) The implications of the findings for problems of cell variant selection and cancer chemotherapy are discussed.

Acknowledgements

I am grateful to Miss Heather Davidson, Miss Pamela Lewis and Mrs Martha McNamara for technical assistance, to Miss Kay Valentine for help with the preparation of the photographs and to Dr R. Goldman for allowing me to reproduce his electrophotomicrographs in Figs. 9 and 10.

REFERENCES

BÜRK, R. R., PITTS, J. D., and SUBAK-SHARPE, J. H. (1968). *Expl Cell Res.*, **53**, 297–301.
FRIEDMANN, T., SEEGMILLER, J. E., and SUBAK-SHARPE, J. H. (1968). *Nature, Lond.*, **220**, 272–274.
FRIEDMANN, T., SEEGMILLER, J. E., and SUBAK-SHARPE, J. H. (1969). *Expl Cell Res.*, in press.
HARRIS, H., WATKINS, J. F., FORD, C. E., and SCHOEFL, G. I. (1966). *J. Cell Sci.*, **1**, 1–30.
LEIBMAN, K. C., and HEIDELBERGER, C. (1955). *J. biol. Chem.*, **216**, 823–830.
MACPHERSON, I., and STOKER, M. (1962). *Virology*, **16**, 147–151.
ROSE, G. G. (1963). In *Cinematography in Cell Biology*, pp. 471–481, ed. Rose, G. G. New York and London: Academic Press.
SEEGMILLER, J. E., ROSENBLOOM, F. M., and KELLEY, W. N. (1967). *Science*, **155**, 1682–1684.
STOKER, M. G. P. (1967). *J. Cell Sci.*, **2**, 293–304.
STOKER, M., and MACPHERSON, I. (1964). *Nature, Lond.*, **203**, 1355–1357.
SUBAK-SHARPE, H. (1965). *Expl Cell Res.*, **38**, 106–119.
SUBAK-SHARPE, H., BÜRK, R. R., and PITTS, J. D. (1966). *Heredity*, **21**, 342–343.
SUBAK-SHARPE, H., BÜRK, R. R., and PITTS, J. D. (1969). *J. Cell Sci.*, **4**, 353–367.

DISCUSSION

Leese: If you take a mixed cell population which has been allowed to grow to the point where cells have been permitted to cooperate through the establishment of cell contacts and then reclone them, is it possible to isolate only IPP$^+$ and IPP$^-$ cells having the same phenotypic characteristics of the parent cell lines or do all cells show varying degrees of ability to take up nucleotide precursors? In other terms, does the information transfer occurring during cooperation represent a permanently acquired phenotypic characteristic, implying the transfer of genetic or epigenetic factors, or is it a temporary phenomenon associated with labile factors eventually lost by the cells but which, for a time, permit the incorporation of purine derivatives?

Subak-Sharpe: This is an extremely important point. As far as our investigations have gone, the new property gained by metabolic co-operation seems to have a half-life and is thus not due to transfer of genetic information which *permanently* changes the "recipient" cell. But I am not yet satisfied that we can completely dismiss the possibility.

Ormerod: If you label the IPP$^+$ cells *before* mixing your populations, do you still get transfer of label?

Subak-Sharpe: We tried to do this. In fact I hoped to detect the transfer of labelled ribosomes in this way. A very few apparently positive cells were observed, but one needs to be cautious because it is easy to be subjective here. The few positive cells could have been the result of ingestion of material from a prelabelled cell which had died.

Burke: I wondered what excluded the possibility that nucleotide is not transferred. If nucleoside phosphorylation occurs at the membrane of the IPP$^+$ cells, on the inside as an active process, what passes freely around between your cells may be nucleotide; you can surely explain your results in this way. If you modified your autoradiographic technique so that you look not at the TCA-insoluble material but at the actual nucleotide (TCA-soluble material), you could perhaps follow the phosphorylated nucleotides and see if they are transferred.

Subak-Sharpe: We have recently done the following experiment which we are trying to repeat under conditions of unambiguously clean labelling. The two types of cells are propagated, one in the presence of carbon (indian ink) and the other of carmine; then they are grown mixed 1:1 for 24 hours. Unmixed controls are treated in the same way. The cells are then removed from the glass in a stream of medium or using trypsin or trypsin-versene, and replated after suitable dilution to prevent further cell contacts. At this point the tritium label is added and autoradiography done a few hours later. As we observed cells which were presumptively IPP$^-$ because they contained carbon particles and yet which had incorporated tritiated hypoxanthine, the nucleotide explanation won't do. Transfer of labelled nucleotide or of labelled RNA cannot explain this result, because the label was added *after* we had separated the cells. This leaves as a possible explanation the transfer of either enzyme or informational nucleic acid or regulator molecules. This experiment has been done once with a clear result but we have had some difficulty in repeating it cleanly. The possibility therefore remains that some of the carbon-labelled cells might have picked up the carbon after separation, and if this happened, we were misled.

Bergel: The question of whether a nucleotide, especially of a purine, is taken up by cells, has been investigated among others by scientists of the

Mayer-Kettering Institute in Birmingham, Alabama, in the course of their work with mercaptopurine ribonucleotide which corresponds to your hypoxanthine derivative (Brookman, 1965). They were mainly interested in the resistance of leukaemic cells to mercaptopurine resulting from deficiency of purine ribonucleotide pyrophosphorylase, like your IPP⁻ cells. The criterion was the therapeutic effect, or its absence; thus this was another kind of measurement from yours. When they esterified the nucleotide phosphate group and diminished its charge, or duplicated 6-mercaptopurine by synthesizing bis(thioinosinic)-5′,5‴ phosphate so that again the molecule had much less charge on the phosphate moiety, it passed into the leukaemic cell, as the resistance was now circumvented. In other words mono-phosphorylated compounds cannot penetrate cell membranes because of the ionic charges of the phosphate groups.

Burke: Yes, nucleotides don't go in. So it is a question of what happens after the nucleoside which goes in has been phosphorylated.

Möller: Professor Subak-Sharpe, you made a distinction between collaboration and a donor–recipient relationship, but all your schemes involve a donor–recipient relationship in which there is transfer of something.

Subak-Sharpe: Yes, but in *both* directions. This is the point. Again, we have a situation where the result of enzymic activity is very rapidly observed in many cells in *indirect* contact with the competent cell.

Jacques: The functional significance of the circular profiles shown on your electron micrographs may be something quite different from the exchange of cytoplasm between neighbouring cells. If they are vesicles, they may be carrying secretory products or endocytosed compounds to or from the intercellular space. If they contain some cytoplasm of the neighbouring cell, they may represent nothing more than the cross-section of a non-vesicular, anchoring, protoplasmic hook.

Subak-Sharpe: I agree with you, and do not place weight on the electron micrographs. They seem to be in keeping with the possibility of an exchange of cytoplasm and membrane but I would not go any further than that.

REFERENCE

Brookman, R. W. (1965). *Cancer Res.*, **25**, 1596.

PATTERN FORMATION AND HOMEOSTASIS

Tom Elsdale

Medical Research Council Clinical and Population Cytogenetics Research Unit, Western General Hospital, Edinburgh

THE single cell from which each human being develops represents, if we disregard for a moment its endowment of genetic information—which is largely unutilized at this time—a level of form and function which is commensurate with the more primitive manifestations of life on the planet. Each revolution in the life cycle of man involves the return to something not too far removed from square one in the organic world. Embryonic development is an autonomous and inevitable process performed without direction from outside, and the pattern-forming methods appropriate to embryos are therefore different from the methods whereby man arbitrarily imposes patterns on the materials at his disposal.

INHERENTLY PRECISE METHODS

A typical man-made pattern-forming machine is the lathe. All of the variables associated with the action of the cutting tool on the object are controlled and the accuracy of the work is related to the extent to which uncertainties in the operation of the machine are banished. On the other hand, some shapes can be formed to any degree of accuracy by the so-called inherently precise methods. These methods have a simplicity and naturalness that contrasts with the essential artificiality of the lathe. Strong (in 1951), who used these methods to fabricate various parts for the Johns Hopkins Ruling Engine, recognized this fact and drew attention to the primitiveness of the inherently precise methods.

Inherently precise (IP) engineering methods are interesting in a biological context (Platt, 1956), because they exemplify in their basic operation, and in terms of man's technology, the pattern-forming situations in nature, where the generation of pattern is the inherent property of the situation and is not arbitrarily imposed from outside.

As an example of an IP method consider the problem of forming a perfectly spherical surface. Take two rough blanks, one concave, the other convex, and of roughly similar radius of curvature, and grind the face of one against the other. The rule that must be obeyed in grinding is that motions in any one direction must not out-number motions in any other, otherwise a bias will be introduced and the final surfaces will not be circular in all appropriate planes. This requirement is satisfied if the grinding motions are random. The requirement could also be satisfied by a carefully ordered and controlled sequence of motions; such a sequence, however, could be regarded in this context as simulating a random process. Given this condition, the convex blank will bed into the concave and the conformity between the two surfaces will increase as grinding continues, until two perfectly spherical surfaces have been generated. This increasing conformity derives from the fact that only two simple shapes show displacement congruence, these being spheres and flats. Displacement congruence indicates the fact that when one perfectly spherical surface is placed over another of the same radius, the two surfaces are in contact at all points and remain so when the joint so formed is moved.

It is interesting to contrast the operation of a simple IP machine such as just described, with the operation of a lathe performing some simple operation such as turning a chair leg.

(1) *Lathe*. Where the pattern is created by a single traverse of the object, half of the object is finished at the halfway stage, the rest is untouched.

IP machine. The pattern is gradually perfected over all parts of the work at roughly the same rate. At a halfway stage all parts of the work are similarly formed.

(2) *Lathe*. The precision achieved in pattern formation is limited by the tolerances to which the machine was constructed.

IP machine. The precision achieved in pattern formation is not limited by the tolerances to which the machine was constructed. The machine may contain no accurately made parts.

(3) *Lathe*. All aspects of the operation of the machine are controlled. Ideally all uncertainties in the operation of the machine would be banished. We shall say that the proper performance of the work requires that *all inputs are controlled*.

IP machine. Whereas certain parts may need to be controlled (in practice, parts held rigid) the energetic input must be random.

(4) *IP machine*. Makes displacement congruent patterns. (The conditions under which displacement congruence is demonstrable can be very limited, and the number of shapes that can be generated by IP methods go far beyond spheres and flats.)

Left-wing students might ponder the thought that lathes are authoritarian machines, whereas IP methods get things done in an atmosphere of participation and accommodation! For the rest: a lathe imposes a pattern on an object, whereas an IP machine is a pattern-forming situation. The primitiveness of IP methods is suggested by considering the ocean beach as such a machine for making spherical pebbles (given homogeneous rocks), and by considering the growth of a crystal in a solution.

I shall now describe a cellular pattern-forming system and show that it can be understood as an IP machine.

THE PARALLEL ALIGNMENT OF FIBROBLASTS IN MASS CULTURES

Fibroblasts in culture have a tendency to align in parallel. I have studied the generation of parallel arrays in populations of human foetal lung fibroblasts grown in plastic Petri dishes. Techniques have been previously described (Elsdale, 1968).

These cells require a rigid or semi-rigid substrate on which they extend themselves by making adhesions. In the absence of such a substrate the cells remain spheres and do not grow. To establish a new culture, an existing one is trypsinized to detach the cells from their plastic substrate, and the suspension of spherical cells so obtained is transferred to one or more new dishes. This procedure involves the destruction of any order in the arrangements of the cells in the dish to be subcultured, and provided the cell suspension consists predominantly of single cells, the arrangement of the cells in new cultures will be random at the start. This feature of the technique is referred to as the random initiation. Following the random initiation, which usually involves setting out a rather small number of cells, well below the number required to cover all the floor space in the dish, the cells recommence growth. After a few days all the floor space will be occupied (confluence) and only after this has been achieved do the cells start to grow on top of one another. In the case of human foetal lung fibroblasts net growth ceases when there are about 8×10^6 cells in a 50 mm dish; this number is about five times that required to comfortably cover all of the dish floor with a monolayer of cells. We say that the cells grow to about five monolayer equivalents (ME = 5).

The tendency for the cells to align in parallel is evident some time before confluence is attained. Many small, independent groups of cells aligned in parallel are observed, and as the number of cells and their density per unit area of substrate increases, these parallel groupings enlarge. There is no evidence for long-range interactions between cells not in contact with one another; the alignments form because cells whose chance collisions bring them into parallel contact tend to stay together. In a culture approaching confluence, the great majority of cells are already incorporated into these parallel alignments, termed groups.

At confluence, the expanding groups, now occupying virtually all the floor space, interlock to give the culture a patchwork appearance. Because the groups originated independently of one another, they vary in size, shape and direction of cellular orientation. When the cells in the adjacent groups have roughly the same orientation, the groups will merge; otherwise, a packing interstice termed a frontier provides a sharp demarcation between the two (Fig. 1).

In the ordinary way these demarcations are rapidly bridged over by cells as growth continues beyond confluence, and complex three-dimensional formations develop, provided that collagen made by the cells is allowed to accumulate. This is another story (Elsdale and Foley, 1969) and does not concern us here. More relevant is what ensues when the accumulation of collagen is prevented by including 60 μg/ml of bacterial collagenase in the medium. The presence of collagenase appears to have no other effect than to inhibit the formation of the three-dimensional structures; the cells remain stretched and elongated, and they continue growing at their normal rate to achieve their customary stationary density of around 8×10^6 cells in a 50 mm dish. In the presence of collagenase all this increase in cell numbers is accommodated within the patchwork pattern of interlocking groups of parallel cells established at confluence. The cells in the groups become very tightly bunched together and there is considerable superimposition of cells, although it is quite possible that most of the individual cells retain some contact by their leading edges with the plastic substrate. The parallel alignment of the cells in the groups remains perfect. As the number of cells increases the frontiers between adjacent groups achieve a striking prominence. They stand out as narrow empty ditches between raised banks of cells, and not until a further rearrangement has brought the cells in adjacent groups into something approaching the same orientation do the frontiers disappear by the merging of the groups (Figs. 2 and 3).

The frontiers mark zones where the leading edges of the line of cells at the periphery of one group confront those of an adjacent group of differently oriented cells. This encounter is contact inhibiting. In sparser cultures inhibiting contacts are transient; in confluent cultures the cells are less free to wander and the frontiers become semi-permanent zones of contact inhibition.

This then is the situation in a newly stationary culture grown up in the presence of collagenase. On further prolonged maintenance these cultures, and those grown without collagenase also, suffer a gradual reorganization, characterized by the disappearance of frontiers and merging of the groups. The number of groups declines and their individual dimensions increase until, ultimately, cultures become organized into single extended parallel arrays of uniform thickness. This reorganization has sometimes remained incomplete after four months, which is the longest time we have maintained stationary cultures. Parallel array in this context has a precise definition: each cell within a parallel array is oriented parallel to its immediate neighbours; parallel arrays consistent with this definition can curve round so that the cells at one end are definitely not parallel to those at the other. These arrays are stable configurations and cultures so organized have no further pattern-forming potential so long as the arrays are left undisturbed.

Turning now to mechanisms, there are three general points to be made at the outset. Pattern formation by cells can be motivated in two distinct ways: first, cells may reproduce in their arrangements a pattern structured in their environment; for example, contact guidance. Second, in the absence of a structured environment cells may create patterns as a result of their interactions.

The Falcon plastic tissue culture dish provides a substrate for fibroblasts that is for practical purposes neutral, without a significant tendency to coerce cells to move or align in one direction in preference to another. The tendency for fibroblasts to align in these dishes according to the description already given requires, therefore, an explanation in terms of cellular properties and cellular interactions.

The next general point to be made is that the tendency to align in parallel so consistently exhibited in fibroblast cultures requires cell movement; the cells must move or be moved into parallel. In the absence of an agent to move the cells passively it is clear that the cells must move into parallel by their own inherent motility. Contact inhibition of movement is an important feature of fibroblast behaviour, and will operate to restrain motility more in dense than in sparse cultures. The observed gradual rearrangement of the cells in dense stationary cultures, however, implies that contact inhibition does not restrain all cell movement, as some have tended to assume would be the case. Time-lapse films reveal that it is only the cells making the frontiers that are semipermanently immobilized by contact inhibition, and that cells within parallel arrays are very much more free to move, but only in directions that do not destroy the parallel alignment. The average displacement per cell per unit of time in dense cultures may be less than in sparse cultures. It is clear, however, that cells in parallel in dense cultures cannot be regarded as non-motile.

Having disposed of these general points, I can now state how I believe parallel arrays are generated from random initiations. My suggestion seeks to explain this phenomenon in terms of three properties exhibited by the cells.

(1) The cells spend most of their time in an extended elongated state. Obviously if this were not the case we could not detect a parallel arrangement at all.
(2) The cells are inherently motile. This assumes not merely that the cells possess motility, but that a cell only ceases to move when it is dividing, or when it is obstructed or otherwise prevented from moving.
(3) Cells, to a greater or lesser extent, mutually obstruct one another's movement. In particular, certain configurations, i.e. at the frontiers, restrain cell movement more than others, i.e. parallel alignment.

Consider a cell in the middle of a dense culture. This cell being inherently motile continuously explores its immediate environment for the opportunities provided for the exercise of motility. Where all cells are doing this, in the course of time they will adopt the arrangement that allows them overall the maximum exercise of their motility. This arrangement in dense cultures is parallel alignment. It follows that once cells have attained this arrangement they have an incentive to maintain it.

HOW GOOD ARE FIBROBLASTS AT RECOGNIZING PARALLEL?

Cells are very small objects, far more susceptible to random fluctuations than we are, and they are flexible. Fibroblasts are in a constant state of agitation, and in a mass culture they are continually being jostled by their neighbours. On the face of it, these conditions seem to favour neither a precise appraisal of nor a precise reaction to their environment on the part of these cells. It would not be surprising if the ability of the cells to recognize the parallel was rather imprecise under these conditions. It so happens that fibroblast cultures provide the opportunity to measure in a quantitative manner the precision of this recognition.

The angle between the orientations of the cells in two adjacent groups separated by a frontier can be fairly accurately measured. If we regard all such angles as acute, the maximum angle is 90°. The minimum angle represents the point at which the frontiers disappear and the groups merge—that is to say, the angle at which cells cease to regard themselves as being non-parallel. This minimum angle is close to 20°, suggesting that the cells do indeed possess but a poor recognition of parallel. However, the alignment within parallel arrays is extremely precise, and the question arises, how do

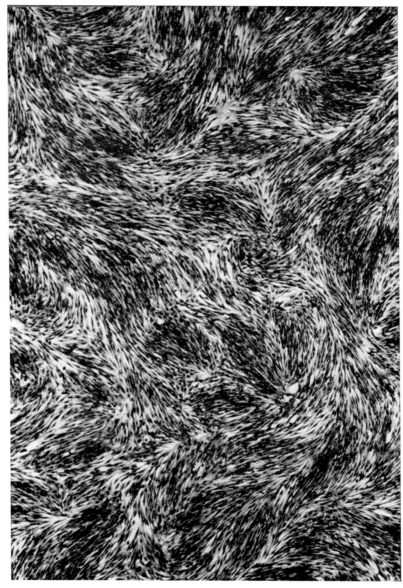

FIG. 1. Human foetal lung fibroblasts in Petri dish culture. Confluent culture. Rough patchwork of groups of cells aligned in parallel. ×62.

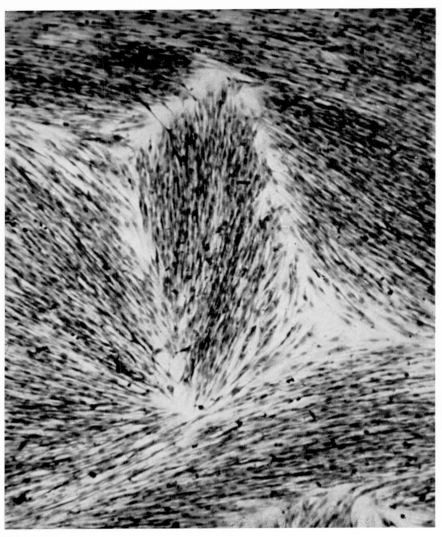

FIG. 2. Human foetal lung fibroblasts in Petri dish culture. Stationary culture grown in presence of 60 µg/ml collagenase. Groups of tightly bunched cells aligned in parallel; frontiers prominent. × 120.

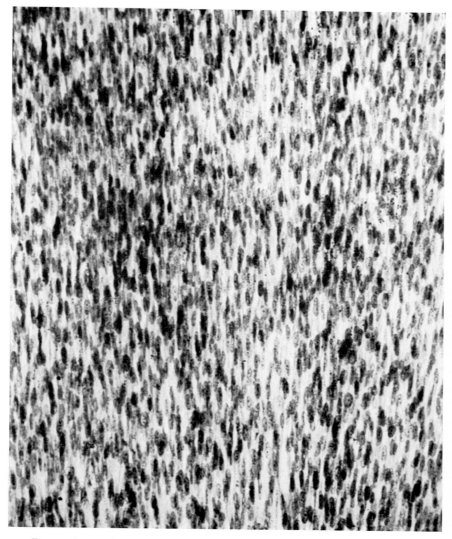

FIG. 3. Human foetal lung fibroblasts in Petri dish culture. Portion of extended parallel array. ×150.

cells arrange themselves so precisely in parallel when their recognition system appears to be inadequate for the task? It may not be necessary to invoke a special mechanism to explain this. Although the cells are not rigid rods, neither are they infinitely flexible. It is reasonable to assume that below 20° cell motility will vary inversely with the angular difference in orientation between cells. If this is so then a cell at the foot of this rising curve will ascend the curve; other things being equal, the steeper the curve the faster the cells will ascend it. One can speculate that a 20° recognition might have been evolved under natural selection to secure the most effective aligning mechanism for cells in the body. The conditions *in vivo* are not sufficiently known to test this idea, but the *in vitro* system suggests why this could be so.

Suppose the recognition of parallel was much better than it is, and that frontiers formed when cells were only 5° out of parallel. (Bearing in mind the consideration raised in the first part of this section, we must assume for the purposes of argument that this is a practicable situation.) The length of frontiers in a newly confluent culture would be greatly in excess of what is observed. Motility in the culture as a whole would thereby be decreased, and as a consequence the aligning process would be slowed down. The result would be an excellent alignment but achieved only very slowly.

Consider, on the other hand, the opposite situation where cells had to be at least 45° out of parallel for frontiers to form. There would be many fewer frontiers in a confluent culture, overall motility would be less restrained, and such frontiers as there were would quickly disappear. However the inclination of the curve relating motility to angular differences in orientation between 45° and zero, might be so gentle that cells ascended this curve only very slowly. The result would be quick but inaccurate alignment.

These considerations suggest that the recognition system could be the lever manipulated by natural selection to secure the best compromise between speed of alignment and accuracy to suit different morphogenetic situations.

"KINETIC" AND "DYNAMIC" VIEWS OF CELL BEHAVIOUR

An enquiry into the science of mechanics involves passing back and forth between kinetic and dynamic views of phenomena (see Lindsay and Margenau, 1957, for example). The kinetic view postulates individual particles with specific properties which govern their interactions, and gives a view of large-scale phenomena in terms of the resultant of many such interactions. The dynamic view on the other hand pays scant attention to minutiae and considers global characteristics such as continuity and compressibility. The fascination of mechanics lies in the way each of these approaches provides its own valid insights into phenomena, and complements the other.

Similarly, in the study of cell behaviour there are "kinetic" and "dynamic" approaches. On the basis of exact studies on the motility of individual cells and an exact description of contact-inhibiting encounters between cell pairs, a notion of how large populations of cells behave can be formed. This view derived from "kinetic" studies alone is likely to be limited because it does not lend itself too well to the construction of a clear mental picture of large-scale phenomena. My own view, starting from observations on mass cultures, is predictably a "dynamic" one. It considers, for instance, that interactions between fibroblasts may be more or less obstructing to cell movement, leaving the finer details vague. The dynamic view is unsatisfactory at the level of detail, but it hopefully compensates by producing a clearer mental image of the behaviour of masses of cells. The IP machine provides an appropriate adjunct in a "dynamic" presentation of cell behaviour, because an intuitive picture of how it works can be gained without attending to details.

THE ALIGNMENT OF FIBROBLASTS IN MASS CULTURES REGARDED AS AN IP MACHINE

The lens-grinding machine we have considered takes rough blanks and grinds them into precise shapes by a process which if continued long enough would grind the parts away to dust. Obviously the parts in the fibroblast system are conserved under movement, and it is the random arrangement of the cells that is eroded during the operation of the "machine" and not the cells themselves. The manner in which work is performed and the internal changes resulting therefrom are clearly quite different in these two situations. However a consideration of pattern generation in these systems shows them to be basically similar, and engenders certain generalizations about inherently precise pattern-forming situations.

The essential feature of an IP machine is that random energetic inputs generate a pattern to potentially infinite precision. Given continued operation of the machine and freedom from outside interference and internal breakdowns, pattern is inevitable.

Fibroblasts are independently motile, being individually energized. The motions of these cells are assumed to be undirected except insofar as they are constrained by features in the environment. A large enough number of motile fibroblasts will therefore constitute a random input for an IP machine.

Clearly in all IP machines there will be a mediation between the shape of the blanks and the arrangement of the parts, or the specific characteristics of the cells, and the pattern produced as a result of the operation of the machine. There is an inevitability vector, which does not itself uniquely determine the ultimate pattern, but indicates the direction of the changes within the machine that take place during pattern formation. The magnitude of this vector is a measure of order. In the lens-grinding machine the inevitability vector is directed towards displacement congruence; in our fibroblast system it is directed towards maximizing the exercise of cell motility. The identification of the inevitability vector is an essential step in constructing an intuitive understanding of how an IP machine works.

The inevitability vector alone does not predicate the outcome of pattern formation. Many different IP machines utilizing displacement congruence are possible, producing a variety of patterns including flats, spheres, cylinders, screws and cogwheels (Strong, 1951). Similarly a variety of cellular pattern-forming systems involving the maximizing of cell motility are imaginable; suppose for instance motility was maximized when cells crossed over one another at right angles. Only when the point of departure or origin of the inevitability vector is taken into account is the course of pattern formation defined.

There are three components therefore to a description of an IP machine. First there are the contextual features—the nature of the blanks and arrangement of parts, the specific properties of the cells and special features of their environment. Second is the inevitability vector. Third is the ultimate pattern.

HOW PREDICTABLE IS THE BEHAVIOUR OF CELLS?

The engineer has two choices open to him in constructing the simple IP machine for grinding spherical lenses. He can arrange for a carefully ordered sequence of motions that simulates randomness, or, assuming that he utilizes an electric motor or some other machine that produces ordered motions, he has the task of disordering these motions before they can be used to randomly energize the IP machine. In either case the energizing of the machine requires contrivance. Turning now to the fibroblast system, it is obvious that no problem arises where there are large numbers of cells, each of which is independently energized. Randomized energetic inputs are thus built into this system. Random inputs may also be built in in more ways than the one just noted.

A fact that distinguishes biology from more exact sciences is that when a biologist repeats an experiment he usually gets a different result; the trends may be the same, but the result is usually different. Biological systems are variable. Where does this variability come from? Consider an investigation reported recently by Merz and Ross (1967). They measured variation and mean clone size in clonal cultures, of human lung fibroblasts. The standard deviation and mean clone size are nearly equal at any given time, and they conclude that the growth of these cells is comparable with the stochastic simple birth process which assumes intermitotic times of random length. Many tissue culturists will have discovered for themselves that when the largest (fastest growing) colony is picked from a dish of diploid cells and recloned, the variation in clone size is not thereby manifestly reduced.

One explanation of this variability is that the cells are displaying rather large responses to probably rather small variations in their microenvironments. This explanation appears to run counter to the fact that in general, living systems are well buffered against fluctuations in their environment. An alternative explanation is that cells are inherently variable in certain aspects of their behaviour.

Cells are small entities and their size is not too far removed from that of objects that display Brownian motion. Our world would be a lot less stable if we found ourselves diminished to the size of cells, and the atoms in the environment remained as now. Wrist-watches would be unreliable and table-tennis impossible.

There is little uncertainty in what cells can do (specificity), for this is dictated by the genetic information, the ultimate repository of order in the cell. Only a small measure of uncertainty is allowed for in the genetic theory of mutation, and in rare mistakes in translation and protein assembly. Much greater uncertainties, significant within periods measured in hours or even minutes, are possible in the direction of motion of cells and in the times they take to accomplish things, to name but two areas.

Concerning the times taken to accomplish things, it is possible that many cells are poorly victualled with certain macromolecules essential to their synthetic processes. The rates at which key processes are performed may vary unpredictably, depending upon how "lucky" a cell is in clocking up the necessary number of chance collisions in particular low-flux reaction spaces. The random element in the growth of Merz and Ross's fibroblasts can be viewed in this light. Consider the function of an organ such as the pancreas. This organ supplies a daily quota of enzymes. This quota could be provided in one of two ways. Each secreting cell might be sufficiently victualled with intermediates that it could be relied on to deliver its fair and equal share of the quota, hour by hour and day by day. Alternatively, the cells may be less well victualled with intermediates, resulting in haphazard production of enzyme. At any one time, some cells may be making a lot, others little, and the fact that today a particular cell did well, might provide no guarantee that it will repeat the performance tomorrow. The mean performance of many cells would remain however quite reliable.

I do not know for sure which method the pancreas adopts but what seems certain is that natural selection will have done the cost accounting and chosen the method that produces enzyme more cheaply in terms of the inputs of order and energy. Certainly the first method would require precise controlling mechanisms if wasteful redundances were to be avoided. The second appeals more to one's biological instincts.

It is interesting to compare fibroblasts with random intermitotic times to fertilized eggs whose first few divisions proceed with an extraordinary regularity. A human ovum is many thousands of times larger than one of Merz and Ross's fibroblasts. From what is known about the constituents of egg cells and the synthetic processes taking place during oogenesis, it is reasonable to infer that eggs are endowed with well-filled, high-flux reaction spaces, and for this reason alone will be larger than other cells. One can speculate that the reason for these well-filled spaces is to ensure that syntheses can be reliably carried through and co-ordinated to a precise time schedule in the interests of securing development of a satisfactory blast-off from the launching pad.

A host of further arguments could be brought forward to support the idea that the behaviour of most cells is subject to short-term uncertainties.

HOMEOSTASIS AND PATTERN FORMATION

The existence of short-term unpredictabilities in cell behaviour could provide for a primitive homeostatic mechanism, serving to buffer cellular systems against outside disturbances. Consider a simple cellular system. If the behaviour of the cells is entirely predictable, then there is no way given by which oscillations within the system initiated by an outside disturbance could become damped. If on the other hand there are short-range unpredictabilities associated with certain aspects of cell behaviour, then the oscillations initiated in these aspects by the disturbance will be combined with a background of chance fluctuations. There will be confusion between the signals due to the disturbance and those generated within the system, with the result that the oscillations will become damped.

An incentive against assuming short-term unpredictabilities in cell behaviour has been the feeling that uncertainties and randomness necessarily stand opposed to order, and that the existence of chance fluctuations in cell behaviour can only serve to combat and sabotage the work of the order-creating processes of morphogenesis. However, the preceding discussion indicates that this view is not obligatory. Random fluctuations in cellular performance can be viewed as contributing to the random inputs that drive IP morphogenetic machinery.

For example the IP aligning mechanism proposed for human lung fibroblasts grown in Petri dishes is driven by an input whose randomness is provided for through the independent energizing of the cells. Within this system components of randomness will not be distinguished. The aligning mechanism would be equally well driven by independently energized cells whose individual behaviour was either free from uncertainties, or subject to chance fluctuations. In the latter case the uncertainties are no longer opposed to order; indeed they are incorporated as participants in the pattern-forming situation. It is possible to imagine a situation where all uncertainties in cell behaviour that might otherwise be hostile to and destructive of order, are institutionalized for constructive ends as inputs to IP morphogenetic machinery. This idea differs from the Turing model in the fact that random disturbances are incorporated into the driving of morphogenesis, not just in the triggering of it (Turing, 1952).

Nowhere is the ability to damp out the effects of outside disturbances better demonstrated than in young embryos. For example, development into a normal larva is possible after as much as one-third of the early neural tube of a frog embryo has been excised. Mutilation of the pattern here does no violence to the pattern-forming processes. It is necessary to envisage an intimate relationship between the homeostatic, regulatory, restorative processes in embryos and the pattern-forming processes; indeed a common base for these processes is a desirable postulate. Such a base can be provided in terms of the utilization of random inputs during inherently precise pattern-forming processes.

If uncertainties in cell behaviour are institutionalized for constructive ends through their contributions to the random energetic inputs driving inherently precise machinery, clearly the destructive potential of these uncertainties is controlled only for as long as the IP machinery keeps running. Note that the parallel arrays of fibroblasts are not static but dynamic configurations, and no fundamental distinction can be drawn between pattern formation and pattern maintenance. The cessation of IP machines means death and chaos, and disintegration follows. It is possible that the ceaseless turnover of bodily and cellular constituents is essential to keep IP machinery running. It is no argument to note that a particular species of RNA for example is unstable, which is merely—if it is true—a statement of fact; the question to be answered is why has nature not taken steps to stabilize this molecule? It is possible that the continuous running of IP machinery commits cells to continuous activity on a variety of chemical fronts.

SUMMARY

It is no more than a platitude to say that in some form or another all the order-creating processes in nature are inherently precise processes driven by a random input ultimately derived from thermal agitations. By restricting the situation and considering cells as the elements in a pattern-forming situation, this idea can be helpful in getting the feel of how cellular pattern-forming situations work.

REFERENCES

ELSDALE, T. R. (1968). *Expl Cell Res.*, **51**, 439.
ELSDALE, T. R., and FOLEY, R. (1969). *J. Cell Biol.*, **41**, in press.
LINDSAY, R. B., and MARGENAU, H. (1957). *Foundations of Physics.* New York: Dover.
MERZ, G. S., and ROSS, J. D. (1967). *J. Cell Biol.*, **35**, 92A.
PLATT, J. R. (1956). In *Symposium on Information Theory in Biology*, ed. Yockey, H. P. Oxford: Pergamon Press.
STRONG, J. (1951). *J. opt. Soc. Am.*, **41**, 3.
TURING, A. M. (1952). *Phil. Trans. R. Soc.*, Ser. B, **237**, 37.

DISCUSSION

Lamerton: If you wish to make a sphere, or indeed a flat surface, by inherently precise methods you have, Dr Elsdale, as you said, to start with a crude template of your finished product. Similarly in the biological situation you would have to start with such a template. What are the biological premises necessary to enable an inherently precise process to produce the complex pattern of the tissues?

Elsdale: There are many different types of inherently precise (IP) machines. The lens-grinding machine was introduced to illustrate the essentials of an IP machine, and situations that include pattern formation or amplification by erosion or accretion are not generally of the same type as cellular pattern-forming situations. As to the premises necessary to enable an inherently precise process to produce a complex tissue pattern, I suggest the situation be analysed into three basic components. The inevitability vector is the key to the intuitive understanding of how the pattern-forming situation evolves, and provides an analogous insight to that provided by the proof in Pythagoras' theorem, for example. The two other components I lumped together as the contextual features; perhaps I now ought to distinguish them.

First, there are the specific properties of the cells—their shape, polarity, deformability, substrate requirements, motility and so on—and any other characteristics that help to define the outcome of cellular interactions. Second, there are the special topological features of the space in which the pattern-forming situation exists. For example, observe that fibroblasts in flat Petri dishes form extended parallel arrays within which all frontiers disappear. Imagine now, in place of a flat dish, the cells had been plated on to the surface of a plastic sphere not so large that an area of a square centimetre or so could be considered as virtually flat. Assume the alignment mechanism operates as well as before. It is now impossible for all the frontiers to disappear; at least one must remain. The topological problem involved here is referred to in the trade as that of stroking the hairs on a spherical dog. This illustrates the principle that a cellular pattern-forming system will not necessarily produce identical patterns in all topological situations. I doubt that this trivial topology is important in embryos.

Vernon: What is wrong with this as an alternative explanation of your final situation? Since the fibroblasts do go together there is an overall attractive force between them, presumably London forces; all you do is to pack them into the minimum volume and all they have done is to maximize these attractive forces. If you take a set of bar magnets, exactly the same thing will happen. The result is inevitable; it wouldn't matter even if the cells were dead, provided the attractive force existed and you imposed a random movement by shaking them.

Elsdale: There is nothing wrong with this explanation and I have suggested as much myself (Elsdale, 1968). It is nice to the extent that it expresses in a simple way the inevitability of the pattern-forming process. However, there is a limitation; if your approach gets too "dynamic", the result is a platitude. We are specially interested, not in how dead cells or bar magnets can be made to align, but how living cells align, because it is living embryos that present us with congenital malformations and living cells that become malignant.

Wolpert: I suspect that Dr Elsdale is really a kineticist in dynamicist's clothing! While agreeing that his approach may be suitable for certain morphogenetic phenomena that involve cell movement I feel that he really is not using the global system at all. All his explanations seem to be really local interactions. This can explain a certain type of pattern-forming process, but there is another class of pattern problems which are inherently incapable of being explained in this way. To give one example, during the early development of the sea urchin embryo about 40 mesenchyme cells come out of the bottom of the embryo and move up the walls to take up a pattern consisting of a ring around the bottom with two branches. If you look at what the cells are doing individually, they are behaving exactly as Dr Elsdale was saying; you cannot predict the movement of any individual cell, pseudopods are being put out—his input—at random; nevertheless the cells take up an inherently quite well-defined pattern and one can explain it by saying that they make the most stable contacts with some sort of pattern or template on the wall (Gustafson and Wolpert, 1967). It is how one specifies this template on the wall which is a pattern problem that I don't think local interactions can explain. The specification of this pattern involves global phenomena which probably require some general mechanism for specifying position in space. This must entail processes quite different from the random ones that Dr Elsdale considered.

Elsdale: I suggest that the pattern on the walls of the sea urchin embryo is either fairly directly prefigured in the construction of the egg, or results from an inherently precise pattern-forming process.

Stoker: One other way in which you could put variability into your fibroblast system is by stopping it at different times. You pointed out that if you take a halfway situation in the culture you actually have a lot of islands of cells. If you stop the system three-quarters of the way through you have a different state, which might help in some of the developmental problems.

Elsdale: You can do this; if you put agarose-stiffened medium on your cells you stop pattern formation, by restraining the motility of the cells.

Subak-Sharpe: To what extent can you influence the pattern at any given point in your system? If you put in a single glass fibre at the beginning of the culture would the cells align in parallel to that fibre? My guess is that they probably would. If you put that fibre in at later stages, at what stage (given that you leave enough time for your system) does the pattern now resist the change imposed from the outside?

Elsdale: Any new feature introduced into the environment of the cells, such as a fibre or a scratch across the plastic, may influence the pattern. We have used this to investigate mechanisms (Elsdale and Foley, 1969).

Stoker: Relating this to homeostasis, it seems to me that Dr Elsdale's system is driving in one direction very strongly, towards a particular situation, but that it doesn't provide for a balanced control from both directions; but you may disagree with me.

Elsdale: I interpret this as asking, how do you get development? The fibroblast system I have described is observed to approach a steady state because the specificity input is constant, owing to the absence of further cellular differentiation. If at some stage a group of fibroblasts were to

11*

change into a different type of cell, thus injecting new specificity into the pattern-forming situation, a new IP machine would be set up locally, pattern-formation waves would follow a different course, and the steady state would be deferred. The system would be to a small degree a developing one.

Embryonic development is impelled by the sequential injection of new cellular specificity as a result of the reading of the genetic information to provide a widening selection of protein molecules. New proteins mean new cells, with new types of interactions, and hence new pattern-forming situations. From the standpoint of morphogenesis herein lies the purpose of the genetic information.

One might imagine an embryo at the outset of development as providing a single IP machine tending towards its steady state. Because, however, of the spatial differences in the distribution of materials, before that steady state is reached new specificities arise locally, creating a plurality of new pattern-forming situations whose increase is the measure of development. Each pattern-forming process contributes to the topological features in which the next in time and next in space operate. One can envisage the mosaic of IP pattern-forming processes in the embryo as being "glued" together in space and time by this dependence.

REFERENCES

ELSDALE, T. R. (1968). *Expl Cell Res.*, **51**, 439.
ELSDALE, T. R., and FOLEY, R. (1969). *J. Cell Biol.*, **41**, in press.
GUSTAFSON, T., and WOLPERT, L. (1967). *Biol. Rev.*, **42**, 442–498.

GENERAL DISCUSSION

Bergel: Having opened Pandora's box at the beginning of the conference, I feel it my duty during this final discussion to put the mysterious and puzzling bits back into the box and look with you only at the clear-cut explicable bits. To carry this allegorical game through to its conclusion, I shall try, with your help, to summarize the distinguishable features of homeostasis. If we equate homeostatic regulation (and the expression homeostatic or homeostasis badly needs an all-embracing definition*) with preservation of life or the preservation of a biological part of the whole, such as cells, tissues, or organs, in its main purpose, namely to stay healthily alive, then homeostasis means some kind of buffering or defence arrangement inside a number of biological systems to defend them against external forces of disorder, anarchy or—borrowing the term from thermodynamics—entropy in the widest sense. These processes include as much the preservation of inherent and inherited patterns as that of the individual as a whole.

Changing from philosophical to more factual matters, one has to allow for the existence of pro-homeostatic and anti-homeostatic forces. The internally or externally threatening forces are manifold: any perturbation, such as injury, viruses, mutagens, carcinogens and other chemicals and drugs, unfavourable temperature, starvation, perhaps even stress and strain. All these disturbances can happen on different biological levels of the organism and so can the countermeasures. Dr Allison mentioned levels distinguishable by time-scales. Professor Subak-Sharpe brought in the three dimensions of space, which are essential from the point of view of growth phenomena. May I suggest that we differentiate between the biochemical or molecular level whose events are rapid, sometimes instantaneous and very tightly controlled; it could be the most protected level and its homeostatic regulation may rest solely on feedback and feedforward mechanisms (see Iversen's Fig. 1, p. 312).

Then follows the subcellular and cellular level with a more sophisticated

---

* Stedman's *Medical Dictionary* (20th edition. London: Baillière, **Tindall and** Cassell): Homeostasis (1) The state of equilibrium in the living body with respect to various functions and to the chemical compositions of the fluids and tissues, e.g. temperature, heart rate, blood pressure, water content, blood sugar, etc. (2) The process through which such body equilibrium is maintained.

See also footnote, p. 105.

homeostatic organization, as these have to maintain "biological equilibria" in a less rigid manner inside certain limits. Tissues and organs are the next level, where for instance hormonal control will watch over homeostatic balances in tissues that are hormone dependent. The homeostasis there may rest for instance on a couple of hormones which possess opposite effects. Normal physiological events such as fertilization can lead to a complete readjustment of the homeostatic situation.

Organisms and population of organisms (for example, cell cultures) must have their own systems which, depending on the specific circumstances, could be very crude or in the case of cell-to-cell relations rather intricate and very much a matter of investigation at present. Of course our main interest has been the arrangements for order on the cellular level. Is it correct, Dr Elsdale, that you consider randomness as a kind of inherent disorder?

Elsdale: Random fluctuations are potentially destructive. There exists however a means by which they can be institutionalized for constructive ends. This means is provided by the inherently precise machine.

Bergel: Would you then agree with this interpretation, that we have not to deal with a "steady state" at all but with a whole range of equilibria, physico-chemical and biological, inside which the cell, the tissue, the organism tries to exist? If the normal limits of deviation are broken, then the buffering or homeostatic arrangements first try to restore the situation but might after that be destroyed and the biological framework will die. The attack against the "limits" could be carried out by some external agent, which can be anything from a bacterium to a carcinogen or even an inorganic chemical sphere which, I think, should be outside our consideration at the moment.

Stoker: I am not sure that it is wise to keep out of the region of pathology. It seems to me that pathological situations are part of the natural order. I wonder whether we are not getting into a rather circular situation if we say that the whole topic of homeostasis can be conceived as a way of preserving life.

Subak-Sharpe: Particularly as, in my opinion, each higher level imposes restriction on the one below it, and the *population* level imposes the restriction that death is necessary at the individual level to keep the population steady. Preservation of life is not an absolute criterion. The continuity of the system is the absolute.

Stoker: The opposing force that Professor Bergel has suggested is disorder, but for any living system there is variability and therefore disorder is essential. So that homeostasis really includes disorder, as he said.

Ormerod: Also, if I take Dr Elsdale's point right, if you are getting order out of disorder you cannot segregate the two.

Allison: May I make a comment on terminology here? I dislike the word "homeostasis". One reason is that the word "stasis" suggests a static situation, whereas we are dealing with a dynamic system. I prefer the word "equilibrium". In genetics for example one has equilibria which are either stable or unstable, or stable within a certain range, beyond which they become unstable. So then one can speak of equilibrating forces, and perhaps disequilibrating forces, that tend to restore the system to, or take it away from, equilibrium.

Wolpert: I have always seen homeostasis as a system which includes some sort of feedback, some loop in the system. That is, there is some sort of sensory monitoring system so that when you disturb the system, the system tends to come back. The context of "equilibrium" doesn't imply this. If you have a ball in the bottom of a bowl and you perturb the ball, it always comes back to the bottom of the bowl. To call this a homeostatic system is to lose what I think is important about homeostasis, and that is the negative feedback.

Allison: In genetic selection for example, if you have a stable equilibrium and deviation from a particular gene frequency, selective forces tend to bring it back. It is the position of minimum potential energy, if you like.

Wolpert: I don't think that the concept of minimum potential energy is the same as that of homeostasis with a negative feedback.

Lamerton: Are we not making our field much too broad? If we consider "homeostasis" in general we are covering the whole of biology, almost the whole of life. But our meeting is concerned with "homeostatic regulators" which I would interpret as factors acting on the cell that either have been isolated or could be isolated.

Subak-Sharpe: But if we wish to discuss homeostatic regulators it is important to recognize at what point they are likely to act, and to do this we need to define homeostasis and to define the different levels of biological complexity at which we recognize that homeostasis occurs.

Vernon: The difficulty of definition arises because people are trying to define homeostasis in terms of particular mechanisms. Professor Wolpert wants a feedback and engineering type of mechanism, which no doubt exists; the biochemists want chemical mechanisms; and so on. And since we don't know the mechanisms of most of these situations and they are not likely to be all the same, we shall not reach a definition that way. A way out is to define homeostasis in terms which are independent of mechanisms, and that is thermodynamically. A simple thermodynamic definition is that homeostasis is the tendency of an organism to resist passage to a state of minimum free energy.

Roe: Before this symposium the term homeostasis had only found general application at the supra-cellular level—that is to say, at the level of tissue or cell growth control. I believe that this restriction of its use should be retained and that subcellular biochemical mechanisms that take part in growth control should be thought of simply as negative feedback systems.

O'Meara: If we try to define homeostasis in relation to all biological systems it is going to be very difficult. If on the other hand we confine ourselves to homeostasis in mammals we have to think in terms of time, as has been pointed out. In the mammal the embryonic state produces an individual who is well formed with definite morphological relationships between one part and another. The process of growth continues until the adult stage is reached. One can morphologically define an individual in whom homeostasis exists when the adult state is reached. That is to say, homeostasis is harmony, and I would not exclude any of the levels of organization from the biochemical upwards. Homeostasis could then be defined as the maintenance of a harmonious and cooperative relationship between the cells, tissues, organs and body fluids and its regulation. A state of dynamic equilibrium exists to make the individual a normal individual.

If this state is to be preserved it must be under some kind of regulation, which we can think of as occurring at the following levels: (1) submolecular level (e.g. electron exchange); (2) molecular level; (3) subcellular level (participation of organelles); (4) cellular level (vital activities of cells, cell relationships, immunological properties); (5) body fluids (immunological, hormonal); (6) tissue and organ level (influence of tissues upon one another, cell proliferation rates in different organs); and (7) population level. This last can have two meanings (*a*) cell population or (*b*) populations of individuals. The cell population influences cell replacement in organs such as the liver in which cell removal or death is a potent stimulus to regeneration. This is constantly exercised in the haemopoietic system. Populations of individuals influence the homeostatic state in units of the population in many ways, the most important perhaps being their genetic influence on a long-term basis and their psychosomatic effects. All these items may be considered to lie within the parameters of the physiological state, but with gross deviations from this state another aspect has to be added: (8) homeostasis and its regulation when a pathological state supervenes.

Bergel: This represents another possibility of subdivision: select the degree of maturity of the biological unit, whether cell, clone, tissue or organ, and use these various levels as subdivisions within this degree of maturity.

Roe: I doubt whether "maturity" is definable. Certain species, such as the rat, continue to change from birth until death. On a commonsense basis one may draw a line between "immaturity" and "maturity" but in a biological sense its position would be rather arbitrary.

Jacques: If our aim is to enumerate the various homeostatic regulators we certainly should distinguish between the several levels of biological organization. The homeostatic mechanisms studied in molecular genetics, metabolism or cellular physiology are usually different. One might even suggest that our survey be extended to pure physical chemistry, on which all the phenomena we have been studying ultimately rest, including cellular growth. For instance, the principle of Le Chatelier which prevails in the inanimate world might be considered as the elementary form of the negative feedback reaction: any alteration in a system in dynamic equilibrium immediately brings the system to react in a manner that will counteract the action.

Möller: Can we look at homeostasis this way? Homeostasis would be those regulating mechanisms tending to keep the interior of the body constant. Examples of this type would be the buffer mechanisms of serum pH and hormonal feedback. Defined this way, however, one would mainly be concerned with steady states within an organism and would omit two important variables: (1) Changes within the organism, which are carefully regulated, but appear because of a demand or a changed environment. Thus the hormonal changes occurring during pregnancy are carefully controlled, but disturb the steady state. Similarly, compensating mechanisms operating because of changes in the environment would be excluded. Antibody synthesis is an example. In these cases steady states are disturbed, but the function is to keep the organism as such "steady" in a changing environment. Besides, the changes are regulated, usually in feedback systems, and, therefore, each change tends to stabilize at a steady state. (2) Differentiation of cells or organisms clearly falls outside the limited definition, although there is no doubt that few processes are so carefully controlled as these. What we are interested in are regulating mechanisms. These could occur at any level (molecular, cellular, individual, population). Homeostasis is just a special case of regulation, namely the steady state situation.

We are all concerned with exceptions to the second law of thermodynamics and are actually studying mechanisms preserving organization.

All homeostatic mechanisms serve the purpose of defence against disorganization, but the term defence is usually used in connexion with intruders from outside. Perhaps the only difference is between external and internal homeostasis—both vital—but external homeostasis is an adaptive, regulated response to environmental changes, internal homeostasis being the classical concept of homeostasis, with the exception that aberrant cell differentiation must be controlled by absolute (suppressing) mechanisms.

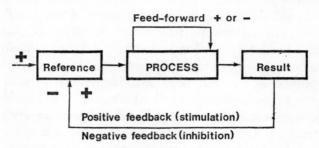

FIG. 1. (Iversen). Generalized feedback system.

Iversen: I have been working with feedback systems of a general nature, which can be summarized as in Fig. 1. The reference value determines the speed of the process if uncontrolled. The result is measured, and a measuring system recognizes the difference between the "correct" value of the result and the actual value. This signal goes back as a feedback which can be positive as a stimulation or negative as an inhibition. Theoretically there is also the possibility of feed-forward, if it is possible to foresee early

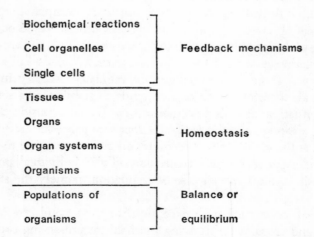

FIG. 2 (Iversen). Levels of interaction: feedback and homeostatic regulation.

in the process that something is developing wrongly. The process may correct itself with a signalling system which is faster than the process itself. It is likely that such regulation mechanisms exist in the nervous system, for instance.

Bergel: Do you think that this applies to all levels?

Iversen: Yes, even on the biochemical levels, and on the organelle and cell levels. But I think it will be useful to call these lower levels simply feedback systems and start referring to homeostasis at the level of interaction between tissues and organs and between organisms, as shown in Fig. 2.

Bergel: In order to help clarify the position we have reached on the definition and function of homeostatic regulators, some of which we think may exist as distinct entities, while others may be only feedback mechanisms or interlinked patterns, I have tried to assemble in a table (Table I)

Table I

SUGGESTED HOMEOSTATIC REGULATORS OR REGULATORS OF CELLULAR GROWTH

Cell-to-cell interactions	Cell-produced factors excreted into intercellular spaces	Intracellular factors	
		Cytoplasmic factors	*Nuclear factors*
Due to: Immune bodies and immunocytes (from spleen, thymus, bone marrow, lymphatic system) Activities of cell membranes	Hormones e.g. growth hormone (from endocrine glands) Thromboplastic agents (from tumour tissues and amnion) —— Wolff factors? (from liver, yeast etc.)	Chalones of: skin granulocytes melanocytes Nerve growth factor Eyecup factor Isoenzymes Interferons, stimulated by: Polynucleotides and polymeric acids Organelles: Lysosomes Mitochondria Endoplasmic reticulum	Operator and regulator genes Repressors and depressors Histones NAD Endonucleases Other enzymes, etc.

the factors we have mentioned and some that we have not. We have first of all *cell-to-cell interactions*, emphasized by Professor Stoker, Dr Forrester and Professor Wolpert. *Outside* the cell borders one finds the *growth hormone* which, as far as one knows, reacts on the whole organism and not on events of the cell itself. *Spleen and thymus effects* on the state of immune bodies are listed as cell-to-cell interactions to indicate their disseminated influence through the lymphatic and circulatory system.

Going into the cell, some of the *immunological events* take place on its surface, from an antigen and antibody point of view. The most important control is exercised by the *cell nucleus* with its genes—operator and regulator ones, histones, repressors, enzymes, RNA, nucleases, NAD. Whether an exact replica of the Monod-Jacob machinery exists in the mammalian cell is not yet clear and developments in our knowledge are needed here. *Subcellular* particles consist of *mitochondria*, with their ATP-ADP equilibrium and their contribution to the respiratory state of the cell, and of *lysosomes* which according to Professor de Duve and Dr Jacques have an essential part to play in homeostasis, clearing out useless debris in the cell.

Distributed over the cytoplasm (bound to some part?) are the *chalones* which if functioning as anti-mitotic agents may counteract stimulatory materials such as hormones and activated enzymes. It is possible that they react more slowly than the stimulating agents (such as NGF), because a suddenly acting brake would be disastrous, while an accelerator should speed up particular reactions instantaneously.

Then there are polymers such as *polynucleotides*, even artificial ones, which have some cytological effects and catalyse interferon formation. Whether *interferons* as such, defending the organism mainly against virus invasions, have other functions of a regulatory character is not quite clear yet. NGF and eyecup factors may have only very specialized functions, although one could imagine that more of their type may be discovered in the future. About promin and retin enough was said in my introduction.

Perhaps this table may be useful as an aid in forming definitions of homeostasis and its regulating systems, particularly with future developments in this field and desirable collaborative efforts in mind. This is where Professor Stoker's comment on the pathological sphere fits in; as Dr Williams has remarked, skin as a morbid subject might be a most appropriate topic for further discussions and studies, with special reference to homeostatic regulation.

Wolpert: I wonder whether we really need to define homeostasis—whether a definition will take us somewhere unexpected or interesting? The important point is what *factors* are involved in control in intercellular relationships in a growing tissue, as Professor Bergel has summarized in his table. To think in terms of factors might be easier than to define homeostasis.

Bergel: Particularly if you will regard the table as an incomplete list only, with many more items soon to be discovered.

Burke: It will certainly be simpler to leave intermolecular and biochemical reactions out of any consideration of homeostasis. The term is not commonly used in biochemistry and it hasn't worried anybody that it

GENERAL DISCUSSION 315

hasn't been used. The intercellular reactions are the interesting ones here. We don't know how the intercellular mechanisms work, whereas the internal cell processes operate by feedback control. I think Professor Wolpert is arguing that we should not be bemused by the sort of mechanisms which we know already when we are looking at new intercellular interactions; let us separate them clearly.

Bergel: I quite agree that one should subdivide the field of homeostasis in the widest sense, but there is danger in ignoring certain aspects, because we don't know yet whether this or that observation is due to *intra*cellular or *inter*cellular events; do we know, for example, how the chalones act?

Jacques: My principal reason for opposing the idea that we limit the field of homeostasis to only part of the continuum between physical chemistry and the dynamics of cell populations is that several homeostatic regulators we have encountered act at two distinct levels of biological organization. Thus the action of some hormones at strategic parts of the *molecular* metabolic pathways is prolonged by their synergistic effect on processes like endocytosis or autophagy, which require a *cellular* organization. Also, spreading on a suitable surface, insulin and serum factors which enhance the endocytic activity of *single cells*, are homeostatic regulators, as Professor Stoker showed us, acting at the level of cell populations.

Bergel: Even if we focus attention on the tissue, organ and organism level only, as Professor Iversen and others suggest, we have to pay attention to the "lower" levels, because there are for instance biochemical deficiencies which lead to disturbances, possibly of a homeostatic nature, and then to considerable changes in tissues, organs and whole individuals.

Roe: Dr Jacques seemed to imply that to distinguish between intercellular and intracellular homeostatic mechanisms would be meaningless because frequently the former (for example, hormones) act intracellularly. This is surely not a valid objection insofar as in the final analysis all control mechanisms are bound to act intracellularly. In regarding a mechanism as "intercellular" or "intracellular" we should take cognisance not only of the location of the site of action, but of the location of the whole system. There are growth control systems which operate entirely within cells; it is reasonable to regard these as "intracellular" or subcellular. There are also systems which are initiated in one cell and operate on or through another. These may be regarded as "intercellular" or supracellular. In my view the two can, and should, be distinguished.

Leese: In essence at each biological level the space–time characteristics of regulatory mechanisms are related to the degree of adaptive biological organization required in the operation of the homeostatic process.

Essentially there is no difference between homeostasis at different levels, and to draw distinctions is both artificial and misleading. The key feature of homeostasis is the ability of the system to adapt in response to changed environmental conditions. There is a spectrum in the space–time relationships according to the degree to which biological apparatus requires to be organized for the homeostatic process to occur. At one extreme there are the millisecond adaptive responses occurring, for example, in changes in the redox conditions in mitochondria or in neural transmission where the biological apparatus is already intact. At the other extreme there are the very long-term effects of evolutionary mutational events on populations mentioned by Professor Subak-Sharpe. At the cellular level adaptive changes in enzyme activity can take place rapidly through interaction between preformed enzyme molecules and small molecular reactants or, on a slightly longer term basis, through enzyme synthesis in response to small molecular effectors. Adaptive changes at a higher level involving cell replication and tissue reorganization occupy yet another region of the space–time spectrum of homeostatic phenomena.

Some of these points have already been raised by Professor Wolpert and Dr Allison but I think it is important to emphasize that little fundamental distinction can be made between any homeostatic processes except on the basis of the degree of biological organization and time involved in the adaptive process.

Subak-Sharpe: At the moment the term homeostasis means many things to many different people; to clarify the situation we shall have to define different levels of analysis. Certain levels of analysis, for example the straightforward biochemical feedback systems, can be dismissed from consideration. As for the remainder, we shall have to define the particular levels of biological complexity at which homeostasis is being considered. If a unified explanation can be given for the observations on systems at any one level, let us state that. From many points of view the most interesting and important levels may well be those of cell and tissue interaction, but to dismiss the whole of the earlier levels would be a grave mistake. But the most inexcusable mistake would be to leave homeostasis as a loose blanket term under whose soft embrace anything can go on.

Bergel: In a completely different field, that of drug addiction, the question of nomenclature and definition has been resolved by adopting the word "dependence" which is to be qualified in the manner Professor Subak-Sharpe suggests; for example, dependence of LSD type or of morphine type. We could use this rather general word homeostasis in connexion with certain specific events.

Subak-Sharpe: Another analogy is the gene, where the early loose definition was found to become valueless with time but then the gene was successfully redefined at different levels of analysis as a unit of function, as a unit of recombination, as a unit of mutation, and so on; the term became again very useful, conveying specific information to most people. The same may have to happen for homeostasis.

Elsdale: An alternative division of the field which cuts right across a division according to the level of regulation is to think firstly of those regulatory mechanisms in the organism which depend upon the presence of "plumbing"—upon the fact that there is an established circulation, that there are individual endocrine organs and target organs; and secondly, those regulatory mechanisms where there is no plumbing, as in embryos. Here the mechanisms are closely tied up with pattern formation.

Wolpert: One of the conclusions of the meeting could be that one should not use the term homeostasis.

Lamerton: There is certainly a big difference between attempting to define homeostasis and defining homeostatic regulators acting in certain situations.

Abercrombie: Homeostasis is really a red herring. The interesting things we have been talking about are control mechanisms in a very broad sense and a lot of them are nothing to do with *stasis*; they are concerned with development.

CHAIRMAN'S CLOSING REMARKS
Professor F. Bergel

With all our papers, discussions and private exchanges of views and the meeting behind us, we should now be able to assess the overall situation and to see if we can answer the questions that were posed at the beginning and during the meeting. These were:

(1) Do cellular homeostatic regulators exist, because, or in spite of, the numerous observations that have been made in many laboratories and discussed during this symposium?

(2) Do we think that the entities mentioned are directly or indirectly involved in homeostasis?

(3) Does the true homeostatic control rest on a multi-membered system or systems, and is it restricted to certain levels of biological organization?

(4) Do any agents among those discussed not belong to homeostatic systems and are they in fact parts of something completely different?

The tentative answers that we seem to have arrived at are as follows:

(1) There appear to exist molecular entities which, if not identical to homeostatic regulators, do exercise such functions in special cells under specific conditions.

(2) It looks as if the integrated action of several factors is necessary to maintain the cellular equilibrium which keeps the cells' development, normality and final ageing processes inside the limits of their inherent nature.

(3) It appears that there may exist several related systems which are more selective with respect to the kind of biological level and type of tissue in which they function. They may rest simply on feedback mechanisms or may promote, slow down or arrest changes due to growth, repair and external perturbations. Functions of subcellular particles were recognized as part of regulating arrangements, but cell-to-cell effects appeared to catch the fancy of many participants as of importance in homeostatic events. This was not specifically discussed inside the framework of the conference's main preoccupation, nor was there an opportunity to concentrate on other pathological states. These matters could form the subjects of further discussions in the future.

(4) It is quite feasible that some of the active principles mentioned have nothing directly to do with homeostatic regulators. That again ought to be re-assessed at a future date.

This attempt to condense into a few sentences a summary of the deliberations makes it clear that a great number of questions raised by the conference have not found clear-cut answers. Many of the utterances rest to a large part on intelligent assumptions that require, as everyone will agree, a great number of systematic research programmes before they reach the level of certainties. It will be a sign of success of the meeting if a number of individuals or working groups remain in touch with each other and perhaps even come to more positive cooperative arrangements.

INDEX OF AUTHORS*

Numbers in bold type indicate a contribution in the form of a paper; numbers in plain type refer to contributions to the discussions.

Abercrombie, M.   55, 72, 99, 103, 271, 273, 317
Allison, A. C.   24, 54, 70, 82, 84, 100, 104, 125, 126, 141, 142, 163, 177, 194, 195, 218, 219, 222, 224, 227, 262, 263, 273, 274, 309
Banks, Barbara E. C.   .   .   .   57
Banthorpe, D. V.   57
Bergel, F.   **1**, 22, **23**, 53, 56, 72, 73, 81, 83, 84, 101, 102, 104, 105, 126, 127, 142, 177, **178**, 219, 222, 225, 226, 228, 263, 273, 274, 289, 307, 308, 310, 313, 314, 315, 316, **319**
Berry, A. R.   57
Burke, D. C.   55, 71, 73, 125, **171**, 176, 177, 178, 179, 289, 290, 314
Davies, H. ff. S.   .   .   .   57
Elsdale, T. R.   27, 163, 220, 272, 291, 303, 304, 305, 308, 317
Finter, N. B.   71, 84, **165**, 176, 178, 179
Forrester, J. A.   163, **230**, 259, 261
Gingell, D. **241**
Gros, F.   23, 24, 71, **107**, 124, 125, 126, 127, 141, 142, 161, 163, 176
Iversen, O. H.   15, 22, **29**, 53, 54, 55, 56, 83, 101, 103, 261, 272, 312, 313
Jacques, P. J.   99, 177, **180**, 193, 195, 263, 273, 290, 311, 315
Johns, E. W.   84, 98, **128**, 141, 142, 228
Kourilsky, P. **107**
Lamerton, L. F.   **5**, 22, 23, 24, 27, 82, 83, 102, 105, 106, 220, 227, 273, 303, 309, 317

Lamont, D. Margaret   .   .   .   57
Leese, C. L.   26, 82, **144**, 161, 162, 163, 164, 288, 315
Marcaud, L. **107**
Mason, J.   **75**, 82, 83, 84, 98, 220, 271
Möller, G.   22, 72, 73, 97, 102, 105, 106, 126, 140, 141, 176, 177, 193, 194, 195, **197**, 217, 218, 219, 220, 222, 223, 224, 225, 226, 228, 260, 261, 272, 290, 311
O'Meara, R. A. Q.   25, 84, **85**, 97, 98, 99, 100, 104, 224, 228, 261, 274, 310
Ormerod, M. G.   178, 217, 289, 309
Pearce, F. L.   57
Redding, Katharine A.   .   .   .   57
Roe, F. J. C.   53, 82, 223, 228, 262, 273, 310, 311, 315
Stoker, M. G. P.   23, 26, 54, 55, 72, 101, 125, 163, 178, 218, 223, 226, 227, 259, 261, 262, **264**, 271, 272, 273, 274, 305, 308
Subak-Sharpe, J. H.   55, 84, 97, 177, 178, 227, 228, **276**, 289, 290, 305, 308, 309, 316, 317
Vernon, C. A.   54, **57**, 70, 71, 72, 73, 83, 103, 162, 218, 219, 227, 304, 309
Wolff, Em. **75**
Wolff, Et. **75**
Wolpert, L.   25, 55, 103, 104, 105, 124, 125, 142, 217, 225, 227, **241**, 260, 261, 274, 304, 309, 314, 317

* Author and subject indexes prepared by Mr. William Hill.

INDEX OF SUBJECTS

Acridine orange, binding to DNA, 140–141
Actinomycin, inhibiting interferon formation, 173
cyclic 3′,5′-Adenosine monophosphate (cyclic AMP), 249
  and chalones, 48, 55
  in cancer cells, 99, 100
Adenylic pyrophosphorylase, 279 et seq.
Adrenaline, as co-factor with chalones, 40, 56
Age, effect on cell division, 7
Aldolase, isoenzymes, 155–157
Alkaline phosphatase, isoenzymes, 147
Allogeneic inhibition phenomenon, 102
Amino acids,
  and Wolff factor, 77, 79, 80
  in histones, 130
Anaemia, pernicious, 228
Antibody synthesis,
  qualitative characteristics, 209–213
  regulation, 197–221
    mechanism, 208–209
    role of feedback, 203–208
  suppressing immune response, 197, 199 et seq., 213
Arabinose, and genetic control in bacteria, 124
Ascites tumour cells, electrophoretic mobilities, 234
Asparagine, and tumour growth, 82
Aspartokinase of *E. coli*, 148–149
Auto-immune disease, 87

Bacteria, replication in, 23
Bacteriophage, repression, 108
Bacteriophage λ,
  chromosomal structure, 109–111, 125
  gene repression, 125–126
  gene transcription in, 24, 107–127
  lysogenic induction,
    kinetics, 116–121
    m-RNA and, 114–115
  repression, 109, 114, 121, 125, 126, 142
  RNA synthesis, control of, 121–123
Biological systems, models of, 15–21, 312–313

Body fluids, role of lysosomes in homeostasis, 185–186
Bone matrix, lysosomes and, 186

Calcium, ion concentration, effect of cell surface, 236–237, 247, 250
Cancer *See Neoplasia*
Carbohydrates, on cell surface, 233
Carbonic anhydrase, isoenzymes, 146–147
Carcinogenesis,
  and immune reaction, 222–223
  early mechanism, 33–34
  lysosomes and, 194–195
  model studies, 21
  plastic film, 272, 273
  theories of, 95, 97
Casein synthesis, involving DNA synthesis, 24
Cell(s),
  alignment, as inherently precise (IP) machine, 298–299, 302
  antibody-producing, 198–199, 209, 217–218, 226
  antigen recognition by, 219–220
  antigen-sensitive, 209, 211
  autophagy, 186–187
  behaviour, predictability, 299–301
  collagen accumulation, 294
  contact between, 26, 241–263, 280, 281
    allogeneic inhibition, 260, 272
    and antigen recognition, 219–220
    and growth inhibition, 264–265, 268, 271, 272
    at molecular level, 248, 259
    control, 249–253
    developmental interactions, 241, 255–257
    electrophoretic mobility and, 238, 244, 245, 247
    immunological response, 254–255
    inhibiting mitosis, 262
    inhibition and paralysis, 250–251, 253–254
    movement and, 241, 249, 251–253
    pseudopodia in, 238, 243, 249, 250, 256
    recognition, 252, 272
    specificity, 248–249, 252, 260

Cell(s)—*cont.*
  cooperators, 284
  coupling, functional, 26, 246–248, 262
  cross-feeding, 281
  culture, regulating systems, 264–275
  digestion of own substance, 186
  division *See under Cell, mitosis*
  duration of cell cycle, change in, 8–10
  dynamic behaviour, 297–298, 304
  effect of interferons on, 166
  electrical communication between, 26, 246–248, 251, 256, 262
  environment, relation to division, 10–12
  functional coupling, 246–248, 251, 253, 255, 262
  functional specialization, relation to division, 12–14, 24, 25, 30
  fusion, and interferon, 172, 177
  growth, 253–254
    action of interferon, 167, 179
    anchorage dependence, 268–269
    contact inhibition, 264–265, 268, 271, 272
    effect of insulin, 269, 274
    interrelationship of control processes, 269–270
    regulation, 102, 313–314
    serum factor dependence, 266–267, 269, 271
  histone fractions in, 131–132
  information transfer, 26–27, 227, 245, 246–248, 255–257, 288
  interferon penetration, 176
  membrane,
    and contact control, 241–263
    as channel, 246–248, 252, 253
    as sensor, 242–243, 252
    as transducer, 243–246, 252, 253, 255
    cytoplasmic interface, 245
    electrical properties, 230–232
    exchange between cells, 286, 290
    mechanical deformation, 245–246, 249, 253
    nucleoside phosphorylation and, 280, 289–290
    properties, 231
    structure, 230
  metabolic cooperation, 276–290
    implications, 286–287
    mechanism, 285–286
  mitosis,
    action of interferon, 168
    and wound healing, 32–33

Cell(s)—*cont.*
  mitosis—*cont.*
    change in time, capacity for, 8–10
    control by chalones, 27
    effect of age, 7
    inhibition, 17–18, 254–255
    mitotic index, 5
    positional information, 26, 27
    rate of, 5–7, 13, 22, 23, 43
    regulation, 22–27, 254–255, 264–270
    relation to environment, 10–12
    relation to function, 12–14, 24, 25, 30
  movement, 295, 296, 305
    and contact, 249, 251–253
  nuclear control, 314
  parallel alignment, 293–296, 304, 305
    self-recognition, 296–297
  pattern formation, 251–252, 291–306
    mechanism, 295
  plaque-forming in spleen, 199 *et seq.*
  population kinetics,
    in relation to homeostasis, 5–14
    of epidermis, 29
  recognition, 220, 252, 272
  surface,
    antigens on, 208
    carbohydrates on, 233
    colloidal properties, 242
    effect of calcium ion, 236–237, 247, 250
    electrophoretic mobilities, 233–238
    in malignancy, 254
    lipid components, 232
    potential and adhesion, 242
    potential energy, 261
    protein components, 232
    sialic acid residues, 234–235
    structure, 230–240
Cellular immunity, 102
Chalones, 3, 314
  action of, 126
  AMP and, 48, 55
  adrenaline as co-factor, 40, 56
  arguments for existence of, 31–34
  assay systems, 40–41
  cancer and, 49–50
  controlling cell division, 27
  definition, 36–37
  diffusion of, 53
  dose response, 42–43, 54
  duration of effect, 56
  effect on DNA synthesis, 46
  effect on protein synthesis, 46
  historical background, 34–37

INDEX OF SUBJECTS

Chalones—*cont.*
  identity, 101, 103–104
  in skin, 29–56
  mechanism of action, 43–49, 55
  production of, 37–39
  purification, 39–40, 53
  specificity, 41–42, 54
Chick embryo, Wolff factors from, 75–84
Chloramphenicol, lysogen induction in presence of, 114–115
Chorion, thromboplastic materials from, 85–100
Chromosomes, damage by lysosomes, 194
Clonal selection theory, 217
Collagen, accumulation of, 294
Cybernetic theory, and homeostasis, 15–21, 103
Cytoplasm, exchange between cells, 286, 290

Deoxyribonucleic acid (DNA),
  binding to acridine orange, 140–141
  hybridization of halves, 116
  in mitochondria, 178
  interaction with histones, 128–143
  of bacteriophage λ chromosome, 24, 109, 125
  purine incorporation into, 278, 279
  ratio to histone in cell, 131, 133
  structure, 135, 137
  synthesis,
    and casein synthesis, 24
    effect of chalones, 46
    inhibition, 46
    rate of, 8, 9
Diphtheria toxins, receptors for, 224
Drugs, and growth, 105

Egg, fertilized, division in, 301
Embryo,
  cell division in, 6, 7
  growth regulation and development, 25, 44, 255–256, 304–306
  resorption of temporary organs, 188
Embryonic tissue, tumour growth factors in, 75–81
Endocytosis, lysosomes and, 185–186, 189, 190
Endoplasmic reticulum, of cancer cells, thromboplastic material in, 91, 99
Enzymes, 107, 144
  adaptation, 144–145
  induction, 107–108 *et seq.*, 144–145

Enzymes—*cont.*
  lysosomal, 190
  repression, 107 *et seq.*, 144–145
Epidermal growth factor, 60–61, 69, 70
Epidermis,
  cancerous, 261
  cell population kinetics, 29
  chalones in,
    assay systems, 40–41
    dose-response, 42–43, 54
    duration of effect, 56
    mechanism of action, 43–49, 55
    production, 37–39
    purification, 39–40, 53
    specificity, 41–42, 54
  dynamic equilibrium, 30
  growth, controlling system, 17–18
  model studies, 36
  maturation growth, 86
  proliferation of, 16–21, 25
  repair, 274
Epithelial cells, division in, 10–12
Epithelial growth factor, 60–61, 69, 70
Erythrocytes,
  electrophoretic mobilities, 233–234, 238, 244, 245
  proliferation of, 13, 22, 24
  specific histone, 137–138
Erythropoietin, 13, 24, 103
Escherichia coli, antigens, 204

Fatty acids, in neoplasia, 91–92, 94
Feedback mechanisms, 106, 107, 309, 312
  regulating immune response, 197, 203–208
Fibrin,
  in neoplasia, 87, 92–93, 97–98
  in plasmacytoma, 89
Fibrinogen,
  formation, in neoplasia, 87
  in oedema fluid, 88
Fibrin-stabilizing factor, in tumours, 89
Fibroadenoma, induction, 262–263
Fibroblasts,
  alignment, 293–296, 304, 305
    as inherently precise (IP) machine, 298–299, 302
    recognition of parallel, 296–297, 305
  behaviour, predictability, 299–301
  clone size, 300
  growth regulation in, 253–255, 264–275, 293–295

INDEX OF SUBJECTS

Fibroblasts—*cont.*
  kinetic and dynamic aspects of behaviour, 297–298, 304
  metabolic cooperation between, 276–290
Foetus,
  isoenzymic lactic dehydrogenase in, 151
Fructose-1,6-diphosphate, 155
Functional coupling, between cells, 26, 241, 246–248, 251, 253, 255, 262

Gastrin, 104
Genes,
  control elements, 124
  in bacteriophage λ chromosome, 109, 125
  transcription, in bacteriophage λ, 107–127
Gene control, histones and, 134, 136–137
Genetic factors,
  in interferon activity, 177
  in isoenzymes, 147, 150
Genetic markers, 285, 286, 288
Genetic variants, 286, 288
Glucose, transport, 157, 159
Glutamate dehydrogenase, isoenzymes, 147
Glycolytic enzymes, multiple forms, 156
Graft rejection, 213, 225
Granulolysis, 187
Growth, 310
  See also Nerve growth factor, Wolff factor etc.
  action of drugs, 105
  contact inhibition, 264–265, 268
  controlling substances, 101–102
  in cell populations *See Cells, growth*
  in developing systems, 25
  interferon and, 167
  invasive, in neoplasia, 87
  maturation, 86, 98
  in neoplasia, 87, 95
  in psoriasis, 86
  regenerative, 86, 87, 102
  regulation of, 44, 313–314 *See also Chalones*
  in fibroblasts, 264–275
  relationship to homeostasis, 85
  theoretical model, 35–36

Hexokinase, isoenzymes, 157–159
Histones,
  and DNA transcription, 141–142
  and gene control, 134, 136–137
  and nucleoprotein structure, 133–135, 137

Histones—*cont.*
  characterization of, 128–132
  definition, 142
  erythrocyte specific, in birds, 137–138
  fractionation of, 128–132
  interaction with DNA, 128–143
  molecular structure, 130, 134
  nuclei without, 142
  ratio to DNA in cell, 131, 133
Homeostasis,
  approach to study, 101–106
  definition, 226–228, 307–308, 309–311, 314, 316
  immune reactions and, 222–228
  levels of regulation, 226–227, 310, 314–315
  model of, 15–21
  neoplasia and, 227–228
  pattern formation and, 291–306
  role of feedback, 312
  space and time relationships, 106, 227, 315–316
  terminology, 309
  types, 227
Hormones,
  and mode of action of chalones, 46–48
  lysosomes and, 189, 195
Hyperplasia,
  and early carcinogenesis, 34
  tendency to, 31
Hypoxanthine,
  incorporation into nucleic acid, 280, 281

Immune reactions,
  and homeostasis, 222–228, 311, 313
  and neoplasia, 222–223, 226
  non-responsiveness, 225
Immune response,
  antibody suppression, 214
  characteristics of, 198–199
  enhancement and, 214
  feedback regulation, 203–208
  regulation of antibody production, 197–215
  regulation of antibody affinity, 197, 210–213
  suppression of, 7S, 202–203, 204, 205, 211, 212
  19S, 199–202, 204, 205, 211, 212
  mechanism, 218–209, 217
Immunological enhancement, 213–215
  neoplasia and, 224
Immunological sympathectomy, 60, 71

INDEX OF SUBJECTS

Immunological tolerance, 217–218, 225
Immunosuppression, 222–223, 225
Inherently precise (IP) pattern-forming methods, 291–293, 303
  alignment of fibroblasts as, 298–299, 302
Inosinic pyrophosphorylase, 276 et seq.
Insulin, and cell growth, 269, 274
Interferons, 314
  and cell growth, 167, 179
  and messenger RNA, 173, 177, 178
  and mitosis, 168
  anti-viral activity, 165, 166, 168, 169, 173, 178
  as possible regulators, 165–170
    biochemical aspects, 171–175
  cell fusion and, 172, 177
  effects of, 167
  formation, 166, 171–174
  inhibition by actinomycin, 173
  mode of action, 167–168, 169
  penetration into cell, 176
  properties, 165–166
  specificity, 167, 177, 179
Interofective system, 105n
Intestine, cell division in, 6, 7, 11
Isoenzymes, 146–159
  distribution patterns, 161–162
  formation, 146–148, 161
  genetic factors, 147, 150
  molecular basis, 146–148
  numbers of, 162
  periodic variations in, 153–154, 163
  physiological significance, 148, 162
  substrate inhibition, 152

Keratinocytes, 30
2-Keto-3-deoxyglucose, and chalones, 101
Kidney, cell division in, 6, 7, 9

Lactic dehydrogenase, isoenzymes, 150–155, 162
Lactose repressor, 142
Lesch-Nyhan cells, enzyme defects in, 277
Leukaemia, and immunological enhancement, 214–215
Liver,
  cell division in, 6, 7, 9, 12
  embryonic, tumour growth factors in, 76, 77
Lymphocytes,
  DNA binding to acridine orange, 140–141
  interaction with antigen, 218–220

Lymphocytes—cont.
  sensitized, destruction of target cells by, 214–215, 260
Lysogenic induction, 108, 125
  of bacteriophage λ,
    kinetics of, 116–121
    RNA synthesis, 121–123
  m-RNA and, 114–115
Lysosomes, 180–196, 314
  and carcinogenesis, 194–195
  and chromosomal damage, 194
  and endocytosis, 190
  and resorption of temporary foetal organs, 188
  functions, 195
  interaction with hormones, 189
  regulation of activity, 189–191
  role in homeostatic regulation, 183–189, 194
  role in vesicular transport, 181, 185, 188–189
  structure and function, 180–183

Malignancy, and homeostasis, 93 See also under Neoplasia
Mast cell tumours, fibrinolytic activity, 98–99
Maturity, concept of, 310–311
Metabolic cooperation between cells, 276–290
Mitochondria, DNA in, 178

Neoplasia, aetiology, 94
  and enhancement, 224
  and immune reactions, 222–223
  cell division in, 7
  cell surface in, 254
  chalones and, 49–50
  clotting properties, 88
  effect of protamine, 98
  fatty acid release from, 91–92, 94
  fibrin in, 87, 92–93, 97–98
  fibrinogen formation in, 87
  growth, 75–83, 87
    and antigenicity, 223
    model studies, 21
  homeostasis and, 222, 226, 227–228
  immunological enhancement, 213–215
  implications of cell cooperation, 287
  lysosomes and, 190–191, 194–195
  malignancy and homeostasis, 93, 222, 226
  proliferating cells, 94
  thromboplastic materials from, 85–100

Nerve fibres, effect of nerve growth factor, 59, 63–65
Nerve ganglia, effect of nerve growth factor, 62–68
Nerve growth factor (NGF), 57–60, 103, 104
  chemical nature, 58–59
  effects, 67–68, 74
    in vitro, 59–60, 62–68
    in vivo, 60, 68
    on neurons, 60
    variation with age, 63, 64
  from snake venoms, 57, 61–68, 71
  immunology of, 60, 71
  level of activity, 61–62
  localization of, 71
  mechanism of action, 72
  nutritional effect, 72, 73
  properties, 62
  protease activity, 70
  purification, 58, 59, 61
  role in homeostasis, 69, 72
Neuraminidase, 235, 238
Neurons, effect of nerve growth factor, 67
Nucleic acid, purine incorporation in cells, 276, 278, 279, 280, 282, 284, 289
Nucleoprotein, structure of, role of histones, 133–135, 137
Nucleotides,
  splitting at cell membrane, 280
  transfer in cooperating cells, 285, 289–290

Oxygen, and isoenzyme synthesis, 153

Pattern, formation,
  and homeostasis, 291–306
  mechanism, 295
Phagocytosis, and pinocytosis, 193–194
Phagosomes, 180, 181
Pinocytosis,
  and phagocytosis, 193–194
  initiation, 243
Plasmacytoma, clotting factors in, 87
Placenta, thromboplastic activity, 89
Polyoma virus, 97, 276 et seq.
Positional information, 256
  in cell proliferation, 26, 27
Positive regulatory mechanisms, 121–122, 124
Postlysosomes, 181
Proliferation, and maturation growth, 86
Proliferation of cells,
  independence of, 87
  in neoplasia, 94

Promin, 3
Prophage, 108
  m-RNA in, 113–116
Protamine, causing tumour regression, 98
Protein,
  synthesis,
    and interferon production, 171–174
    effect of chalones, 46
Psoriasis, cell maturation in, 86
Purine, incorporation into nucleic acids in cells, 276, 278, 279, 280, 282, 284, 289

Recognition systems, 126–127, 219, 224–225, 241, 243, 246, 247, 248, 252, 254–255, 260, 272–273
Regeneration,
  local reactions, 31
  of tissue growth, 86
Relaxation times in biological systems, 105–106, 227
Repressor protein, 109, 114, 125, 126, 142
Retin, 3, 101
Ribonucleic acid (RNA),
  and interferon formation, 171–174
  purine incorporation in, 278, 279, 289
  synthesis,
    inhibition by histones, 136
    in lymphocytes, 141
    temporal control in phage, 121–123
  messenger Ribonucleic acid,
    and interferon, 173, 177, 178
    bacteriophage λ specific, 111–112
    in prophage λ, 113
    species identification, 111
    synthesis, by bacteriophage λ, 108
Ribonucleic acid polymerase, 137, 143
Ribosomes, interferon type, 168, 173, 176

Salivary gland, nerve growth factor from, 58, 68
Sarcoma, clotting properties, 88
Serum antibodies, regulation, 210–211
Sialic acid residues on cell surface, 234–235
Skin,
  cell division in, 10
  chalones of, 29–56
  effect of carcinogen on, 33–34
Snake venoms, nerve growth factor from, 57, 61–68, 71
Spermatozoa, isoenzymic lactic dehydrogenase in, 151, 153

Tetanus toxin, receptors for, 224
Thrombin, in neoplasia, 89
Thromboplastic materials, from human tumours and chorion, 85–100
Thromboplastin, in neoplasia, 89
Thyroid,
  cell division in, 9
  control mechanism, 103
Tight junctions, 246, 251, 261
Tumours *See Neoplasia*

Vacuomes, 181–182
Vesicular transport, lysosomes and, 181, 185, 188–189

Vipera russelli, nerve growth factor from, 61–68, 71
Viruses, and interferon, 165, 166, 168, 169, 173, 178

Wolff factors, 75–84
Wound healing,
  and mitotic activity, 32–33
  hyperplasia in, 31
  local reaction, 31

Z 200, organ culture of, 73 *et seq.*
Z 516, organ culture of, 75 *et seq.*